Potentiating Health and the Crisis of the Immune System

Integrative Approaches to the Prevention and Treatment of Modern Diseases

Potentiating Health and the Crisis of the Immune System

Integrative Approaches to the Prevention and Treatment of Modern Diseases

Edited by

Avshalom Mizrahi

The Israeli College of Complementary Medicine
Tel Aviv, Israel

Stephen Fulder

Consultancy and Research Services
Clil Village, Israel

and

Nimrod Sheinman

The Center for Mind–Body Medicine
Kadima, Israel

Springer Science+Business Media, LLC

Library of Congress Cataloging-in-Publication Data

Potentiating health and the crisis of the immune system : integrative approaches to the prevention and treatment of modern diseases / edited by Avshalom Mizrahi, Stephen Fulder, and Nimrod Sheinman.
p. cm.
"Proceedings of the Third Dead Sea Conference on Potentiating Health and the Crisis of the Immune System, held October 20-24, 1996, in Zichron Yaakov, Israel"--T.p. verso.
Includes bibliographical references and index.
ISBN 978-1-4899-0061-6
1. Alternative medicine--Congresses. 2. Health--Congresses. 3. Medicine, Preventive--Congresses. 4. Immunity--Congresses. I. Mizrahi, A. II. Fulder, Stephen. III. Sheinman, Nimrod. IV. Dead Sea Conference on Potentiating Health and the Crisis of the Immune System (3rd : 1996 : Zikhron Ya'akov, Israel)
[DNLM: 1. Immunity--congresses. 2. Immune System--congresses. 3. Health--congresses. 4. Nutrition--congresses. 5. Behavioral Medicine--congresses. QW 540 P861 1997]
R733.P68 1997
615.5--dc21
DNLM/DLC
for Library of Congress 97-16693
CIP

Proceedings of the Third Dead Sea Conference on Potentiating Health and the Crisis of the Immune System, held October 20 – 24, 1996, in Zichron Yaakov, Israel

ISBN 978-1-4899-0061-6 ISBN 978-1-4899-0059-3 (eBook)
DOI 10.1007/978-1-4899-0059-3

Originally published by Plenum Press, New York in 1997
Softcover reprint of the hardcover 1st edition 1997

10 9 8 7 6 5 4 3 2 1

PREFACE

With all the enormous resources that are invested in medicine, it is sometimes a mystery why there is so much sickness still in evidence. Our life span, though higher than at any time in history, has now leveled off and has not significantly increased in the last two generations. There is a one-third increase in long-term illness in the last 20 years and a 44% increase in cancer incidence, which are not related to demographic issues. In some modern countries, the level of morbidity (defined as days off work because of sickness) has increased by two thirds in this time. Despite $1 trillion spent on cancer research in 20 years, the "War On Cancer" has recently been pronounced a complete failure by the U.S. President's Cancer Panel.

Evidently we still have a long way to go. The goal of "Health for All by the Year 2000" as the World Health Organization has put it, is another forgotten dream. As ever, the answer will be found in breaking out of the old philosophical patterns and discovering the new, as yet unacceptable concepts. The problems of medicine today require a Kuhnian breakthrough into new paradigms, and new ways of thinking. And these new ways will not be mere variations of the old, but radical departures.

This book, and the conference upon which it was based, is part of a search for these new pathways. The Third Dead Sea Conference is a forum set up by pioneers in health theory. It is intended to provide an opportunity to exchange ideas and benefit from cross-fertilization of an interdisciplinary gathering. It is held every 2–3 years in Israel, a land that has so often provided new thinking and new teachings for the Western world. It will be obvious from a review of the contents of the book that the thinking is broad, revolutionary, and courageous. The theme is the main challenge that we face today—the burden of chronic disease and the failure of our ability to cope with the health challenges of the modern world.

The solutions are highly varied and often original: deep nutritional changes, a more vitalistic view of the human being, holism, the energetic view of health embodied in Oriental medicine, multidisciplinary and systems attitudes in research, the inner healer, etc. To some extent there is too great a variety of ideas here. But this is partly a consequence of the diverse group of original thinkers brought together from many parts of the world.

The editors hope that the chapters here offer the reader some food for thought, some new ideas and new visions within a health field that at times seems exhausted. They also hope that the chapters open a Pandora's box of new questions that can be raised at further meetings, in research projects, and in discussions. With that in mind, the editors would like to warmly thank all the contributors to this volume, and also all those who helped to arrange the Third Dead Sea Conference, and, of course, all the participants who helped to make it a great success.

Avshalom Mizrahi
Stephen Fulder
Nimrod Sheinman

ABOUT THE DEAD SEA CONFERENCES

The concept of the Dead Sea Conferences (DSC) forum was formulated in 1989 by a group of health professionals who now constitute its Organizing Committee. The aim of the forum is to establish a base for international conferences in Israel to explore the many aspects of health and medicine. The DSC is advised and supported by an International Advisory Committee and is in informal alliance with several health, medical, scientific, and holistic organizations worldwide (Conferences forum address: P.O.Box 11868 Tel Aviv 61116, Israel).

The goals of the DSC forum are to encourage a better understanding of health and medicine. The DSC forum emphasizes the following:

1. The enhancement of body–mind–spirit, by encompassing the widest variety of possible approaches, philosophies, and traditions of healing.
2. A better understanding of the connections between medical modalities and research to bridge the gaps between them.
3. An exploration of the holistic approach to health, including the treatment, prevention, and optimization of health.
4. An exchange of ideas, theories, and findings that search for integration among biological, physical, behavioral, and medical sciences, to serve as a tool for a better understanding of the complexity of health.
5. An understanding of the scope and effectiveness of behavioral interventions.
6. An exchange of ideas for the establishment of education programs that facilitate the teaching of integrative, holistic, and multidisciplinary approaches to health and medicine.
7. Promoting and developing international health strategies and policies aimed toward achieving a healthier world.

ORGANIZING COMMITTEE

Prof. Avshalom Mizrahi, Ph.D.—"Dreamer," Founder, and Chairman
Prof. Zvi Bentwich, M.D.
Dr. Stephen Fulder, Ph.D.
Dr. Nira Kfir, Ph.D.
Mr. Yossi Morgenstern, D.N.A.T.
Dr. Nimrod Sheinman, N.D.

The Dead Sea Conferences are organized in Israel, in informal alliances with:

The American Holistic Medical Association
British Holistic Medical Association
Japan Holistic Medical Society
The Institute of Noetic Sciences, USA
The Scientific and Medical Network, UK
Forum Health Care, The Netherlands

THE 1st INTERNATIONAL DEAD SEA CONFERENCE was held in Israel, on the shores of the Dead Sea, March 1–5, 1992. The title of the conference was: The Interaction between Western and Eastern Medicines.

THE 2nd INTERNATIONAL DEAD SEA CONFERENCE was held in Israel in Tiberias, on the Sea of Galilee, November 28–December 2, 1993. The title of the conference was: The Anatomy of Well-Being.

THE 3rd INTERNATIONAL DEAD SEA CONFERENCE was held in Israel in Zichron Yaakov, on the Carmel Mountains, October 20–24, 1996. The title of the conference was: Potentiating Health and the Crisis of the Immune System.

CONTENTS

Potentiating Health and the Crisis of the Immune System

Integrative Approaches to the Prevention and Treatment of Modern Diseases

UPDATING FLEXNER*

An Expanded Approach to Health and Healing

Elliott S. Dacher

Fellow, The Institute of Noetic Sciences
Practicing Internist
3 East Pasture Way
Gay Head, Massachusetts 02535
Telephone: (508) 645-9058; Fax: (508) 645-3109; Email: esd@healthy.net

> The real voyage of discovery rests not in seeking new landscapes, but in having new eyes.
>
> —*Marcel Proust*

ABSTRACT

Although there is general agreement that our current approach to health and healing is undergoing substantial change, there has been a lack of critical discussion regarding the extent, character, and direction of change. It is important for us to articulate the values that we would like expressed in a reconfigured approach to health and healing, and to ensure that our efforts to initiate change are aligned with these values. An over view of western history suggests that the post-modern world view is characterized by four essential values: multidimensional realism, intentionality, holism, and personal authenticity. A healing model, the Whole Healing model, and two clinical programs which express these values are presented.

In 1910 Abraham Flexner in his seminal report to The Carnegie Foundation For The Advancement of Teaching offered a series of recommendations that fundamentally altered the course of medical education in the United States. Based on the teaching model at the Johns Hopkins Medical School, he recommended that medical education be directly linked to university based clinical and laboratory sciences. This suggestion became the

* The content of this paper has been in part drawn from the following sources: *Whole Healing*, Dacher, E S, Dutton, New York, 1996. *Towards a Post Modern Medicine*, The Journal Complementary and Alternative Medicine, Vol: 2, Issue 4, December, 1996.

foundation of our current training programs. But in the same report Flexner also warned us that training in these sciences would in itself be insufficient. The practitioner, he stated, requires further skills, ".. a different apperceptive and appreciative apparatus to deal with other more subtle elements. .... One must rely for the requisite insight and sympathy on a varied and enlarging cultural experience." Neither in science nor in the affairs of men and woman is our culture similar to that of 1910.

Confronted with the complexities of lifestyle and stress-related degenerative diseases, addictive disorders, anxiety, depression and their physical counterparts, dissatisfaction with the over use of pharmacological and interventionist therapies, a rising antipathy with professional arrogance and authority, the rise of self-care initiatives, feminism, holistic and ecological perspectives, along with a growing demand for high level health and wellness conventional medicine has finally reached its limitations. There is now a broad based consensus that change is necessary and desirable. Much like in Flexner's times, our enlarging cultural experience is forcing us to reconsider our century old approach to the critical issues of health and healing.

In the United States individual states are increasingly providing licensure to new categories of health practitioners, medical schools are offering programs on holistic healing, the Office of Alternative Medicine is funding research on complementary therapies, insurance carriers are beginning to offer reimbursement for these therapies, the Internet is overflowing with information and informal dialogues, and the state of Washington has passed legislation mandating that "Every health plan after January 1, 1996 shall permit every category of provider (chiropractors, acupuncturists, naturopaths, etc.) to provide health services or care for conditions included in the basic health care services (offered by the health plan)..."

But the current pace of change has allowed both practitioners and the general public little opportunity for reflection and evaluation. As a result there has been a lack of significant discourse in regard to the extent and the direction of change. Motivated by very real concerns yet conditioned by old patterns of thought, fired up with enthusiasm and hope yet compelled by complex professional and financial interests, and carried along by a seemingly unstoppable momentum, we simply assume that our current initiatives are taking us in a beneficial and innovative direction. As a result, we have failed to ask the critical questions whose answers can either reassure us about our current efforts, or cause us to reconsider them. Consider these two simple but basic questions:

- What perspectives do we wish to see expressed in a reconfigured approach to health and healing?
- Do our current initiatives reflect and support the development of these perspectives?

AN ENLARGING CULTURAL EXPERIENCE

The first question, one that deals with the articulation of a newly emerging world view, must be considered in the context of our unique historical moment. Today we find ourselves living in an extraordinary in-between time, a sort of gap in time that has been created by the decline of our previously unquestioned optimism and faith in the 500 year tenure of modernism, and the slow and as yet uncertain emergence of a new post modern viewpoint. As practitioners and individuals in search of a more meaningful approach to health and healing, we rarely concern ourselves with these larger cultural movements, issues we usually leave to historians, social scientists, and philosophers. Yet at times of great transformation we cannot afford to do so. Only to the extent that we can accurately comprehend the historical forces that are driving and shaping our times can we effectively embrace and empower these forces rather than oppose them with potentially misguided efforts. With this in mind it becomes incumbent on us to inform our efforts with an understanding of our extraordinary historical moment, an understanding the can enable us to best answer the two questions we have proposed.

As westerners our cultural history can largely be traced to Hellenistic Greece. This meeting of the ancient mythological world and the emerging world of rational inquiry gave rise to an extraordinary culture which sustained, for a brief period of time, a precarious yet highly creative balance between sensory and intuitive knowledge - the seemingly opposing perspectives of analysis and synthesis. Among this culture's many achievements was the rise of Aesclepian medicine, an amalgam of the rudimentary elements of a scientific medicine, the practice of hygiene, and the active invocation of the imagination and spirit.

Several centuries later this union of rational and intuitive knowledge was sundered apart with the rise of the monotheistic Christian mythos. Faith and scripture replaced rational inquiry as the primary route to knowledge and truth, and faith, revelation, salvation, and healing became one. Independent intellectual exploration of the human experience and the natural world was surrendered to the absolute authority of the Church. But the rise of a monochromatic perspective in either personal psychology or history always empowers its counter-balancing force which in time forces the decline of the previously dominant perspective. It is in this manner that the dominance of the Christian era eventually declined, giving way to the Copernican revolution and the modern era, an extraordinary epoch which was to last for 500 years, and is only now beginning its decline.

Initiated by Copernicus and completed by Kepler, Galileo, and Descartes this paradigm shift engaged the western world in a compensatory, yet equally monotheistic world view, one that was sensory-based, factual, and mechanical. This powerfully pragmatic perspective has been highly successful in elucidating the mechanistic aspects of nature, but it has left us with a disenchanted and devitalized world, one that is devoid of meaning, spirit, and faith. We have deconstructed metaphysics, leaving ourselves with no encompassing vision of life, transcendent or immanent. We have deconstructed the individual, turning the immense experience of humanity into a mechanistic collection of biological parts. We have alienated ourselves from the natural world, preferring to control and manipulate it rather than to learn from and be nurtured by it. We have delegitimized the poetic, imaginative, and aesthetic realms. And as a result, we are losing our centuries old faith and hope that an objectified and technological understanding of life acquired through reason alone would provide us with a progressive and endless improvement in the quality of our lives.

The recognition of the limitations and dark side of the modern world view has placed us in a time of great uncertainty, one that is simultaneously filled with an unlikely mixture of personal and social nihilism and an unfettered optimism. The consequences of the former can be seen in social decay, personal despair, and the of loss meaning and hope, and the consequences of the later can be seen in a momentous surge of creativity, pluralistic thinking, and hopefulness. It is in this gap between world views that we live. And both our disillusionment with the existing medical model, and our efforts to revitalize and reconfigure our approaches to health and healing emerge from this pregnant historical moment. To understand this circumstance is to comprehend that the changes we must now envision are fundamental rather than cosmetic, and as much compelled as chosen.

Through the mist of the uncertainty of our times the elements of a post modern world view and post modern medicine are slowly becoming evident. This new perspective, unlike those preceding it, will most assuredly be polychromatic, allowing for the inclusion of multiple perspectives while denying the dominance of any single perspective. And the individual, within the framework of community, will rediscover his or her unity with life while simultaneously maintaining an authentic sense of autonomy and transcending the alienation imposed by the modern world view. For the purpose of seeking guidance in our efforts to reconfigure our approach to health and healing within the context of the larger cultural movements of our times I would like to set forth four perspectives that now appear to be critical elements of the emerging post modern world view, perspectives that must be central structural supports for a post modern medicine.

MULTIDIMENSIONAL REALISM

The modern world view requires that reality be objectified, sensory-based, impersonal, measurable, quantifiable, opaque, and collectively experienced and validated. The post modern view of reality is far more complex. It rejects the view that reality is limited to the receptive capacities of our five senses, and validates the legitimacy of non-sensory, intuitive knowledge. Reality is extended, revitalized, and personalized. It is seen as a multidimensional amalgam of sensory and non-sensory knowledge. This amalgam denies the extremes of a purely sensory-based universe, and its antithesis a radical subjectivism.

INTENTIONALITY

The modern world view postulates that all phenomena are caused by unchanging universal laws that exist independently of human consciousness. In essence, causality is seen as physically based and upward in its direction. The post modern perspective validates and legitimizes the causal nature of consciousness which is individually willed and downward in direction.

HOLISM

The modern world view assumes the distinct separateness of subjective and objective, external and internal, material and immaterial, mind and body, reason and intuition, and man and nature. At its essence, irrespective of its extraordinary accomplishments, it is an alienating perspective. The post modern viewpoint accepts the value of an analytic methodology within a larger context that comprehends and honors the unbroken unity of all life.

PERSONAL EMPOWERMENT

The modern world view assigned great importance to the individual and his/her rights, an ideal that too often was degraded to a self-indulgent, egoistic, and aggressive quest for power and material gain accompanied by personal alienation. The post modern world view revitalizes and deepens the meaning of individualism by asserting the significance of the individual search for authenticity through self-knowledge, legitimizing the shift in authority from belief systems, institutions, and professionals to the individual (a shift in authority from external to internal), and recognizing that authentic individualism and personal empowerment comes into being in the context of relationship and community.

These four perspectives underlie, inform, and drive our current process of cultural change. Together they constitute the historical imperative of our times, an elastic world view that unlike previous religious and empirical perspectives is both pluralistic and inclusive. These essential perspectives characterize the post-modern world, and as such they can most appropriately guide and define an approach to health and healing that is unique to the needs and character of our times - an approach that I have called post modern medicine.

A SHIFT IN WORLD VIEW OR MORE OF THE SAME

With these new perspectives in mind, let's consider our second question: Are our current efforts expressing and supporting these perspectives, the viewpoints that characterize fundamental change? Consider the following recent attempts to expand our ideas about health and healing.

- John Travis, M.D. opened the first wellness center in the late 1970s in Mill Valley, California. Influenced by Halbert Dunn's book *High Level Wellness,* Travis' concept of wellness sought to expand our ideas about health beyond the customary focus on preventing and curing disease to include a concern for the promotion of well-being. Health and healing were seen as a personal affair, a psychosocial process of education and lifestyle change.
- The idea of holism, first described by Jan Smuts in his 1920s book *Holism and Evolution,* was revived by individuals and practitioners seeking a broader vision of health and healing. As a concept, holism expressed the view that life at all levels is organized as a unity. Although reductionism had been successful in explaining the mechanistic workings of nature, it was increasingly seen as a limited and partial approach to knowledge, an approach which distracted us from a more comprehensive and ecological view of the human condition that offered a more meaningful, vital, and enchanted view of nature.
- In the 1980s alternative and complementary practices began to emerge as a further expression of the rapid changes in our ideas about health care. Naturopaths, chiropractors, acupuncturists and others sought and achieved state licensure, and began the initial steps towards full integration into the mainstream of institutionalized health care, a process aimed at achieving conventional acceptability and consensual validation. The Office of Alternative Medicine was established at the National Institutes of Health to examine the efficacy and appropriateness. of these diverse approaches to health and healing.

Each of these initiatives were honest attempts by sincere individuals and institutions to bring change to an entrenched health care system, one that no longer seemed effective in dealing with present day problems and sensibilities, and was at odds with the emerging post modern viewpoint. Let's examine the results of each of these efforts.

The idea of wellness was rapidly integrated into our culture. But as it entered the mainstream of our cultural life, and particularly when it was integrated into existing health care institutions, wellness was reduced to four physically based issues: nutrition, smoking cessation, fitness, and stress management. Its fundamental emphasis on personal development, and its psychosocial framework and values were largely jettisoned, and with its assimilation into the larger culture it was reshaped until it more resembled traditional preventive approaches, packaged as generic commodities carrying the "wellness" label, than the dramatic shift in perspective envisioned by Travis.

The idea of holism suffered a similar fate. As a philosophy, holism evolved as a counterforce to atomistic and reductionistic perspectives. Sixty years after Smuts defined this concept, his vision was reduced to packaged commodities that could be bought and sold with labels such as holistic medicine, holistic dentistry, and so on. And further, it became a marketable credential that was self-applied by a diverse group of practitioners who confused humanism and an expanded repertoire of remedies and practices with holism. And even in the case of those practices that evolved from a more comprehensive framework, the "holistic" components rapidly receded in importance, or were completely discarded as they were secularized and reduced to disease-oriented treatments. As we are discovering, this is the cost of integrating into and accommodating to the institutional structures of mainstream health care whose perspectives are solidly embedded in the traditions of the modern world view. It is the price of cultural acceptability and third party reimbursement.

Alternative and complementary approaches to health and healing, however valuable in diversifying our treatment options, have similarly failed to significantly alter our existing world view. Conventional and alternative practitioners, irrespective of their rhetoric or intention, generally use their specific expertise to prescribe techniques, practices, drugs, or supplements for the purpose of repairing or fixing an abnormality. The professional defines the approach solely within the context of his or her professional domain, and the prescribed treatment is external rather than internal. The individual is a more or less passive recipient of the therapeutic process gaining little in the way of personal insight or additional self-healing capacities. Because all of us are conditioned to turn to authoritarian structures and external remedies at times of adversity, we often demand and easily accommodate to the treatment model. There are always individual practitioners whose practices reflect a substantial shift in perspective (both conventional and alternative healers can access holistic principles within their traditions), but this remains an individual prerogative in distinction to a cultural shift.

The answer to our second question is now apparent. With few exceptions these and similar efforts ultimately failed, usually in their implementation, to explicitly and consistently express the perspectives that characterize a post modern medicine, perspectives that most of us would agree with and ones that would take us in the direction of fundamental change? So why has this happened? The answer is clear. Old perspectives and parochial interests are powerful and enduring. They silently and effectively reshape our efforts to more or less conform to existing conventions, incorporating and reshaping them until they accommodate to the assumptions of the existing world view. Because each of these initiatives explored new approaches and perspectives, and in this way succeeded in expanding existing perspectives and stretching our imagination, they have been useful endeavors. But so far they have failed at fundamental change. As a result, these initiatives have ultimately fallen within the hegemony of the existing values, perspectives, and practices, falling far short of taking us in the direction of a post modern medicine. Wellness became prevention, holism became an empty word, and alternative approaches became alternative treatments. The powerful influence of the existing world view subtly but surely changes us before we can change it, and our efforts fall of short of animating the perspectives of the emerging post modern viewpoint. Overcoming ourselves and our deeply conditioned and often unconscious assumptions is a difficult task.

POST-MODERN MEDICINE: THE WHOLE HEALING MODEL

I would like to propose a model that responds to our contemporary personal and cultural imperatives, to the character of the post modern world view. We begin by considering the full range of healing already available to us, the four distinct healing systems that we can identify as part of our personal experience. Each of these systems has its own frame of reference, operating principles, characteristics, practitioners, practices, and research methodologies. I call the four systems Homeostasis, Treatment, Mind/Body, and Spiritual. At any one time, most of us are using one or two of these systems. That is what we are taught and what we are accustomed to. Used individually, these systems are limited and partial approaches to healing. Taken together they form a whole healing system with a flexibility, adaptability, and comprehensiveness that cannot be accounted for by the mere sum of the individual systems.

The figure below illustrates the natural nesting of the four individual systems to form a single system. However, living systems do not come in parts; it is only our capacity for abstraction that makes it seem this way. And although the descriptions of the separate and distinct healing systems can stand apart, human life, as we know it, is dependent on the presence and interaction *all* of these four systems.

Each of these systems draws upon a different aspect of consciousness, operates through a unique mechanism and process, and applies its resources towards achieving a particular aspect of health. As we add one system to another we will notice an expansion of consciousness, an increasing number of available resources for healing, and an enlarging sense of what it means to be healthy. We will discover that health is many things: a well functioning mind and body, the capacity to recovery from disease, the development of personal autonomy, and the progressive achievement of wholeness.

THE HOMEOSTATIC HEALING SYSTEM

In 1929 Walter Cannon, the famed physician-physiologist, described the most primary and basic healing system available to us, the homeostatic system. This inborn system

Figure 1.

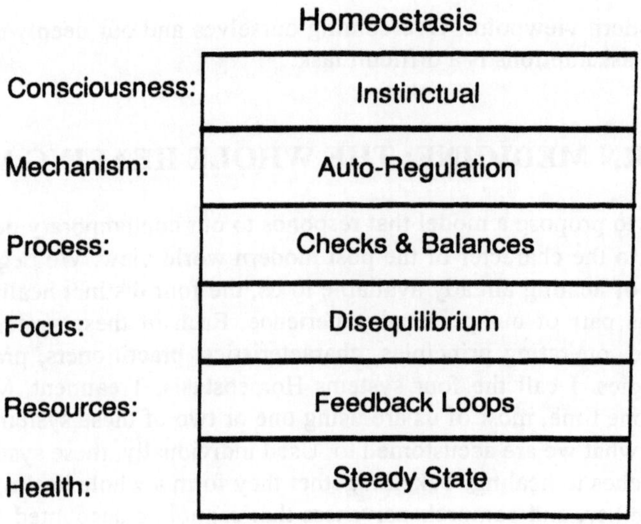

Figure 2.

of internal physiologic checks and balances which evolved over the millennia of human development makes it possible for us to automatically respond to internal states of disequilibrium with immediate, reflex-like physiologic corrections. As a result, body temperature, fluid and mineral balance and other automatic activities are at all times kept in balance. In this way homeostatic healing contributes to our health by maintaining a constant internal environment, a necessity for life.

The homeostatic healing system developed into its present form through a progressive accumulation of checks and balances designed to respond to the disruptive effects of internal and external stresses on our physiology, stresses that would tend to shift our system towards imbalance. However, because it takes approximately 1 million years to accomplish a 5% sustained change in the human biologic system, our homeostatic system is far more suited to the life of our ancient ancestors than it is to the more recent and dramatic changes in lifestyle and environment that characterize urban life. As a result, the homeostatic system, a relatively fixed and unchanging system, is often poorly adapted to the lifestyles, practices, and environments of our day-to-day lives; our nutritional choices, exercise patterns, physical environments, and above all, our stress levels. This mismatch of our inherited natural protective mechanisms and the realities of a twentieth century lifestyle have resulted in significant limitations and deficiencies in the effectiveness of this system.

For example, consider the human stress response which evolved as a quick on-and-off reaction to the abrupt appearance of physical danger. In modern times, lions no longer appear suddenly in the bush only to disappear shortly thereafter. Our modern "lions" take the form of worries, fears, and anxieties which constantly activate the stress response. Worse, unlike our ancient ancestors (or the newborn infant), we, through conscious intervention, can block or avoid the natural response which is to escape from or avoid stressful and dangerous situations. For intellectual reasons, we often choose to remain in stressful circumstances so that our stress response is ironically activated. When this happens, the normal protective response, which once insured survival, is unable to effectively respond to the new realities of our lives, and the result is the development of acute and chronic stress related diseases. What once insured the survival of life, now threatens our survival.

We can best support and enhance its effectiveness by providing the environment, nutrition, physical activity, and social relationships that most approximates the circumstances under which it developed. To a large extent these activities fall under the label of "prevention." In a sense, we are attempting to prevent a malfunction of this system by giving it what it needs. Perpetual mental stress, high fat, processed foods, and a sedentary lifestyle are creations of urban life and do not support homeostasis.

To remedy the deficiencies of a fixed and too often maladaptive homeostatic healing system, we have developed "treatment" models whose purpose is to step-in and restore normal function when homeostasis has failed. Treatment practices, conventional and alternative, increasingly draw upon man-made interventions which, unlike our automatic protective mechanisms, are flexible and can respond to changing conditions. To an extent we can say that our capacity to design treatment systems that augment nature's natural mechanisms reflects our progress as humans. One can also say it is an indication of how far, for better or for worse, we have removed ourselves from nature.

THE TREATMENT HEALING SYSTEM

Treatment, in its various forms, is the dominant model of healing in western culture. The figure below illustrates the major characteristics of this system. It is activated by our reaction to the signs and symptoms of illness, and works towards repairing abnormalities through the use of external resources such as drugs and surgery. At one time or another each of us will use the resources of the treatment system to address the inevitable adversities of living.

To treat is to apply a process to a problem with the intention of resolving it. In the case of biomedical treatment the process usually consists of the use of external agents such as drugs, surgery, or physical therapy. Other forms of treatment may come in the form of vitamins and other supplements, biofeedback, relaxation techniques, body work, energy work, chiropractic, acupuncture, and a host of other practices. We activate treat-

	Treatment
Consciousness:	Reactive
Mechanism:	Repair
Process:	Reductive
Focus:	Disease Categories
Resources:	Drugs/Surgery
Health:	Restore Function

Figure 3.

ment when we seek assistance from a health care practitioner as a reaction to the appearance of a symptom, or the presence of overt disease, an indication of the breakdown of the homeostatic healing system. The initial complaint is routinely followed by the requisite testing, establishment of a diagnosis, and the prescription of therapy according to the particular practice, a therapy which is usually directed at a specific body part. Decisions are made by the health professional, and treatment is exclusively dictated by the type of disease. Treatment is generally tailored to the specific disease rather than to the unique characteristics and needs of the specific individual within whom the disorder expresses itself. The advice given to physicians by the famed internist Sir William Osler is well taken here. He said "It is better to know the person that has the disease, than the disease that has the person." The treatment system usually works in the reverse order.

It is currently fashionable to call many interventions and techniques holistic, suggesting they aim at something other than complementary approaches to treatment. The facts tell a different story. However well intentioned a practitioner may be, a practice strategy that has one person (usually considered the expert) doing something to someone else most often falls under the treatment system as we know it. Although the intellectual intention of a practitioner and a client may lean towards holism, the practice and its impact may be quite different. The effort, irrespective of the rhetoric, is usually directed towards repair and restoration of function leaving the individual without further empowerment or enhanced personal skills and resources.

As with the homeostatic system, an understanding of the treatment system, its assets and limitations, demonstrates the need for a still further approach to healing. That is, exclusive use of a treatment system, by its very definition, is incapable of *meaningfully* including psychological and social aspects of our lives. As we move from homeostasis and treatment to mind/body and spiritual healing we are moving from an automatic system that is inborn and a treatment system that is culturally imposed to systems of healing that rely on our consciousness and intention. Unlike the first two systems, mind/body and spiritual healing offer us the capacity for self-regulation, and in doing so, a more direct and personal involvement in our health.

THE MIND/BODY HEALING SYSTEM

Unlike the physical context of the homeostatic and treatment systems, the mind/body and spiritual healing systems evoke a very different view of the human condition. These systems call upon the capacities and qualities that characterize human life: consciousness, intention, will, creativity, faith, love, and compassion. In these systems we deal less with parts and increasingly focus on wholes. The mind/body and spiritual healing systems operate through "downward" causation, the process by which higher levels of human organization, the mind and spirit, effect changes in cells, tissues, and organ systems by re-organizing the whole.

The mind/body healing system relies on personal responsibility and self-motivated effort. It requires the development and use of personal skills and capacities—physical, psychological, and psychosocial—that can help us connect mind, body, and spirit, and the development of the capacity for self-regulation. In contrast to homeostasis, which operates automatically, and the various forms of treatment, which are applied in response to the appearance of disease, the mind/body healing is proactive and intentional. Its focus is on personal attitudes and lifestyles, and the skills that are necessary for healthy relationships, conflict resolution, and personal growth and development, central factors in development

Mind/Body Healing

Consciousness:	Intentional
Mechanism:	Self-Regulation
Process:	Developmental
Focus:	Person Centered
Resources:	Mind and Body
Health:	Autonomy

Figure 4.

of stress-related degenerative disorders, and the critical components of a health promotion program.

The full potential of this system is developed over time as a result of our choices and our efforts. It is neither automatic, like the homeostatic system, nor culturally imposed, like the treatment system. We have a choice, to develop its potential or not. It is a person-centered system rather than a disease-centered one. Mind/body healing is concerned with psychological development, personal transformation, and mastery, to the extent possible, over the activities of the mind and body.

This aspect of healing bases its scientific legitimacy on the emerging research in the field of psychoneuroimmunology. The discovery that the inter connectedness of our thoughts, feelings, images, and biochemistry is mediated through a mobile neuropeptide messenger system, a series of natural chemicals that transfer information between the mind and body, has provided us with an understanding of the biochemical pathways that link the mind and the body.

The change in focus from diagnostic categories to issues of personal attitudes, life-styles, and psychological development alters the relationship of the health practitioner to his patient. It becomes more of a partnership in contrast to the hierarchical relationship that characterizes the treatment system. The intent of the mind/body system is more educational than therapeutic and a health practitioner serves more as an educator and coach. It is at this level of healing that we most profoundly see the shift from outer aids to inner resources, a shift that is intentional rather than compelled.

As with the homeostatic and treatment systems, the defining focus of the mind/body healing system, self-regulation, self-awareness, and psychological development, accounts for its contributions as well as its limitations. This system approaches, but fails to fully consider the spiritual aspects of human experience which transcend and extend the boundaries of personal development. To convey a holistic and intuitive understanding of the living experience that in itself is healing, the spiritual healing system comes into play.

THE SPIRITUAL HEALING SYSTEM

Although the spiritual experience is singular in nature, there are many paths to it and different names for it. It can arise quite suddenly, through prayer, devotion, love, compassion, meditation, music, dance, art, and nature, it may last a few precious brief moments, or at times, for longer periods. It can also evolve slowly over a lifetime of study, practice, growth, and development. In the latter instance, it is as if small islands of understanding expand and coalesce over many years to provide a more comprehensive awareness and understanding of the whole of life.

The spiritual perspective is a way of understanding life that provides meaning to our day-to-day lives and the capacity to experience wholeness. Spirituality sees wholes rather than parts and patterns rather than details. When we are guided by this perspective, life seems to make sense, everything is in its place, and we feel balanced and connected to life. This deeper sense of self and nature is satisfying to the soul and spirit. It can have profound effects on personal attitudes, values, relationships, and unresolved conflicts, and as a consequence, it can influence biochemistry and physiology. We call these effects on the mind and body spiritual healing. We experience it in silence as well as in nature.

Unlike the preceding healing systems, spiritual healing results from a way of *being* rather than an active doing. Whereas mind/body healing results in an increasing psychological understanding, spiritual healing conveys wisdom and a persistent sense of oneness with life. In its quest for wholeness spiritual healing relies on intuitive knowledge, an often unused aspect of our consciousness. It operates by conveying to us a unifying and integrated vision of life, and a sense of meaning, purpose, and coherence.

As we explore the four healing systems as a single, fully integrated system, we can discern qualities that are related not to the individual healing systems but to the healing system as a living indivisible whole. It is somewhat like the relationship of individual notes to the melody found in a musical composition. The notes are the elements of the mu-

	Spiritual Healing
Consciousness:	Intuitive
Mechanism:	Integration
Process:	Unifying
Focus:	Myth/Symbol
Resources:	Consciousness
Health:	Wholeness

Figure 5.

sic, yet the music is much more. It is the sense of rhythm, the juxtaposition of the notes with the spaces between the notes, a certain context, movement, significance, and meaning. We sense, feel, and intuit something that is alive and developing in contrast to the static non-vital nature of the individual notes. We find in the musical composition a dynamism and vitality that is not a property of the notes by themselves, but only emerges through their organized interaction. And so it is with the whole healing system. Each of the individual systems taken alone is static, stationary, and devoid of the dynamism of life. But taken together we find a fully integrated, ever moving, flexible, adaptive, and organic process. It is all *one* carefully orchestrated process that comes alive only as we consider the living whole.

INITIATING CHANGE

So how do we assure fundamental change? How do we align ourselves with the future? First, we must clearly articulate the perspectivesthat we choose to assert, then carefully design and embed them into innovative models and programs, and finally, measure their success by their demonstrated capacity to foster these perspectives. To accomplish this goal these perspectives must gain priority over our conditioned thoughts and actions, and our parochial professional interests. The changes that will result from such an effort will not be an accumulation of new ideas and practices that are subtly but assuredly reshaped to resemble the past, but rather a fundamental revision of our approach to health and healing.

Because the central perspectives of the post modern world view and post modern medicine—multidimensional realism, intentionality, holism, and personal authenticity—are activated, animated, and validated through direct personal experience guided by an inquiring consciousness, *the individual* (in contrast to professionals and institutions) *becomes the essential focus and primary healer of post modern medicine. Health and healing—its character, symbols, and metaphors—become personal issues, ones that are uniquely defined and orchestrated by each individual.* In a sense, the individual is the healer, the healee, and the healing. Directly engaged in the historical process of actively integrating and living a new world view, the individual himself is transformed. This transformation is marked by a shift in authority from professionals and institutions to the individual, a shift that is accompanied by an expansion of personal consciousness and capacity.

It follows from this that initiatives that result in an expansion and extension of professionalism and its monopoly over knowledge, conventional or alternative, expropriates power and possibility from the individual and runs counter to the values of a post modern medicine. Practitioners and their therapies will remain an important component of a person-centered post modern medicine, but not a dominant one. They will be a valuable resource to individuals who are actively engaged in composing their lives, defining their personal visions of health, and learning from and responding to life's adversities.

Because we are still living in the gap between world views, we can only catch glimpses of what the full flowering of post modern medicine will look like. Yet there is much to gain from these glimpses. Let's consider two programs: the *Dean Ornish Lifestyle Intervention Program* and the *Planetree Hospital Unit and Consumer Education Program.* In examining each of these programs we can measure them against the perspectives that will characterize a post modern medicine.

In 1977 Dr. Dean Ornish began to explore an alternative, non-pharmacological approach to atherosclerotic heart disease. The central elements of his program included a low fat diet, meditation, yoga, exercise, and psychological counseling and support. When I visited this program in 1990 I had an opportunity to join an evening meeting and to speak in some detail with several of the participants. What most impressed me was the extent to which these individuals had become empowered in the pursuit of their own healing. They had developed a repertoire of new skills, resources, and capacities, gained insight into their lives and relationships, cultivated a more expansive understanding of health and disease, learned to make conscious and self-directed choices in a complex and pluralistic universe, and accomplished each of these goals within the context of a community. As a result of these experiences the participants extended the scope of their personal autonomy, expanded consciousness and self-knowledge, and created new options, each of these valued outcomes was built into the ongoing program. By transforming their approach to health and healing they had simultaneously transformed themselves.

The goal of Ornish's program, as I view it, is to support the personal growth and development of the participants so *they* can assert their primary role in the healing process, recovering from illness and promoting vital and healthy lives. Ultimately, the professional fades into the background, and the individual and his or her experience becomes the central factor in health and healing. This is not a treatment program in the way we have previously conceptualized treatment. I'm not quite sure what to call it, but my sense is that it expresses the values we have discussed, and contributes to the creation of a fundamentally new and effective approach to health and healing, one that begins to engage the central elements of the post modern view.

Let's look at another example, the Planetree model. In 1977 Angelica Thieriot, an Argentinian, was hospitalized during a visit to San Francisco. Although impressed by the technology she was appalled by her hospital care. As a result of this experience she approached the chief of medicine at the Pacific Presbyterian Medical Center in San Francisco with the idea of creating a model program, a program that would respond to the needs of the individual by supporting personal autonomy.

In 1981 the first Planetree Health Resource Center opened. This consumer library was designed to assist individuals in acquiring up-to-date medical information that would enable them to be active and informed participants in the healing process. The center maintained a library, subject files on conventional and alternative health care, access to the National Library of Medicine's search service, selected bibliographies, and listings of national and local organizations and support groups.

	Homeostasis	Treatment	Mind/Body	Spiritual
Consciousness:	Instinctual	Reactive	Intentional	Intuitive
Mechanism:	Auto-Regulation	Repair	Self-Regulation	Integration
Process:	Checks & Balances	Reductive	Developmental	Unifying
Focus:	Disequilibrium	Disease Categories	Person Centered	Myth/Symbol
Resources:	Feedback Loops	Drugs/Surgery	Mind/Body	Consciousness
Health:	Steady State	Restore Function	Autonomy	Wholeness

Figure 6.

In 1985 the first Planetree hospital unit was established. In each of the patient's rooms, the colors, lighting, carpeting, and other details were specifically designed so that the healing needs of individuals could be met. The patients had full access to their medical records, and were encouraged to add their observations, feelings, and responses to their files. The new unit provided kitchen facilities, flexible visiting hours, and a health educator. Alternative practitioners were permitted within the hospital setting, and patients had the option of wearing their own clothes, robes, and pajamas. In what is for most individuals a highly vulnerable circumstance the Planetree program focused on enhancing personal autonomy, expanding the individual's knowledge and capacities, and allowing for a pluralistic approach to the healing process.

When I visited the Planetree hospital unit and consumer library the difference was clear. I did not feel I was visiting a treatment facility, but rather a healing center, one that was focused on the individual. Patients could leave the hospital more informed, aware, resourceful, empowered, and autonomous, a unique experience in health and healing. This project is another example of how postmodern perspectives when designed into the core of a program can support the emergence of a fundamentally new kind of medicine.

Each of these programs is a first step in the right direction, a movement away from the limitations of an exclusively reductionistic and professionally-centered treatment program. In each instance the advances of modern science are not discarded, in fact they are honored and then integrated into a larger, post modern world view. The result is the emergence of a post modern medicine, or what I have also termed whole healing

Thomas Kuhn, in his seminal book *The Structure of Scientific Revolutions,* said: "The transition from a paradigm in crisis to a new one .... is far from a cumulative process, one achieved by an articulation or extension of the old paradigm. Rather it is a reconstruction of the field from new fundamentals, a reconstruction that changes some of the field's most elementary theoretical generalizations..... When the transition is complete, the profession will have changed its view of the field, its methods, and its goals." It is time that we step back and begin to speak about fundamentals, about the perspectives and models that define our lives and our work. Such a conversation will surely assist us in creating and successfully implementing the fundamental changes that are now awaiting us.

2

THE NEW SCIENCE OF HEALTH

Robert Lafaille

International Institute for Advanced Health Studies
Kardinaal Mercierlei 26
B-2600 Antwerp, Belgium

SUMMARY

Health promotion and preventive medicine developed tremendously during the last decades. A lot of initiatives and policy measures were taken. Not always on sound scientific bases. The call for a "new science of health" is an invitation to develop a more solid basis for health oriented action. In this presentation I will give an overview of developments, draw attention for weaknesses and challenges, and explore practical implications of this new way of thinking.

1. INTRODUCTION

In the mid eighties Dr. Ilona Kickbusch, at this moment Director of the Division of Health Promotion, Education and Communications of WHO in Geneva and her collaborator Dr. Helmut Milz, at that moment WHO consultant, asked me to bring together a group of pioneering researchers who would be able to think through the possibility of a "New Science of Health". In 1989 a first symposium was organised in Louvain-la-Neuve on this topic and since then more initiatives emerged to built further at the foundations of a solid new science of health. We face a lot of difficulties. Many research (such as the risk factor model) in the field of health has very weak epistemological and empirical foundations, but at the same time there's an urgent need for action.

The new science of health* is emerging from different traditions. It is based on values which emerged in various contexts, such as grass-roots movements, the self-help movement, developments within academic medicine, democratisation processes in society, contemplementary medicine, the women's movement etc. The new science of health is very clearly linked to a new emerging worldview. The major characteristics of the new science of health are:

* I use here the singular is an overall indicator of a future oriented movement, in according with the book published about this topic (Lafaille & Fulder 1993). Other authors or traditions (like the German) prefer to use the plural "health science", stressing that not a single science will bring the new ideas, but that a collaborative movement is going on.

Potentiating Health and the Crisis of the Immune System
edited by Mizrahi *et al.* Plenum Press, New York, 1997

- A deep conviction that people to a very high extent are the author of their life, which includes their health and illness. Life is a journey into the unknown. Education has to prepare for that by teaching tools which can be used during this journey.
- Focussing on health, instead of on disease and pathology (what was called by Antonovski "the salutogenic approach").
- Focussing on prevention instead of on clinical practice. Everything that can be prevented is better than organising endless chains of cure and care.
- Focussing on cultural, social and political determinants of health instead of limiting the focus of attention to the body. All system levels (atoms, molecules, genes, tissues, organs, consciousness, social relationships, groups and institutions, societies, etc.) have to be taken into account for promoting health. Especially the health services and health policies have to be reoriented quite drastically (see the WHO Charter of Ottawa).
- The new science of health will be multi- and transdisiplinary in nature and integrate new developments in different fields and disciplines.
- It will focus on the study of complexity, instead of "instant" treatment procedures. It rejects dualistic ways of thinking.
- It will include subjectivity, instead of excluding life and culture. Biographical research is a necessary part of it.
- Operating when there is no disease at all, instead of waiting until someone is ill.
- Promoting the maximalisation of all human potentials. This also means the maximalisation of self-healing forces, vitality, placebo-effects, human values, the experience of the fullness of life (incl. sexuality), etc.
- Encompassing human experience (of which free choice is a part) as an integrated part of health. The new science of health is (and should be!) more than any other science in a very high degree "an experiential science". Scientists in general tend to forget that.
- The science of health is non-invasive. It uses personal choices, awareness, meditation techniques, etc. to reach its goals.
- Self-care techniques and its dissimination through all societies is a main goal to reach self-healing, self-determination, freedom and empowerment of people. Professionalism which serves only the self-interests of professional groups is in contradiction with the basic values of helping. As a consequence of a different value system, and other aims for professionals, the new science of health implies redefinitions of the role of the medical doctor or therapist.

Now, I will present first of all an overview of some theoretical traditions which contibute to the grow of a new science of health, and then I will give some overview of major results.

I will use a classification here, which I elaborated at length in the book "Towards a New Science of Health" (Lafaille & Fulder 1993). This kind of classification is a first step towards a more solid foundation of the new health sciences and a preparation towards theoretical integration on a meta-levels.

2. PARADIGMS IN THE HEALTH SCIENCES

All science tries to make theoretical reconstructions of reality. For centuries, philosophers and scientists thought that real knowledge would consist of one ultimate reconstruction of reality. Science, therefore, was conceived as an entreprise to discover this single reconstruction. It was an adventurous search for "truth" or "the laws of the universe". However it gradually became clear that there could not be a one-to-one relationship between the scientific reconstruction of reality and reality itself. It became clear too that one's observations are not only "influenced" by reality, but they also have influence on reality itself. By observing reality this reality can be changed. This is especially true for the human sciences. The observer is included in the system of observation. The same observations may generate an endless series of interpretations; and reversed, every theoretical interpretation/model may generates its own observations. Scientific distance is therefore relative, and dependent upon socio-historical processes.

The following four-field matrix is proposed to classify the different types of current scientific thinking in the different health sciences. We distinguish a *biomedical* -, an *existential-anthropological* -, a *systems* -, and a *culturological* paradigm. There is a degree of "Wahlverwantschaft" between every paradigm and certain types of research, methods of observation, epistemological assumptions, conceptions about causality, or more generally, about the kind of relationships between empirical phenomena, and political and policy issues (figure 1).

In the literature a distinction is often made between a so-called "medical" and "holistic" model. The medical model only applies to the first matrixfield. The holistic one to the three other matrixfields. A four-field classification is more accurate than a dual one to describe the many differences in theoretical viewpoints and diminish the danger of unnecessary polarization (good-bad oppositions). There is an increase, from top to bottom, of the number of factors which are simultaneously investigated (increase of complexity and synthetic power of theories). In the two upper matrixfields, theories are strongly directed

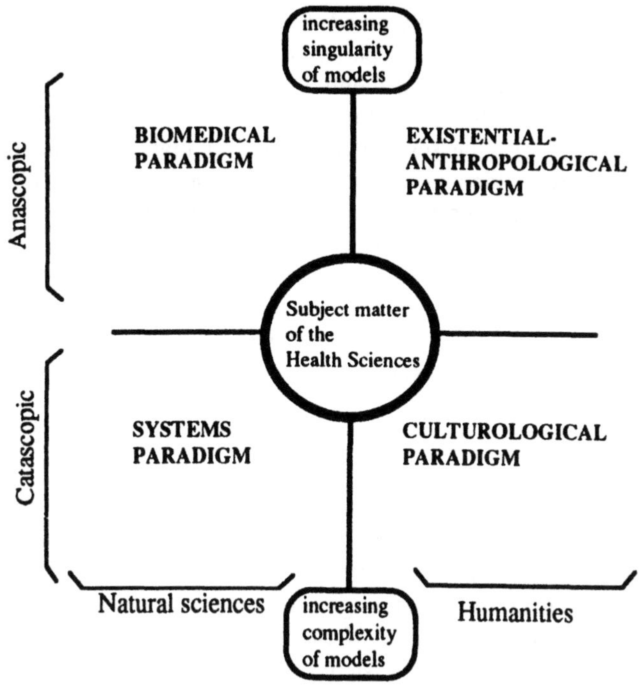

Figure 1. Contemporary paradigms in the health sciences.

to the individual (or his body), in the two lower fields the influence of social relationships and society as a whole receive much more attention. To clarify this distinction, we will use the proposed terms by Zijderveld (1966) anascopic and catascopic. The term "anascopic" refers to scientific forms of thinking which strongly focus upon the individual (or its smaller components such as his/her body, an organ, tissues, cells, etc.) and looks from that point of view toward bigger wholes; the term "catascopic" is used when wider contexts (groups, society, etc.) are taken as a main frame of reference and the individual behaviour is studied and explained in relationship to this broader context. The theories in the two fields at the left side are characterized by a positivistic, causal, quantitative and operational style of thinking; the theories at the right side have a more historical, interpretative signature and are much more open to symbols, emotions, meaning in life, etc. The left side is grounded in the natural sciences, the right side in humanities. In principle, one could add to Figure 1 a more historical dimension, and also integrate in the figure different concepts about methodology, but for our goals here, the four-fould matrix is sufficient. Below, I will describe each part of this scheme at greater length.

A. The Biomedical Paradigm

The central characteristic of this paradigm is the notion that health and illness are the result of the influence of objective factors (determinants). Illness becomes materialized and a "biological fact". It is defined in terms of tissues, organs, cells, nerves, bloodstream, etc which structures and evolution are governed by the laws of the natural sciences (physics, chemistry, fysiology, etc.). The biomedical paradigm is based upon retionalism, mechanism and empiricism. We can further distinguish within this paradigm another two subtypes. A monocausal subtype (which attempts to find one cause or one class of (biomedical) determinants) and a multifactorial subtype. This extension of the paradigm to its multifactorial subtype is certainly an advance, but nevertheless it is still based on all the original postulates of this positivistic paradigm.

The biomedical model is very influential in modern medicine, which it dominates nearly completely. With the help of this paradigm new insights were generated which led to a much better control of illnesses and a successful high-technological treatment of many diseases. The narrow-sideness and some dysfunctions caused a lot of problems to which different social movements reacted (anti-psychiatric movement, self-help groups, feministic movement, etc.). Especially its underlying dualistic concept of man has been criticized which implies a division between an objective body and a subjectively experienced corporality. Experiential aspects are then declared to be non-specific factors and treated as epi-phenomena. Further, on a more epistemological level, one has warned against the danger of atomism or causal thinking in terms of a too narrow concept of causality.

B. Systems Paradigm

The systems paradigm grew out of simultaneous developments in various scientific disciplines or their fields of application (cybernetics, information theory, the neurosciences, the computer). In this paradigm the concept of a system has a central place. A system can simply be defined as a set of elements between which a specific, non-random pattern of relationships, exists. A system can be separated from its environment; the environment is defined as everything which does not belong to the system but holds a relationship to it. The relationship between system and environment can be described in terms of input and output. It is worth noting that not every input generates an output and that the period of time between input and output can be very long. The system concept is not only applicable to physical phenomena, but encompasses an area of phenomena (inclusive language) as vast as possible.

There is a difference between open and closed systems. In the case of closed systems the final position of a system (equilibrium) is quite totally dependent upon the starting position such as in the case of certain chemical reactions. This is not the case in open systems. In such a system the position of equilibrium is much more dependent upon the characteristics of the whole system instead of being dependent upon specific changes of its internal elements. Examples are the composition of the blood or the temperature of the body. This tendency of open systems to maintain a certain state of equilibrium is called homeostasis.

In general, a system can be divided into subsystems that are mutually dependent and that can be conceived as environments to each other. Sometimes a clear hierarchy between these subsystems is assumed in which functions and processes which are situated at one level use the possibilities and characteristics of other levels. In the systems paradigm the simple causal scheme (stimulus-response) has been elevated to a more complex concept of causality and interdependency between levels, in which cybernetics and its feed-back mechanisms are operating.

More recently, there has been a shift of attention toward building theories about open systems in which free will and individual choices and less rigid concepts of environment are involved. There have been significant developments in the epistemological foundations, the mathematical and statistical aspects in linking the different system levels to each other and in situations of disequilibrium.

C. The Existential-Anthropological Paradigm

The existential-anthropological paradigm refers to a systematic body of theoretical concepts which underlie many intellectual schools or traditions. Representatives of this paradigm can be found in the traditions of psycho-analysis, existentialism, symbolic interactionistm, symbolic medical anthropology, modern psychotherapeutic schools, phenomenology, etc. A certain area of witness literature about health and illness can also be regarded as part of this paradigm.

By "existential" we mean that the living experience is seen as very important. That is, meaning, structures of significance, consciousness, human values, interpretations of reality and emotions, are recognized as important constituent parts of fundamental knowledge. A qualitative "humanistic" approach is proposed. This leads in this paradigm to an emphasis on the uniqueness of the individual and his world of living experience. A basic assumption is that the human subject is creator of the world in a mutual dependent way with other people. In the field of medicine and health science this implies an appreciation of the individual biography and the significance of life events, crises, emotions, etc.

D. The Culturological Paradigm

The culturological paradigm is allied to the existential-anthropological paradigm by emphasizing human experience, but differs from it on core elements. The focus on the individual is broadened to include the societal and cultural roots of health and illness. By enlarging the focus it has a certain affinity with the systems paradigm. The cultural paradigm has its roots in the critique of culture, the sciences of history, macro-sociology and social and cultural anthropology. It is less developed and less systematic than that of the biomedical paradigm or systems theory. Nevertheless many scientists are searching in this direction.

The culturological paradigm stresses, not so much the discovery of universal laws, but that reality is a process. It stresses the variability of phenomena. Historical comparative research is therefore a favourite element of theory building. Comparisons are not only made between populations, but also cross-culturally. The culturological paradigm aims to cultivate a critical-reflexive attitude within the everyday practice of research and looks for ways to integrate more explicitly the social-cultural influence.

Investigating the relationship between culture and behaviour has a high priority in this paradigm. Culture refers to existing and operating values, conceptions, norms, targets and expectations in society. These have mainly a dynamic character (patterns) and are collectively shared by people. The relationship between culture on the one hand and health and illness on the other hand is complex, dialectical and multidimensional. Culture influences the definitions of health and illness and the concrete situations in which these labels are applied, namely the concrete processes of ascription of the ill-role, perceptions of the causal determinants of illness and health, the incidence and prevalence of illnesses, the experience of being ill and the way to deal with it, coping with emotions and pain, the collective and individual patterns in relation to care in health and illness, and life-style itself. Critical objections have been raised towards the spread of preventive measures because they could legitimize processes of social disciplining. Healthy life-style programs (such as yoga, macrobiotics, health education programs, transcendental meditation, etc.) can be seen as an indication of the increasing individualization process and the lack of meaning and spiritual values in contemporary societies .

3. DIVERSITY OF HEALTH ADVICE

Every paradigm generates its own definitions of health and disease, scientific and observational procedures, theoretical models of interpretation (theories), treatment procedures, preventive measurements and conceptions about the health care system and social and political policy issues (figure 2).

The matrixfields in Figure tow refer to the paradigms of Figure 1. Matrixfield 1 is the biomedical paradigm, Matrixfield 2 the existential-anthropological paradigm, Matrixfield 3 the systems paradigm and Matrixfield 4 the culturological paradigm. One practical problem is that the advice according to each paradigm may be radically different or contradictory. For example, disciplinary measures(Matrixfield I) might be in contradiction to the principle of self-realization (Matrixfield II). This diversity will increase in proportion to the development of new theories. There is already a surfeit of health advice which hinders a real and good choice. The same problems were uncovered by research on self-care techniques (Lafaille, et al. 1981–1985; 1984). There are so many possible self-care techniques that one needs considerable expertise to cope adequately with this bulk of information. Neither criteria of scientific research, nor leaving this problem to personal choice, are sufficient conditions for a good solution. In proportion to the development of new theories these centrifugal forces will increase. At this moment already we can see a mer à boire of health advice which hinders a real and good choice.

On the level of the foundations of a new science of health, we face a lot of problems. From the other hand, scientists are working on it, and I am convinced that for a lot of problems we will find solutions within a time span of 5 to 10 years. Especially I men-

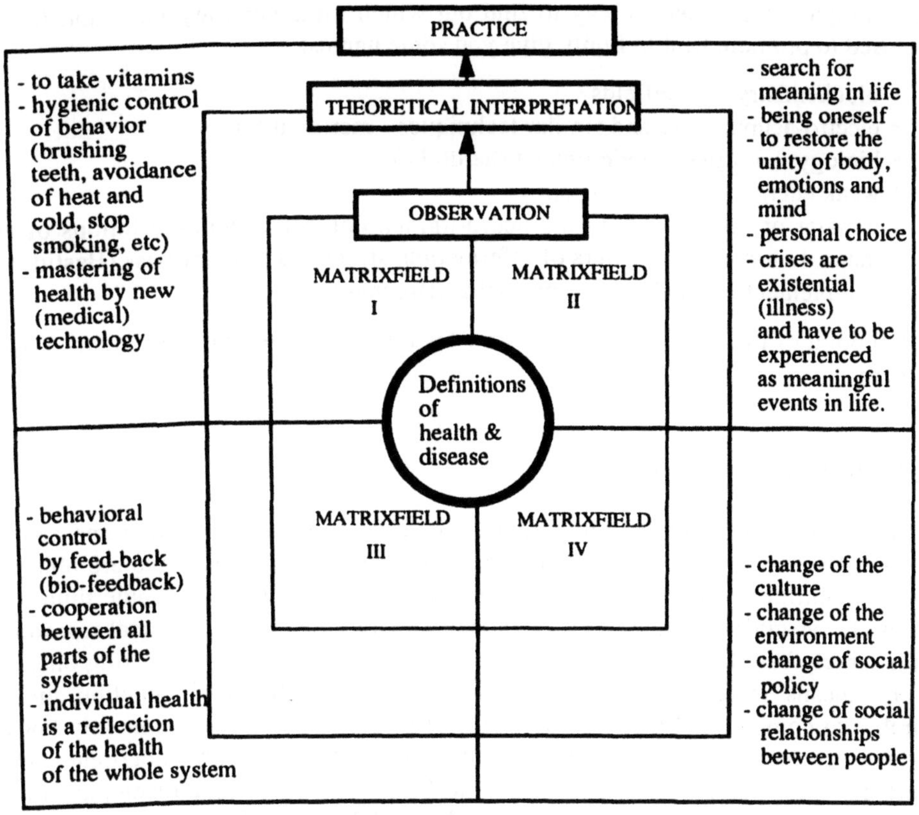

Figure 2. Relations between paradigms, observations and health practices.

tion here the need for meta-models who are able to integrate various views and theoretical traditions and the need for an appropriate biographical methodology. I will speak about this last point in another lecture. I'm quite optimistic, although the circumstances in which people have to develop the new are mostly not very comfortable. Many researchers are outside the academic institutions, because the current academic subculture there blocks a lot of developments.

4. SOME RECENT CONTRIBUTIONS

I now will turn to another point. If a new science of health rises up, what kind of concrete research could be a part of it? I will give you a brief overview and mention some examples. It will be clear that, although I invest a lot in detecting new developments, my possibilities are limited and will always to a certain extend be determined by personal taste.

Weak Energies Research

One of the most promising developments in medicine and the health sciences is—according to my view—the study of weak, subtle energies. In various scientific fields research projects are under its way to find out which influence they have and how they can be used to promote health. I only give you some names:

- biochemistry - forcefields
- electro-acupuncture and similar techniques - meridian forces
- biology – morphogenetic fields (Sheldrake)
- weak energies emissions (Popp)
- transfer of information in the neuro-immunological system - dilutions of Hanemann, non-material carriers of information, structures of emptiness (Bastide)
- complementary medicine – aura fields (Brennan)

A special point of interest here is if, and how far weak energies can be influenced by consciousness.

Research into Complexity

Science becomes more and more able to investigate situations of extreme complexity. Especially through the computer and statistical software packages we are able to investigate the influence of tens of variables and their interactions at the same time. Heath advice is always given in situations of great complexity. The current classical research tradition is largely based on simplicity of (causal) relationships (like comparising only averages and means instead of systematically working out structural equation models). Research into complexity will bring us also more insight in what is commonly known - and which very commonly is condemned with a kind of denigratory undertone - as placebo effects. Since for certain illnesses the placebo effect is very high, medicine will have to develop in the future a very effective and skilful "placebo" effect - how contradictory this may sound for many health professionals at this moment. Placebo-medicine will be holistic in nature. It will encompass at least the following levels (Figure 3).

There is still a very open space in the market here. This type of medicine will be fully scientific, since the effectiveness of its interventions could be proven by empirical research.

Research into Life Style Changes

Especially epidemiology has offered a lot of material through which it becomes clear how great the impact of life style on health is. This opens the opportunity to ameliorate health by changing one's life style. Healthy life style programs are the favourite instruments here. Dozens of programs exists, some of them have proved to be effective by scientific research. To illustrate some of the advantages of healthy life style programs, I will use yoga; a topic on which I carried out some research. I found the following effects (with this remark that I want to put aside the question of control groups, because this would lead us to a too technical discussion) (Figure 4).

The black dots give the percentage of psychosomatic complaints at the beginning of practicing yoga (first measurement during the first three months of practicing yoga), the open dots refer to the percentage of psychosomatic complaints after one year of practicing yoga. The results show clearly that the number of persons without complaints increase (from 25 to 50 %), and that as an average the % of complaints decrease. The differences are statistically significant.

Biological variables

• psychoneuroimmunological reactions on transfer of informations • body-mind interactions • transfer of (weak emission, subtle) energies • learned conditional reflexes

Patient variables

• type of illness (incl. psychiatric disorders) • suggestibility • the personal reaction to the stimulus of the placebo • learning and socialisation • expectations about treatment • characteristics like age, sex, intelligence • patterns in and stage of inner development • anxiety • faith and hope • personality traits • mind-body relationship • somatization • victim/participatorship in life • defence mechanisms • motivation • meaning • guilt dynamics • unconscious processes • self-concept • level of experienced stress • dreams, inner images, fantasies • conditioning of the intake of certain pills (intake rituals) • positive expectation in relationship to the substance • attitude vis-à-vis medical authorities • loneliness

Situational variables

• treatment procedures • size, colour and shape of tablets or capsules • social and physical context of treatment (e.g. clinic, GP practice or laboratory) • general life conditions of the patient • influence of other patients(exp. influence of treated on untreated patients).

Patient-Physician variables

• attitude of the physician towards his patient (interest, warmth, friendliness, neutrality, lack of interest, rejection, hostility, etc.) • empathic abilities of the therapist • transfer of attention, sympathy, status, etc. • redefinition of social roles • competence and ability of suggestive influencing by the therapist • authority of therapist; social prestige • (general) therapeutical competence • physician's expectation of improvement, and more general his attitude towards treatment (faith, belief, enthusiasm, conviction, commitment, optimism, positive and negative expectations, scepticism, disbelief, pessimism, etc.) • dominance - submission • conviction about the effectiveness of treatment • elicitation of catharsis • the theoretical frame of reference of the physician and his belief in it • the factual interaction between physician and patient • the physician's attitude towards his own competence: anxiety, feeling of guilt and inadequacy • communication of reassurance about fundamental fears and uniqueness "I'm allowed to be who I am".

Social relationships in which the patient lives

• quality of the partnerrelationship and family life • crises in relationships • loss of partner, family members or friends • reactions of others on the treatment; acceptance of rejection of the sick role • quality of social and work relationships • quality of friendships • environmental changes

Figure 3. Overview of determining factors in the placebo effect.

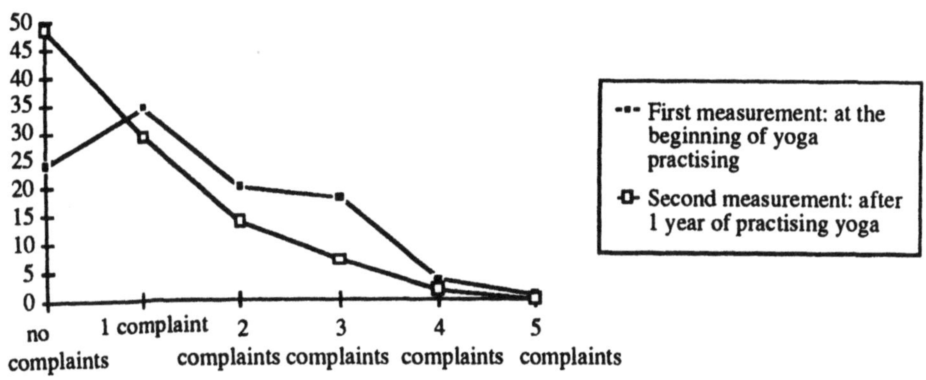

Figure 4. Evolution of psychosomatic complaints due to yoga practising (year 1986–1987).

Research into Meaning in Life

There is a lot of evidence that meaning in life and certain inner developments have a great impact on health and illnesses. Unfortunately, for various reasons the research in this field is still not very advanced. Especially we lack a solid biographical methodology to observe and investigate these themes (Lafaille & Lebeer 1991; Lafaille, Lebeer & Mielants 1995; Lafaille & Wildeboer 1995). Inner development is to a certain extent related to inner images, which show themselves in dreams, fantasies, imagery, meditations. Its precise relationship to developmental processes is still largely unclear, but investigators with experience in this field don't doubt the presence of such a relationship. The best way to explain what an inner process is, is to show it with an example. I take the example of one of my friends, which is described by Weil in his scientific bestseller *Spontaneous Healing*. It is the case of Mr. Terayama, a former cancer patient, who cured from his cancer in accordance with a deep spiritual crisis. The example shows an inner development which started as a person who is "victim" of his illness. The illness is experienced as something that comes from the outside, on which he cannot exert any control. Later on the illness was the carrier of a process of liberation of authority and a passway to deep spiritual experiences. The key of the process is that he went through a deep, in his case a spiritual crisis, and simultaneously the cancer disappear. More of these cases are described in the medical literature. I do not want to go into a discussion about the aetiology of cancer here. There is still a lot unknown. But it is an interesting line of thought. It leads to new hypotheses which should be proven by scientific research. I also want to make the following remark. If these kind of processes do not play a part in cancer, in many other illnesses they do. So the new science of health will have to deal with these topics.

Research into Body Awareness

Since two decennia more and more research shows the influence of psychosocial factors in many diseases. I will focus here on coronary heart diseases which is a serious treat to health especially during mid life. Social isolation, loneliness, depression, etc. are recognized as harmful risk factors for these diseases (Lynch 1977 & 1979; Frasure-Smith et al. 1993). On a parallel route besides this aetiological research, intervention studies were carried out, which showed that certain types of life-style modification could ameliorate the health status of patients with coronary heart disease. According to research of Ornish, it would be even possible to reverse the coronary stenosis. Intervention programs which proved by scientific research to be effective are Soon, Ornish, Van Dixhoorn, Frasure-Smith, Thoreson. A meta-analysis of 15 different studies showed that these kind of interventions are significantly effective.

I will expand here on the method of Van Dixhoorn as an example of this kind of approach. He developed a new type of treatment in the field of cardiac rehabilitation which consists of a combination of relaxation and breathing techniques. The intervention consisted of six individual treatment sessions of one hour, in which the patient was taught relaxation and breathing exercises. Also some guidance on how to cope with tensions in daily life was part of the program. This kind of intervention was added to the regular rehabilitation program of patients in the cardiac rehabilitation department of a general clinic. Now this method is one of the three rehabilitation programs which have been recognized and adopted by the Dutch National Heart Association. This method has been fully described (so far only in Dutch available).

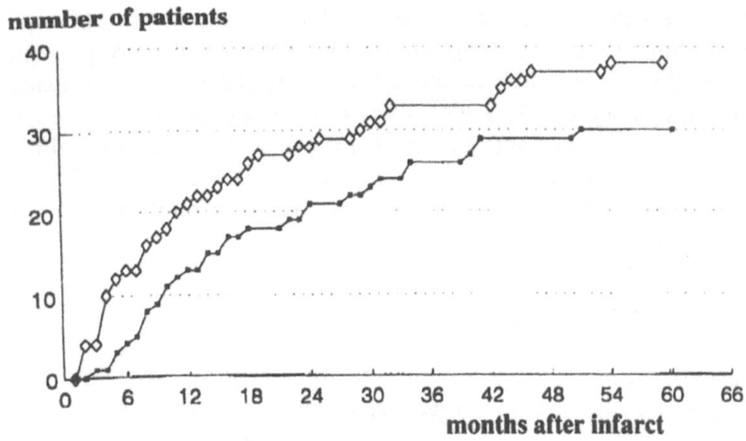

Figure 5. Cardiac rehospitalisation and number of deaths. ● = relaxation + bodily exercises; ◇ = bodily exercises alone.

One of the findings of the research was that this method reduced significantly the number of cardiac rehospitalisation days, especially during the first two years (van Dixhoorn, 1987), especially for angina pectoris. There showed to be significantly lesser open heart surgery – especially during the first year. The amount of cardiac mortality decreased slightly (not statistically significant), but the moment of death shifted to a later date. In average 42.8 months in the experimental group compared to 24.1 months for the control group.

The economic gains were calculated—and the results are very supportive for the developed method. More people went back to work, medical consumption decreased, especially for the more expensive care as hospitalisation and heart surgery. An estimation shows that for the totality of the 156 patients in the study 288 200 Dutch guilders were paid for treatment in the experimental group, against 576 950 guilders for the control group; or 7212,- guilders against 3879,- guilders for each person in the experimental group. A reduction to 54% of the original costs (figure 5).

At this moment, further research is needed to evaluate the preventive potential of these kinds of methods.

Research into Chronic Diseases

As is widely known the amount of chronic diseases will increase significantly in the population. Here there lies a space open for innovative therapies and prevention. By the following example, I will show that in this field much more is possible, than one generally accepts. My friend and colleague Jo Lebeer investigated exceptional development of brain damaged children. Building further on the work of Feuerstein, Pëto, Doman, etc. he developed a new method consisting of new therapeutical interventions as well as a supportive system for the parents to reach optimal development of these children. I could give examples of possibilities here in different fields. Other applications certainly can be developed by the use of similar principles. At least in the following areas is much more possible than is classically offered: cancer, rheuma, high blood pressure, authism, neural pains, coma, etc. Also want to mention the research of Prof. David Aldridge and his team at the Witten/Herdecke University in Germany who developed a method to bring people out of coma using a combination of breathing and music therapy. For many patients this is an incredible solution. Further research is needed to determine the precise category to which this method applies. Very briefly I also want to mention the field of chronic pain. For chronic ill, this is a very important domain. A lot can be done in this field. It mostly asks for intensive guidance. A kaleidoscope of innovative treatment methods is collected by a workgroup, sponsored by WHO. It's has been published through the excellent edition of Annette Kaplun (1992) from Switzerland.

Research into Healthy Cities

WHO formulated some years ago the idea of a healthy city. Now there is a world wide network of cities who engage themselves to realise this idea. At our Institute in Antwerp we founded this year an industrial think tank. Health scientists and captains of industry meet there in order to discuss and to cooperate about new technologies which are in concordance with the principles of the Charter of Ottawa and the health policy as promoted by WHO. Although many policy makers mostly hold a not very optimistic view about opportunities to create a healthy environment, I am convinced that this pessimism—they call it very often "realism"—is not very appropriate. It reflects more the current difficulties to turn down certain power games in society, than that it has any technical foundation. I am very optimistic that we are able to develop in the short run technologies that save the ecological environment to a very high degree. I give only two examples to support my optimism:

- The Welbi-oven is a new type of incinerator for waste (developed by Mr. Marc Keersmaekers and he has already taken out a patent on it). It has the following characteristics. A new type of drum oven in which oxygen is injected. This augments the efficiency of the combustion of waste. By the recycling of the fume gas energy is used which would otherwise be lost. The emission of the combustion process is practically for 100% clean. Almost absolute avoidance of the emission of dioxin. At least measures will be reached low beyond the values of the ovens which are operating at this moment. It has a tremendous temperature limiting effect, although the combustion performance is still increased, which leads to a serious reduction of costs. At this moment a new version is under study, which will optimalise further the combustion process.
- On the first of January 1997 a new control method for catalysators will be installed in the European Union (the so-called "four gas tester"). Together with new types of catalysators, it then will become possible to reduce the emission of polluted materials with ± 95 %. On Internet we found that Mercedes Benz is already able to built a car who works on hydrogen. Television shows races with cars working on solar energy. In our industrial think tank, we are working on an overall plan, a kind of combination of policy measurements and new technologies, which should reduce pollution by cars almost to 100%.

I am strongly convinced that when politicians and the population really choose for it, every town with average characteristics can be made a "healthy town" in a very short time. The technology is available, can be prospected, can be developed in a very short time. And with "short", I mean within the limit of a 5-year period. New types of evaluation research—e.g. with the help of urban health planning—allow to register very accurately the health state of a town, detect developments and allowing local policy to adjust their policy to that. In the field of non-industrial pollution, it is more a question of making proper choices, than waiting for scientific or technological innovations. In my part of Europe, I still can observe a lot of resistance to make these necessary choices. The background of this resistance is clear, although there is no alternative. The survival of our planet is at stake. One also has to consider the following. In the near future national and local areas will compete not only in the field of the classical economic goods, but also in regard to the costs to maintain an ecological equilibrium. Areas who make these choices first, may be later ahead in this field of new economic competition.

5. CONCLUSIONS

I could mention much more research and topics. But I guess the material I have given is sufficient to show that a new science of health is growing. Many researchers explicitly or implicitely are engaged in this process. There is a lot of resistance from the professional world by defending traditional classical ideas. Nevertheless, if the new research is in real connection with reality, this means if really biographical changes, life style changes, healthy cities etc. bring more health to people, then we cannot doubt who will be the winners in the long run. Scientists have to be honest and modest in relation to reality. In a process of change many times it is difficult to determine what is real valuable and what not. So many ideas are formulated. Many of them are very speculative. There is nothing wrong with that. It's a signal of creativity. But only those ideas who have a strong connection with reality will survive. This is true for old and for new insights. I am personally very hopeful about the future of the new science of health.

6. ABOUT THE AUTHOR

Robert Lafaille, sociologist, used to lecture in general sociology and sociology of welfare at the University of Tilburg and was staffmember of the Institute of Family Medicine of the University of Antwerp. Was visiting professor at the Fachhochschule Magdeburg. He now coordinates the International Network for a Science of Health. He received his Ph.D. from the University for Humanistic Studies in Utrecht for a dissertation on healthy life-style programs. His main areas of research are health theories, healthy life-style programs, history of health and life-style, methodology of biographical research, self-help and self-care techniques, meditation and theories of social problems.

7. REFERENCES

Beck, U.(1986) *Risikogesellschaft. Auf dem Weg in eine andere Moderne*, Frankfurt: Suhrkamp.

Capra, F, (1975) *The Tao of Physics*, Berkeley: Shambhala.

Capra, F, (1986) Wholeness and Health, in *Holistic Medicine*, vol.1, p.145–159

Crawford, R.(1980) Healthism and the medicalisation of every day life, in *International Journal of Health Services*, vol.10, nr.3, p.365–388.

Diemer, A.(1978) *Elementarkurs Philosophie. Philosophische Antropologie*, Düsseldorf.

Dijksterhuis, E.(1989) *De mechanisering van het Wereldbeeld. De geschiedenis van het natuurwetenschappelijk denken*, Amsterdam: Meulenhoff.

Elias, N.(1969) *Uber den Prozess der Zivilisation. Soziogenetische und psychogenetische Untersuchungen*, 2 vol., Bern and München: Francke Verlag.

Elias, N.(1990) *Ueber sich selbst*, Frankfurt am Main: Suhrkamp.

Europäische Monographien zur Forschung in Gesundheitserziehung(1984) nr.6, Wien: Bundesministerium für Gesundheit and Umwelt Schuz.

Foucault, M.(1971) *Madness and Civilization: a History of Insanity in the Age of Reason*, Routledge, London.

Foucault, M.(1973) *The Birth of the Clinic: an Archeology of Medical Perception*, New York: Vintage Books.

Foucault, M.(1976) *Mental Illness and Psychology*, New York: Harper.

Göpel, E.(1987) *Lebensmodelle und ihre methodischen Konsequenzen für die Gesundheitsbildung*, Materialien des Oberstufen-Kollegs, Bielefeld: Oberstufen-Kollegs, Universität Bielefeld.

Hampden-Turner, Ch.(1981) *Maps of the Mind. Charts and concepts of the mind and its labyrinths*, New York: Macmillan.

Helman, C.(1984) *Culture, Health and Illnesses. An Introduction for Health Professionals*, Bristol/London/Boston: Wright.

Hoefnagel, A.H.J.M.(1977) Systeembenadering en sociologie, in: Rademaker en Bergman (red.), *Sociologische stromingen*, Het Spectrum/Intermediair.

Hurrelmann, K.(1988)*Sozialisation und Gesundheit. Somatische, psychische und soziale Risikofaktoren im Lebenslauf*, Weinheim /München: Juventa Verlag.

Ingrosso, M. (ed.)(1987) *Della prevenzione della malattia alla promozione della salute*,Milan: Franco Angeli.

Jantsch, E.(1986) *Die Selbstorganisation des Universums. Vom Urknall zum menschlichen Geist*, München: Carl Hanser Verlag.

Kaplun, A. (1992) *Health Promotion and Chronic Illness. Discovering a new quality of health*, Copenhagen: WHO Regional Publications, European Series, No.44.

Kickbusch, I.(1989) *Good planets are hard fo find*, WHO Healthy cities papers, nr. 5, Copenhagen: Fadl.

Kleinman, A.(1980) *Patients and Healers in the context of Culture*, Berkeley: University of California Press.

Lafaille, R. and Fulder, S. (1993) *Towards a New Science of Health*, London: Routledge.

Lafaille, R. and Hiemstra, H.(1990) The Regimen of Salerno, a contemporary analysis, in *Health Promotion International*, vol.5, Nr 1, p.57–74.

Lafaille, R. et.al. (1981- 1985) (eds.), *Zelfhulptechnieken* (Self-Care Techniques), loose-leaf edition, 3 vol., Deventer/Antwerp: Van Loghum Slaterus.

Lafaille, R. (1996) The Placebo: Mysterious Forces or Investigating Complexity, in Høg, E. and Olesen, S.G. (Ed.) *Sudies in Alternative Therapy 3. Communication in and about Alternative Therapies*, Copenhagen: INRAT/Odense University Press, p. 127–148.

Lafaille, R.(1984) Self-help as Self-Care, in S.Hatch and I.Kickbusch (eds.), *Self-Help and Health*, Copenhagen: W.H.O. , p.169–176.

Lafaille, R.(1991) *Programma's voor Gezonde Leefwijze. Een aanzet tot een culturologische en synthetische analyse* (Healthy Life-Style Programs. A Contribution to the Development of a Culturological and Synthetic Analysis), Antwerpen: International Institute for Advanced Health Studies.

Lafaille, R., and Lebeer, J. (1991) The Relevance of Life Histories for Understanding Health and Healing, in *Advances*, vol.7, nr.4, p.16–31

McKeown, Th.(1979) *The role of medicine. Dream, mirage or nemesis?*, Oxford: Basil Blackwell.

Niehoff, J.-U. en Wolters, P. (1990) *Ernährung und Prävention. Köpergewichte - ein Beispiel präventionstheoretischer Probleme*, Berlin: Wissenschaftszentrum Berlin für Sozialforschung.

Noach, H. (1992) Conceptualizing and Measuring Health, in Badura, B. & Kickbusch, I. (Eds.) *Health Promotion Research. Towards a New Social Epidemiology*, WHO regional publications, European series, no. 37, Copenhagen: WHO.

Parsons, T. (1951), *The Social System*, New York: Free Press of Glencoe.

Peeters, H. (1978) *Historische gedragswetenshap. Een bijdrage tot de studie van menselijk gedrag op de lange termijn*, Meppel/Amsterdam: Boom.

Robes, J. (1988)*De gezonde burger. De gezondheid als norm* (The healthy Citizen), Nijmegen: SUN.

Schipperges, H. (1985) *Homo Patiens. Zur Geschichte der kranken Menschen*, München/Zürich: Piper.

Verbrugh, H. (1978) *Paradigma's en begripsontwikkeling in de ziekteleer*, Haarlem: De Toorts.

Verbrugh, H. 1983, *Nieuw besef van ziekte en ziek zijn. Over veranderingen in het mensbeeld van de medische wetenschap*, Haarlem: De Toorts.

Vries, M. De (1985) *Het Behoud van Leven* (The Preservation of Life), Utrecht: Bohn, Scheltema & Holkema.

Weil, A. (1992) *Spontaneous Healing*, New York: Knopf.

3

HEALTH AND HEALING

Developing Our Potential

R. Newman Turner

1 Harley Street
London W1N 1DA, United Kingdom

ABSTRACT

Health has been defined as the realisation of the capacity for maximum enjoyment of life with or without infirmity. In spite of having greater physical, biochemical, and spiritual resources to enable people to attain and maintain health, the burden of chronic disease and the crisis of the immune system continue unabated.

An examination of the reasons we become unwell suggests it is no longer sufficient to confront the agents or symptoms of disease or even to facilitate the inherent self-healing mechanisms. With so much now capable of overwhelming these it has become essential to examine and develop ways in which the healing response can be potentiated.

While the means to enhance healing now exist at the biochemical, physical, and psychological levels perhaps we also need to develop our own potential as healers. But how?

This question is fundamental to the theme of the Third International Dead Sea conference.

INTRODUCTION

The Dead Sea Conferences were founded on the belief that health is the birthright of every individual and that the means of achieving it should be available to all without restraints other than those of sound clinical judgement. Implicit in the spirit of integration which runs through these conferences is the participation of exponents of many disciplines which all share a respect for the natural environment of mankind, both within and around the body, and a recognition of the inherent self-healing mechanism, the *vis medicatrix naturae.*

Throughout this series of conferences we have explored many of the key issues which underpin our theme of integration and, by taking another look at some of them, I hope to show that we must not only understand how to use the tools at our disposal but, perhaps, examine how we might develop our own capacity to apply them more effectively.

To take up the theme of this particular conference, I also intend to look at the idea of potentiating health. Is it necessary? Can it actually be done?

HEALTH

We first need to decide what we mean by health. Our individual aspirations may vary enormously. For one person health may be the ability to take a pain-free stroll in the park each day; for another it may be the ability to run a marathon. This is best expressed by the World Health Organisation definition:

> 'Health is the ability of each individual to enjoy the maximum capacity for enjoyment of life with or without infirmity'

So why are we still so far short of the WHO objective of 'health for all by the year 2000'? It is not just a matter of providing health care to under developed nations but the state of well-being of even those who have abundant resources. In spite of having greater physical, biochemical, and spiritual resources to enable people to attain and maintain health, the burden of chronic disease and the crisis of the immune system continue unabated. The remarkable technical resources of the late 20th century may be a mixed blessing. The dazzling benefits they have brought may have blinded us to the inner resources with which nature has provided us.

There has been a tendency for all of us in health care, both complementary and mainstream, to become like the occupants of the Tower of Babel - all shouting our wares from our own window without much idea of what is going on around us. On the other hand we cannot ignore the fact that it is technology which has enabled us to tune in to and explore, in various ways, the very nature of some of our inner resources. We can measure the waves of the brain's activity under various states of consciousness; we can photograph the aura or energy fields around the body; and we can tease out the most intricate detail of our genetic blueprint.

In spite of the march of technology and the dominance of scientific materialism, there is probably more widespread awareness of the spiritual dimensions of our lives than ever before. (It has to admitted that one reason for this is its dissemination through the books and instruments of mass communication which are part of the technological revolution.)

The problem is that there is still a great divide between the humanistic and the technological aspects of health care. One of the aims of the DSCs is to find ways of bridging this gap.

WHY WE BECOME UNWELL

If we examine the reasons why people become unwell it is clear that there are tremendous forces massed against the inherent self-healing capabilities of our bodies.

We come into the world with a genetic endowment which is estimated to account for up to thirty percent of our potential for well-being in life. The burden of pre-natal factors such as our parents health and lifestyle, nutritional status, and environment, are superim-

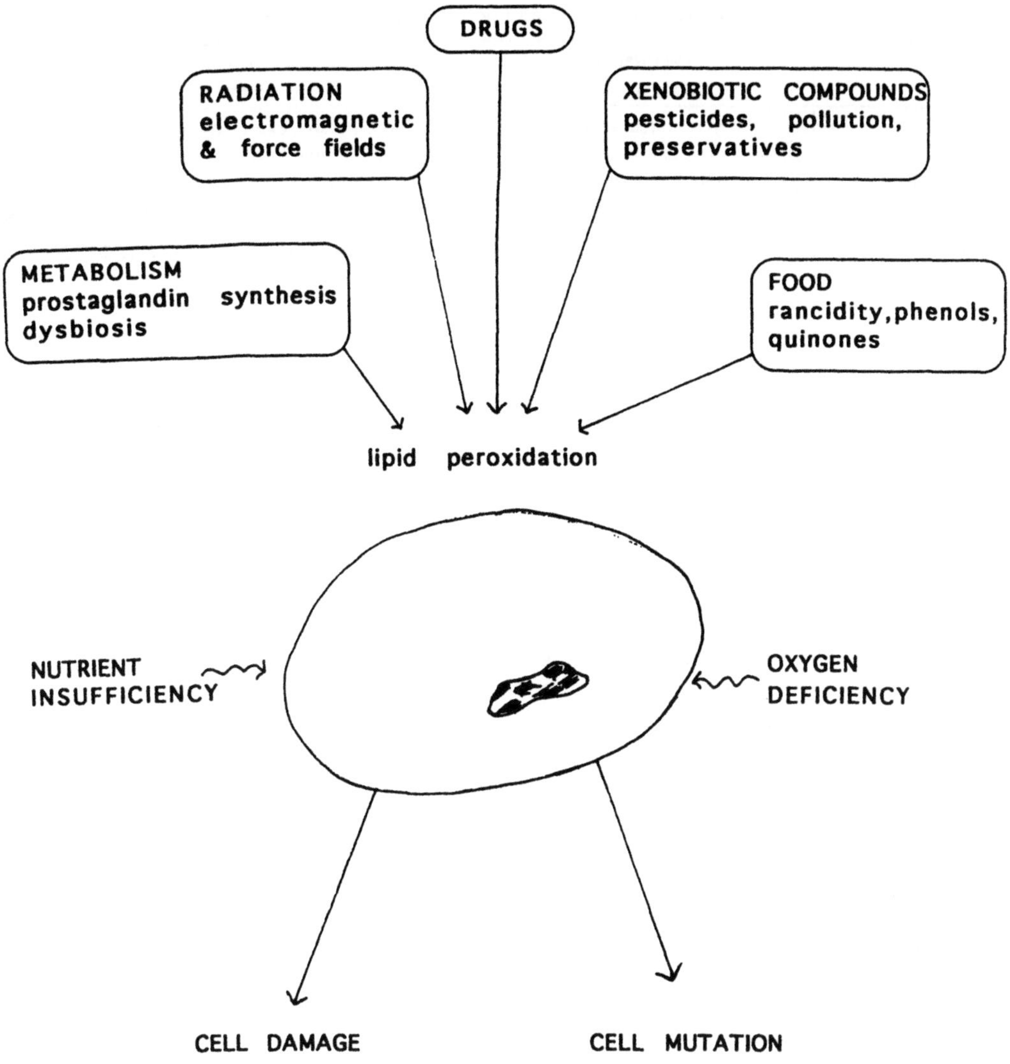

Figure 1.

posed on any genetic susceptibilities. After birth we are exposed to a wide range of challenges, both external and internal, which contribute to the process of mesotrophy the 'slow but imperceptible decline in the health of the cell'.[1]

Much of this comes from the burden of free radical compounds to which our modern lifestyle exposes us. Free radicals are ubiquitous in the environment and within our bodies. Some play a vital part in energy production, others can form highly toxic compounds. Pollution, electro-magnetic fields, chemicals in food and medicines, are all sources of these compounds which, if not adequately neutralised by the detoxification mechanisms of our body, may attack the lipid membrane of our cells causing damage or mutation[2] (See figure 1).

Health care which deals only with the consequences of these causative factors without supporting the individuals ability to deal with them will only generate a downward

Table 1. Facilitation and potentiation procedures

	Facilitation (homoeostasis)	Potentiation (heterostasis)
Structural	Correction of mechanical obstacles to function by osteopathy, chiropractic, etc.	Hydrotherapy, exercise—increase hormonal and immune function
Biochemical	Providing or modulating nutritional requirements for healthy metabolism	Orthomolecular nutrition, instigating Law of Mass Action—stimulates metabolic function
Emotional	Insights and resolution of conflicts	Visualisation—potentiates immune response

spiral into chronic disease. We, therefore, have to contend with an underlying vulnerability to disease which calls for more than the restoration of self-healing processes.

THERAPEUTIC PROCESSES

Therapeutic action can be either confronting or catalytic. Confronting procedures can often be life-saving in their short-term speed and efficiency. Catalytic measures, though, are also essential to survival in the longer term. One has only to think of the necessity for support of the immune system in the face of many of today's illnesses.

The discoverer of the General Adaptation Syndrome, Hans Selye, described the process of self-regulation and healing as *homoeostasis* and the potentiation of that process as *heterostasis*.[3] Therapeutic action can, therefore, be passive or facilitating, that is allowing homoeostasis, or active and potentiating as in heterostasis.

Many complementary therapies aim to facilitate the self-healing of the body by removing obstacles to its action or providing the nutrients or other raw materials it needs. There is also evidence that a number of natural procedures can actually enhance the defensive and healing capabilities of the body. A summary of homoeostatic and heterostatic procedures is given in table 1.

In considering the healing process we must not overlook the power of the placebo. This is greater in the more impressive procedures such as acupuncture or surgery, the latter having produced significantly better results than the oft-quoted thirty percent benefit. In the now notorious (by todays ethical standards) comparison of internal mammary artery ligation and sham operation for angina all patients did equally well.[4] Clearly placebo is a factor which we should embrace as part of our therapeutic armamentarium.

DEVELOPING POTENTIAL

An important part of therapeutic power lies in us as practitioners. How can we release our potential as healers? Most of you are likely to be actively engaged in therapeutic work or some activity which serves it directly or indirectly. To achieve this required, in many cases, a great deal of application, allegiance to an academic tradition, even adherence, at times, to outdated dogmas. Is this always the best preparation for the task of healing? Essential though they may be there is a danger that rigid adherance to the rules and rigours of our training and disciplines may hinder our own ability to assist patients effectively.

Many of us were pitched into work necessitating the application of those principles without much time to think about anything beyond reduction of the immediate problems with which we were faced. Even those of us whose education encouraged us to think and act more holistically often found we were compromising those principles to meet the demand for short-term relief. In mainstream medicine compromise and confrontation are essential to save life. There is not much scope for working with the wisdom of a natural healing process that has been significantly challenged and overwhelmed. Nevertheless, it is still worth bringing something of that indefinable spiritual dimension to the most practical aspects of our work. It can only enhance it.

EPISTEMOLOGY OF SUBJECTIVITY

This means being able to let go of the dogmas and traditions of our scientific training. At his keynote address to the second DSC, held in Tiberias in 1993, Dr Willis Harman made an eloquent case for what he called an 'epistemology of subjectivity' in science.

> 'It will be more participatory in recognising that understanding comes, not alone from being detached, objective, analytical, coldly clinical, but also from co-operating with or identifying with the observed and experiencing it subjectively. It will involve recognition of the inescapable role of the personal characteristics of the scientific observer including the processes and contents of the unconscious mind.'[5]

What Willis was arguing was that it is possible to sustain objectivity whilst allowing the mind to be opened to the wider realms of experience. It is a matter of maintaining rigour whilst allowing our creative and intuitive faculties to expand from the firm base of scientific precision.

There is a case for what, in some circles, has been called 'fuzzy logic', which is, effectively, keeping our feet on the ground while allowing our minds to go into free association. We can also act within the more structured framework of lateral thinking. In naturopathy and functional medicine the diagnostic emphasis is on lateral rather than linear thinking. We continue to explore within the analytical paradigm but widen our database to take account of peripheral influences. It is, in essence, the approach of Significance Diagnosis, which evaluates factors that influence the patient's ability to cope as much as the immediate disease signs and symptoms. We diagnose the patient rather than the disease.

There may, however, be situations in which we need to abandon the strict adherence to a formal framework in order to allow the mind to adopt a more creative and intuitive mode - to be a little 'primitive'. It is not hampered by the rigours of analysis but can be receptive to the intuitions on the wilder side of human reason which we would normally screen out. But having once allowed them free expression they must be subject to the rules of logic. They can be considered as part of the scheme of things and either integrated or discarded in the light of the evidence we have before us.

Although a certain detachment is necessary when working with patients I am not suggesting that transcendental states are necessary to achieve this. It does help, though, if we are able to make time to get in touch with our inner being by whatever means we feel appropriate, whether it be meditation, prayer, or simply communing with nature. This is just one aspect of a creed we might all work by as healers.

AN ETHOS OF HEALING

I would like to propose an ethos of healing by which all of us engaged in healthcare would wish to live and work. These are the qualities - too often neglected - which I believe we can all cultivate to assist our patients more effectively and contribute to the advancement of the healing arts.

- **Respect**
  - for the self-healing capacity
  - for individuality
  - for the beliefs of others
  - for our environment
- **Responsibility**
  - for our actions
  - for our fellow creatures
  - in nurturing
- **Receptiveness**
  - to new ideas
- **Curiosity**
  - a spirit of enquiry
  - a willingness to explore
  - a commitment to rigorous research
  - a readiness to review
  - a willingness to question our own assumptions
- **Flexibility**
  - the willingness to change our beliefs in the light of
  - experience and evidence
- **Culture**
  - contact with literature and the arts and, where possible,
  - active participation
  - laughter
- **Insight**
  - self-knowledge and awareness of 'the inner intimation of things' (Krishnamurti)

These are qualities with which I am sure no one would disagree. The difficulty may be in allowing them a place in our lives. At least if we can keep them in mind I feel sure that our efforts to facilitate and potentiate health will gain a dimension which might have repercussions far beyond the narrow confines of our own clinical environment.

SUMMARY

The theme of this 3rd International Dead Sea Conference is 'Potentiating health and the crisis of the immune system'. I have attempted to show that the challenges to our self-healing (homoeostatic) mechanisms are now of a magnitude which requires the stronger natural therapeutic measures to potentiate the healing response (heterostasis).

In order to nurture our own potential as practitioners an Ethos of Healing is proposed by which we might all endeavour to live and work.

REFERENCES

1. Kollath,W. (1950) Uber die Mesotrophie, ihre Ursachen und Praktische Bedeutung. In: Grote,LRR.,Kollath,W. eds. Ernahrungswirkungen. (Schriftenreihe d. Gansheits-Medizin. Band 3) Hippokrates-Verlag, Stuttgart.
2. Newman Turner,R. (1996) Free radicals and disease - the toxaemia hypothesis. Complementary Therapies in Medicine 4: 43–47
3. Selye, H. The Stress of Life. McGraw Hill, New York. 1976
4. Dimond, EG, Kittle, CF, Crockett, JE (1960); Comparison of internal mammary ligation and sham operation for angina Am.J.Cardiol. 5:483–486.
5. Harman,W. (1993) An epistomology of subjectivity: its importance to personal and global well-being. (Abstract) 2nd International Dead Sea Conference, p20.

LIMITS OF SPECIALIZATION AND INTEGRATED APPROACHES

Herbert Pietschmann

Institute for Theoretical Physics
University of Vienna, Boltzmanngasse 5
A-1090 Vienna, Austria

ABSTRACT

The two opposites "specialization" and "integrated approaches" are taken as an example to explain the subtleties of a dialectic situation. In order to overcome the either-or relation in a possible synthesis, it is necessary that both sides make themselves aware of advantages as well as disadvantages of their own approach. A model is suggested to help in the dialectic process aiming at the synthesis.

1. INTRODUCTION: DIALECTIC SITUATIONS

Modern science has developed a method based on a "frame of thinking" which has proven to be very fruitful when applied to the description of matter in space and time[1]. This frame of thinking is based on the axioms of logic and of experiment. The axioms of logic (dating back to Aristotle) require the *unique definition* of all notions and the *absence of contradictions* in all statements.

In the 17th century, the scientific method has been based on the axioms of the experiment, which require restriction to *reproducible* events, *quantification* of all results and *analysis,* that is to say the reduction of complex phenomena, which we meet in everyday life, to simple elements (atomistic thinking is a most prominent example). After this was accepted, the second half of the 17th century brought about the first great success of its method together with an amendment of the axioms of logic by the requirement, that only *causal reasoning* should be used (i.e. tracing similar events to a common cause, like gravity as the reason for the attraction of massive bodies).

Due to the tremendous success of this "scientific rationality" based on the frame of thinking, those aspects of human society, which cannot be included within this frame, have been more and more neglected. The most important example is a situation, in which contradictory statements can be made where both sides have some truth in their view.

These situations shall be called "dialectic situations" and they form the central point of my considerations.

Whenever a dialectic situation comes up, it is not advisable to make a decision in the sense of either-or. In so doing, the truth which is contained in one of the two sides would thereby be lost. On the other hand, our logic tells us, that there is no other alternative. (Aristotles axiom of the excluded third, tertium non datur). Therefore, it is necessary to change the basic attitude in such a way, that a completely new level of insight or consideration can be reached. This can be achieved only by a learning process of both sides. This learning process can begin only when both sides accept the fact, that neither they possess the whole truth nor is the other side totally wrong.

To accept the necessity of learning, change or growth, is seldom easy and often even hurts. Therefore, we have developed many strategies to avoid such a genuine learning process by compromising between two dialectic sides. Before we enter into our central topic, let me illustrate this by another example: interdisciplinary work. It is my claim, that this has also developed into a dialectic situation for the following reason: On the one side, it is generally acknowledged that interdisciplinary work (research, discussions, conferences or mere contacts) are a necessary goal of our time. On the other hand, what I would call genuine interdisciplinarity, requires the communication of two distinct fields on an equal level. Since each field has its own language, its own tradition, its own paradigm, such interdisciplinary discourse soon runs into misunderstandings and possibly even mistrust, unless both sides undergo a learning process to understand the other side and thereby re-interpreting its own field in a deeper sense. Consequently, what we observe on a large scale today, are two possible compromises which avoid this learning process: the first is called interdisciplinarity, but I claim it is only faked, when one discipline acts as a supporting discipline for another one, for instance when physicists use their radiological methods to determine the age of sources or tools for historians. The second possibility of such a faked interdisciplinarity can be met in so-called interdisciplinary seminars, when various disciplines talk about one and the same subject without communicating with each other. What I would call genuine interdisciplinarity is the result of a dialectic process: the representatives of two distinct special fields try to understand the language of each other without necessarily entering into this field in the sense of active participation. To give an example: when a physicist hears the word "energy" from, say, a biologist, he or she should neither assume that the biologist has the same notion of "energy", nor should he or she claim that any other notion is simply wrong. In trying to understand the meaning of the same word in another field, he or she will recognize the limits of the domain of applicability of his or her own notion.

Let us now turn to our main topic: the dialectic situation between specialization and integrated approaches. It is quite obvious, that—on a first glance—they form an either-or-relation: we can either require specialization or integration, at least not both simultaneously. It is also quite clear, that modern times have chosen specialization as the path for progress in almost all fields of science, in particular also in medicine. But as time progresses, more and more people become aware of the fact that integrated approaches may become equally important. Therefore, the problem comes up how to find a third alternative, a synthesis between the two approaches. In order to possibly reach such a synthesis, it is necessary but not sufficient to make oneself aware of both the advantages and disadvantages of each of the two sides. This has to be done by the representatives of each side for their own case. We shall therefore now take each side and analyse it in this respect. Only afterwards shall we try for the next step: a possible synthesis.

2. SPECIALIZATION

Let us recall the historical routes of specalization in our culture. Aristotelian logic requires (in its first axiom, the theorem of identity) that every notion is uniquely defined. It also gives the rules for such a definition: a notion is defined by identifying the next higher level in the hierarchy of notions and the specific difference to the other notions on the same level. Hence the mere logical structure of notions suggests the hierarchy as the optimal ordering principle. It is inherent to hierarchical structures, that each level differentiates (specializes!) the position of its immediate superior. Thus any hierarchical order, in which the elements on one level communicate via the next higher level (and not directly with each other) leads to specialization. We find this phenomenon also in the structure of scientific disciplines (including medicine), in large organizations (be it educational, economic or otherwise) as well as in the structure of our laws and many other areas. Let us notice in passing, that the division of labour, which brought about tremendous progress for humankind, can also be considered as a kind of specialization according to hierarchical structures.

In the 17th century, Aristotelian logic has been amended by the method of science using the experiment as a criterion for the validity of general statements. It cannot be stressed enough, that this meant to give up everyday experience and direct observation as the criterion and replace it by the artificial construction of an experiment. To illustrate this, let us recall that according to Aristotelian physics, heavy bodies fall fast, lighter bodies fall slowly and firelike "bodies" (for instance smoke) do not fall but rise. This corresponds to everyday experience and direct observation. In modern physics since Galileo Galilei, it is replaced by the law of gravitation, which states that all bodies fall equally fast. No one can directly observe the prediction of this "law of nature", it has to be tested by experiments which are organized in such a way, that they can be extrapolated to the case of vanishing resistance (i.e. vacuum). In other words, the everyday world which we meet in our true life is too complicated for the derivation of natural laws. It has to be disected (analyzed) into simple elements in which these laws hold. Although this is not immediately comparable to specialization, the analytical method of science complements the afore-mentioned logical specialization. Together, they are one of the driving forces for progress in scientific understanding of our world.

Let us now try to evaluate merits and limits of specialization.

2.1. Advantages of Specialization

We have already mentioned *division of labour* as one of the positive results. It allows for the development of expertise and experts, who know every little detail in a limited field. We can find hardly anybody in our complex technological society, who has not benefited from the specialized knowledge of experts in one or the other case whenever serious problems have come up.

Hierarchical structures are so constructed, that, as an ideal, they are *free from contradictions.* Therefore, it is always possible to distinguish between right or wrong whenever a decision has to be made. It may need an expert to find out what is right and what is wrong, but the possibility always exists at least in principle, when there are no contradictions in the system. Thus we can have very large organizational structures which are extremely efficient and fast in decisionmaking. It has to be noted, however, that *right* (or wrong) is not necessarily the same as *true* (and false). In actual life, we all know situations in which we could not achieve a true goal, in spite of the fact that every single decision has been right. (In medicine, there is the common folklore characterizing such a situation by "operation successful but the patient is dead"). But even in a purely technical field such as aeronautics, there are examples where a pilot has saved his plane (and the life of all passengers) by deliberately making a "wrong move" in an unexpected critical situation. But all this leads us to the next sub-paragraph in which we look at limitations and disadvantages of specialization.

2.2. Disadvantages of Specialization

Let us take up the difference between "right" and "true". Typically, right (or correct) is a statement which can be formally proven. A formal proof (for example in mathematics) is completed, when it has been shown that there are no contradictions between the statement and its suppositions and that all notions are well-defined. (For more details see for example ref. 1). However, as we have pointed out above, a correctly proven statement may become untrue when it is applied to a concrete situation in real life. Let me just illustrate this by some typical examples: a mathematically proven statement is for example, that it is impossible to divide an arbitrary angle into three equal parts. However, this is proven under the supposition of "construction" which is defined as limitation to the use of certain instruments (ruler and circle). In practice, there is absolutely no problem to do this, either with different instruments or simply by lifting one of the above mentioned restrictions[2].

In physics, the first law of Kepler, that the orbits of planets are ellipses, can be proven from Newtons law of gravitation, however, under the assumption that there is only one planet. In the real world, no planet whatsoever has a genuine elliptical orbit.

In human relationships, it is very often necessary to convey a truth by deliberately using incorrect statements (in particular when the truth is shocking and has to be transmitted in small steps).

Finally, we have started our article by mentioning dialectic situations, in which two contradictory statements both contain part of the truth. In these situations, it is hardly meaningful to talk about right or wrong at all. In short, the truth of a statement cannot be evaluated without incorporating all the connections to the entirety of the situation at hand, in other words, a specialist or an expert will come up with right decisions but in his capacity of expert he is not able to see whether they also correspond to the truth of the situation. Specialization is *blind for connections*, decisions are made *without holistic view*, they rest on the assumption that all contradictions are necessarily errors. The chinese taoist philosopher Dschuang-dsi has put this in the words: "With an expert you cannot talk about life, for he is bound by his doctrine"[3].

3. INTEGRATED APPROACHES

Let us now turn to the second aspect: integrated approaches; a cooperation with specialization cannot possibly be reached, unless both sides are truly aware of its advantages and limitations, as I have tried to point out. It is tempting, to simply take advantages of the other side as disadvantages and vice versa. However, in so doing, we would suppress the essentials of a dialectic situation, in which the counterpart cannot simply be reached by negation. Let us therefore look more closely into the goals of integrated approaches.

3.1. Advantages of Integrated Approaches

Integrated approaches aim at the totality of a given problem in the sense that no part of it should be considered totally unimportant. If contradictions between various parts come up, they are therefore not considered to be errors or mistakes, nor are they hindrances for the holistic view. On the contrary, the contradictions may be viewed as essential source for development into a new position, a new view of the given problem.

"Decisions" in the integrated approach aim at a consensus between all participating parts. They are therefore not "ruling in" and "ruling out", but try to "incorporate".

3.2. Disadvantages of Integrated Approaches

We have said above, that in the integrated approach contradictions *may* be viewed as essential source for development. However, there cannot be any doubt that some contradictions are simply errors or mistakes. If in such a case the contradiction is taken to be essential, it turns out to be nothing but a hindrance for the decisionmaking process. Thus, the integrated approach cannot immediately attack a given problem, it has - in a kind of preface-action - first to decide which of the contradictions are simply errors and which of them are necessary in order to incorporate all parts of the problem. Hence there are two kinds of possible conflicts: those which stem from essential contradictions and lead to a growth process and those in which there is disagreement whether a given contradiction is a mistake or essential. This may lead to very cumbersome decisionmaking processes, it may take a lot of time and it may even waste time.

In some cases, it is necessary to go into all the details of a very small part of a big problem, in other words it is necessary to contact an expert. Though these problems are better attacked with the specialization approach, it may again be a source of conflicts and disagreements, in which case experts are necessary and in which case the broader view is more helpful.

For those of us (and I gather that it is the majority) who feel more at home in the specialized approach, the disadvantages of the integrated approach are exactly the reason for generally abandoning it. However, this leads us now to the next chapter, an attempt at a synthesis of both approaches or at least some kind of equilibrium.

4. A POSSIBLE SYNTHESIS

Let me reiterate, that any coexistence between the two sides specialization and integration, not to speak of a possible synthesis, cannot be achieved, unless both sides truly accept the fact that they neither possess the full truth nor is the other side totally wrong. Each one has to be aware of its own advantages and disadvantages. Only through this self-reflexion is it possible to generate the need for the other side as a complement. In order to become aware of inhibitions and hidden prejudice, I have designed what I call the **H**-model.

4.1. The H-Model

Advantages and disadvantages of each side are shown on the two vertical bars of an **H** in fig. 1. The horizontal bar of the **H** should be understood as the balance between the two sides. The question to be asked on both sides is: where do we accept the disadvantages in order to benefit from the advantages of our own side? Whenever the disadvantage of one side become unacceptable, the **H** has to be thrown into balance again by accepting the other side. To be more specific: when the elimination of all contradictions leads to a too narrow view, integrationalists have to be asked for help rather than experts. Whenever

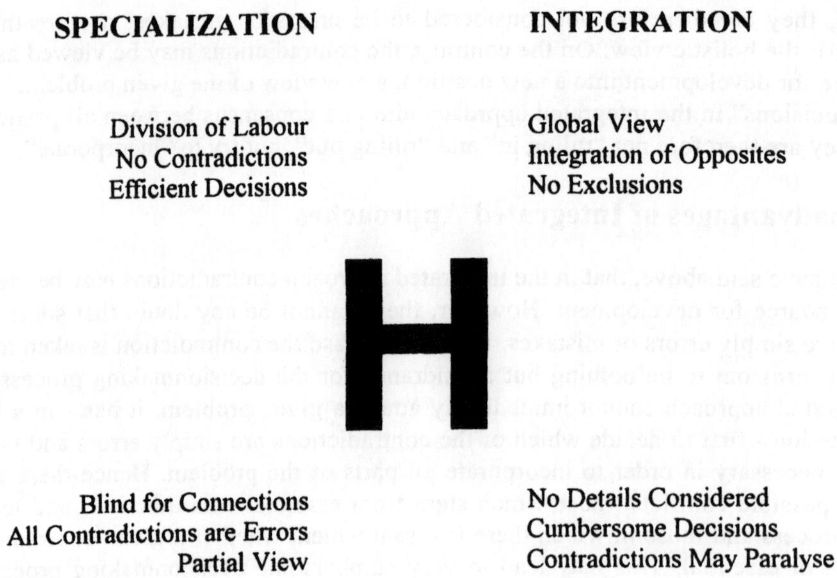

Figure 1. The **H**-Model for a balance between two dialectic partners, here Specialisation and Integration. Advantages and disadvantages are given on each side along the vertical bars. The horizontal bar symbolizes the balance.

the number of contradictions becomes so large that it blocks all further movement, experts have to be asked to analyse the problem instead of integrationalists with their holistic view.

4.2. Beware of HX-Confusion

The balance between two dialectic opposites is always fragile because of the following reason: we all have a natural tendency to suppress our own limitations, but we are glad to observe and indicate those of others. Therefore, it is much easier to reject the opposite side because of its disadvantages, rather than require its help because of ones own limitations. If this happens on both sides—and it usually does rather fast—we arrive at a state which I call **HX**-confusion (Fig. 2). Each side points out its own advantages and fights the disadvantages of the other. In this way, the own limitations and the advantages of the other side are never considered and instead of the **H** we arrived at an **X,** which by its crossing bars indicates the fierce fight. Once this state is reached, the only way out is a separation between the two opposites. It is necessary, that - possibly by a third party - each side reflects again its own advantages and limitations in order to be able to arrive at a new "**H**", a new balance between the opposites.

5. SUMMARY

We have discussed a dialectic situation, in which the two opposites "specialization" and "integrated approaches" cannot be viewed in an either-or relation, but should be brought into balance or possibly even a synthesis. The **H**-model should serve as a help for both sides, to first make themselves positively aware of advantages and limitations of their

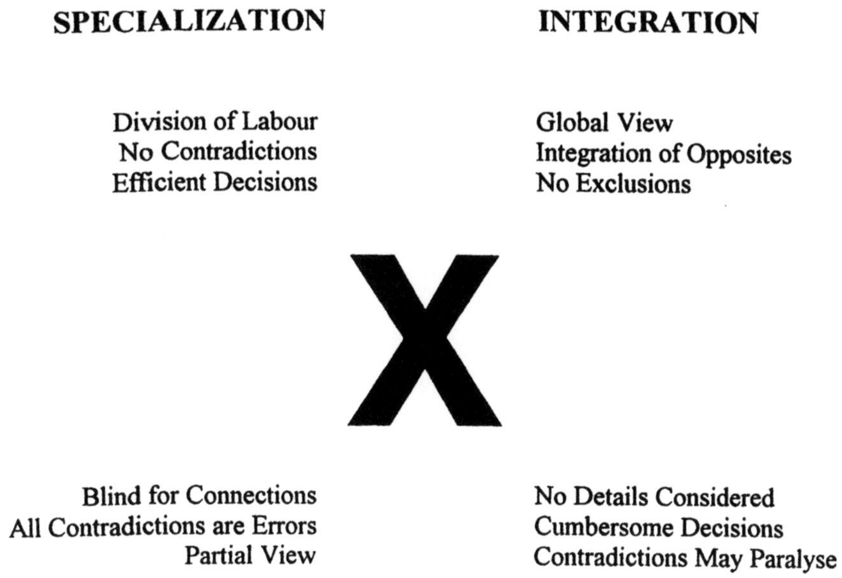

Figure 2. A case of **HX**-Confusion in which each side, instead of remembering its own limitations, rejects the other side because of its disadvantages.

own point of view. Only afterwards is a connection (by the horizontal bar of the **H**) possible. Even this connection can always get out of balance, it forms an unstable equilibrium which needs energy input in order to be kept alive. The greatest danger for the survival of such a balance is the "**HX**-confusion", in which each side rejects the other because of its disadvantages without recognizing the own limitations.

I would like to emphasize, that specialization and integrated approaches formed only an example, but that the **H**-model and the **HX**-confusion is a much more general phenomenon which can be applied to many other dialectic opposites. In particular, in medicine, scientifically oriented approach and alternative method (holistic medicine, homoeopathy, acupuncture, ayurvedic medicine and the many others) form such a dialectic pair. It can be arranged with advantages and limitations of both sides again on a big **H**. It is quite obvious, that in this case the **HX**-confusion is the presently dominant situation and it would be very beneficial for medicine and probably for humankind as a whole if it could be overcome, that is to say that the **X** could be exchanged by an **H**.

6. REFERENCES

1. See for example: Pietschmann H., Phänomenologie der Naturwissenschaft. Springer Berlin 1996
2. One possible method is described in: Pietschmann H., Die Sicherheit der Naturgesetze, in: Eranos Jahrbuch 1986, (Ritsema R., Editor) Insel, Frankfurt 1988, p. 89
3. Dschuang-dsi. Das wahre Buch vom südlichen Blütenland. Diederich, Düsseldorf 1977

5

THE HEALTH CONSEQUENCES OF COMPETING CONVENTIONAL AND ALTERNATIVE DEFINITIONS OF HEALTH

Stephen John Fulder

Consultancy and Research on Biomedicine
Clil, D.N. Oshrat, Israel 25233

1. ABSTRACT

Conventional medicine defines health as either absence of disease, or various utopian definitions, such as "a state of complete well-being". The definitions of the absence of disease is the basis of all therapeutic interactions in modern medicine. These definitions have led to a frustrating lack of familiarity with health itself and its manifestations. It can have drastic effects in practice on our experience of medicine and of illness. Various other views have arisen in response to the overly mechanistic view of the human being. In particular, the salutogenic view of Antonovsky sees health as a social phenomena. This has led to the field of health promotion, with mixed results.

Alternative medicine defines health more openly and more vitalistically. It is based on a rich and ancient source of experience into the nature of health and the process of becoming more healthy or more sick. Most alternative therapies see the individual and his life's journey as the key to health or sickness. health itself is definable only in relation to context – the individual at that moment in his life. There is also an element of the 'will to be well', the self-healing powers, which are respected and affirmed. Models of health and disease are more in tune with subjective experience. there is therefore a strong biographical aspect both to treatment and to research in alternative medicine.

The alternative definitions are more vitalistic and as such run counter to the mechanistic world view. This can create a real struggle for patients, for example in the treatment of chronic diseases. The mechanistic view would propose toxic treatments which target specific body processes. The vitalistic view will propose health building attitudes and activities, such as a more harmonious, and balanced life. The patient will often try to do both together and this can create conflicts. Besides, alternative medicine can itself at times be mechanistic, such as in chiropractic or formula acupuncture, and the reverse is also true, such as in the typical caring and listening family practitioner. It is helpful to explore these themes so that both patients and practitioners can learn how to negotiate within the various models of healing available in our current society.

Potentiating Health and the Crisis of the Immune System
edited by Mizrahi *et al.* Plenum Press, New York, 1997

2. INTRODUCTION

Conventional medicine can be said to have been the dominant medical system in the post-industrial world for not much more than 150 years. Prior to that, although differential diagnosis was the main diagnostic method used, physicians had to compete with traditional practitioners and bonesetters, and apothecaries with Galenicists. Traditional and ancient medical concepts such as the four humours, the elements, the *vix medicatrix naturae*, and crasis/dyscrasis (i.e. that health is based on inner and outer balance) only went out of fashion during the early part of the last century.[1] In its short history, modern medicine has proved to be so apparently effective, and so well adapted to the industrial world view that it gave the impression that indigenous, ancient or traditional medicine had no validity, and was well-nigh extinct. In fact, this was not so. It clearly existed in the East and the Third World, and in hiding in western culture, where it took a defensive cultist posture in the face of modern medicine's self-confidence.

However in the last 20 years there has been a radical renewal of interest in, and use of, traditional or alternative medicine. So much so, that we are reentering a period in which scientific medicine and its services share and compete for customers with alternative medicine, within a pluralistic national medicine.[2] Indeed the British Medical Association's new report acknowledges that alternative medical systems are full systems, that they are here to stay, that doctors must learn about them even at undergraduate level, and if a doctor wishes to study them, he or she must undertake a full course of instruction.[3] Alternative medicine is becoming available on the National Health Service throughout the UK.[4] The scale of the current use of alternative medicine is not always appreciated. Surveys have shown that in Europe roughly a third of the population have used alternative medicine.[5] and roughly the same proportion in the USA.[6] Polls of doctors have shown that 3/4 of British General Practitioner trainees and nearly half of those in practice want to learn one or more alternative medical techniques,[7] that virtually all doctors in primary care want alternative medicine to become fully part of national medicine.[8] In addition, where doctors work with complementary practitioners, complementary medicine is highly popular.[9] Demand outstrips supply.[10] These developments may have profound implications on the concepts of health, which will be examined below. The implications arise from a growing familiarity within society of alternative explanatory systems concerning health and disease.

2.1. Common Features of Nonconventional Medicine

Nonconventional medicine is an aggregate term for a variety of ancient or traditional medical systems in their modern forms. They include the specialities in Table 1 which are grouped according to broad similarity.

Some of the above modalities are complete medical systems with their own diagnostic and therapeutic methods based on a unique, global and self-consistent theory of health and disease (e.g. acupuncture, herbalism, homeopathy and naturopathy). Others, (such as the physical and manual therapies) are subsidiary techniques. Practitioners of these subsidiary methods do not consider themselves to be first-call primary care practitioners. Compatibility of alternative theories with conventional medical theory also varies. The physical therapies such as chiropractic, medical herbalism and to some extent naturopathy utilise essentially conventional diagnosis together with concepts of disease which are different from, but understandable by, conventional science. For example chiropractic is based on conventional anatomy and physiology, but extends knowledge of the pathogene-

Table 1. Therapeutic modalities of complementary medicine

Ethnic Medical Systems
Acupuncture and Chinese Medicine
Ayurveda and Unani Medicine
Manual Therapies
Reflexology
Chiropractic
Osteopathy
Alexander Technique
Massage Therapy
Therapies for "Mind-Body"
Hypnotherapy
Psychic Healing
Radionics
Creative Therapies
Anthroposophical Medicine
Nature Cure Therapies
Naturopathy
Hygienic Methods
Non-allopathic Medicinal Systems
Homeopathy
Herbalism

sis and treatment of musculoskeletal problems (such as "adhesions" and "subluxations") into subtle areas that are regarded as invisible and unproven by conventional medicine. Herbal medicine recognizes and uses conventional descriptions of disease such as eczema. However it chooses medicines that affect the supposed deeper imbalances (e.g. that allergies originate in part from inadequacies in the liver housekeeping functions) as well as treatments that attempt to restore proper local tissue function. On the other hand homeopathy, naturopathy and Oriental medicine employ different concepts of disease, based on an alternative world-view which is not easily translatable or compatible with scientific medicine. It has been difficult or impossible to map these systems onto conventional medicine or vice versa. Even in China, where strenuous efforts have been made, there is still no agreement on whether acupuncture meridians or Oriental viscera such as "kidney" correspond to any known anatomical structures.[11]

2.2. Characteristics of Alternative Medicine Relevant to Descriptions of Health

There are certain basic features of the practice of alternative medicine that involve another view of health from that implicit within modern medicine and modern life. I would like to review some of these features here and then discuss their implications for our perspectives on health.[12] Not all therapists will employ these concepts. However they represent the foundation of authentic alternative medicine, are laid out in the texts and taught in the colleges, even if the therapy has compromised towards the biomedical model during the current struggles between various competing systems. For example many acupuncturists find themselves called upon to focus on symptoms more directly at the expense of the slower restoration of energetic balance, because of the expectations of patients who are conditioned by modern medicine to expect a fast restoration of comfort.

2.2.1. Self-Healing Is Paramount. The in-built natural healing process is respected and recruited during treatment, although it is not necessarily understood. Resistance is improved by preventive measures. Particularly in Oriental and Ayurvedic medicine, a considerable proportion of traditional practice is devoted to the restoration of vital force and self-healing energy. For example in Western herbal medicine there is very frequent use of a category of herbs called 'alteratives' or 'blood purifiers'. These are herbs such as echinacea (*Echinacea purpurea*), cleavers (*Gallium aparine*), burdock (*Arctium lappa*), sage (*Salvia spp*) and myrrh (*Commiphora mol-mol*). They are used during the treatment of most acute and chronic infections and inflammations, along with fasting or special diets and nutrients, and other herbs to promote circulation of lymph and body fluids, all of which is intended to awaken a more powerful immune response and encourage long term immune function. Other natural remedies such as garlic or thyme may be used as natural antimicrobials, but they are not the primary tools. This is also a fundamental position of naturopathy and homeopathy.

2.2.2. Working with, Not against, Symptoms. Symptoms are a guide in the journey to a cure. They are managed, not suppressed. For example the daily ebb and flow of a symptom such as headache may be used by an acupuncturist or homeopath as a guide to the course of treatment of deeper problems with organ function. For example a migraine-type headache could be seen by an acupuncturist as arising from overactive liver metabolism (liver 'fire' or 'yang'). Real treatment involves an adjustment to the propensity of the liver to create inflammation, not merely relief of symptoms by analgesia. The type and location of the headaches (frequency, severity, vertigo, sharpness, one-sidedness, etc.) is constantly monitored throughout the treatment as a guide to the effectiveness of the draining of the liver's excess. Accompanying symptoms, such as nausea, may be an indication of the energetic state of other organ systems such as the spleen.

2.2.3. Individuality. Each person's condition is different, has arisen from different reasons, against a different constitutional background, and requires a different path for treatment. Decisions are personal and individualistic, not statistical. One of the indications of the richness of the medical system is the development of a typology with which individual differences in health, disease and response to the environment can be understood. For example the constitutional picture in Ayurvedic medicine is a highly detailed art, which integrates thousands of characteristics of body, skin, personality, habits, etc. are defined in terms of *Vata* ('airiness'), *Pitta* ('fieriness') and *Kapha* ('wateriness') integrated. This establishes an individual's susceptibilities, strengths and weaknesses, and guides both prevention and treatment. Western (Thompsonian) herbalism by contrast does not make extensive use of constitutional differences, and modern medicine ignores it completely unless there are inherited pathologies.

2.2.4. Integration of Human Facets. Individuals are regarded 'holistically' in diagnosis and treatment. There is less *a priori* division between Mind-Body-Spirit or Environment-Society-Individual. 'Holism' is just one of the approaches that may or may not be incorporated within a therapeutic modality. It is not a medical system in itself, although the term 'holistic medicine' is sometimes loosely used. Alternative medicine is essentially more holistic. In homeopathy and Oriental medicine, for example, emotional, psychological and behavioural signs are always included in diagnosis. This is rather less so in naturopathy, herbalism and the manual therapies, but even here holism is often applied as an approach of an individual practitioner. For example naturopaths may encourage relaxation and imagery along with diet and herbs to treat high blood pressure. Or osteopaths and chiropractors occasionally explore the psychosocial stresses that may give rise to a repeated musculoskeletal problem in a certain patient.

2.2.5. No Fixed Beginning or Ending. There is no defined or determined state of illness where treatment must begin and wellness where treatment must end. Such points are defined contextually. One patient may require assistance to reach a state of well-being and accommodation to his cancer. Treatment will finish when this is achieved, although in conventional terms he is still seriously ill. Conversely, another client may be treated so as to improve his energetic balance and condition or vitality. He may seek treatment with Oriental medicine to cope with an addiction, an energetic dullness, convalescence or even to improve *Shen*, i.e. to bring light to his eyes. Treatment in this case, in conventional terms, is of a healthy person.

2.2.6. Conformity to Universal Principles. Remedies are discovered and employed in conformity to patterns of relationships (such as yin/yang) between all living creatures and their environment. These patterns are often subtle and involve energetic rather than material phenomena. For example *Ch'i* in Chinese medicine is a tangible but invisible vital force which operates continually as the basis of all function. In Oriental medicine it is sensed and utilised in much the same way that modern man would sense and also utilise gravity. Despite the fact that *Ch'i* is so universal, it is enormously elaborated as an explanatory principle to describe detailed changes in function, e.g. constrained, stagnant, wild, deficient, excess, etc. of liver, spleen, kidney, etc. This is in contrast to conventional medicine, which derives from reductionist science. Therefore processes in the body are examined as discrete entities, unconnected to basic forces and elemental qualities. Consider again the use of *Ch'i* in the diagnosis of 'Excess Liver 'Ch'i' rising' for a migrainous headache, in contrast to the conventional medical view of 'overstimulation of local vascular serotonin receptors at the pain site'.

2.3. A Meta-Model for Complementary Medicine

It is possible to summarise many of the features of complementary medicine in a descriptive, qualitative manner. The basic axioms or concepts have been given above. When put together a working model would look something like this. The human being moves through his personal time and space, which is mapped out by biological, psychological and biographical events. He is in constant dynamic relationship with his internal, familial, physical and social environment, requiring constant adjustment and adaptation. This is a learning process for both body and mind. Some health related behaviour is always required so as to harmonize this relationship, and if the disharmony increases beyond a subjectively determined limit, it becomes time to seek professional assistance, for example, if the seasons change and the person feels chronic pain in the joints. Alternative medicine functions: a) To monitor the extent of the disturbance or distortion in the psychobiological field. b) To understand the inner and outer constitutional picture of the person, and the special susceptibilities of the person to his environment. c) To evoke appropriate self-healing capacities. d) To provide remedies for both mind, body and even spirit so as to restore health in an individualistic and holistic manner, and to combine this with appropriate instruction so as to reduce susceptibility in the future. e) To halt and reverse disease processes and repair damage. If the distortion is too great, the damage too severe, the disease acute or highly infectious or life-threatening, the condition becomes more suitable for the stronger methods of conventional, allopathic medicine. In such cases the patient is usually referred by the complementary practitioner or self-referred to allopathic health care systems.

2.4. Some Conceptual Elements in Modern Medicine

The mechanistic basis of modern medicine has grown out of an early Christian view that it is important to study nature as a sacred task, a recognition of God's work. However as the sacredness within the Church declined in the Post-Rennaisance period, the observation of nature became a valued tasks in its own right, as a part of control over nature and its forces. Nature could and should be tamed, and it was man's duty to his (God-given) intelligence to do so. Taming nature meant understanding its laws. And so science was born and with it scientific medicine.

The basis of medicine thus became the search to understand objective processes going on in the body and mind, and if they have gone astray, to put them right. Gradually this objective relationship to the body as mechanism, led to all kinds of extreme positions. For example, the letting of blood, repeatedly, was justified by a bad-blood, mechanistic theory, which was held to even though it should have been clear that in many cases it was killing the patient. Arabic medicine, which was more holistic and naturopathic in orientation, never having lost its Greek parentage, looked on such bloodletting by European doctors with horror.

Concern for processes, and mechanism, the need to find objectively verifiable laws, and to prove them to be absolutely real within nature, and not a figment of the 'unreliable' human mind, and the growing sense of power over nature, led eventually to the modern medicine of today, which is so highly mechanistic that patients are generally viewed as no more than carriers of symptoms. The human being is a machine, the mind is an elaborate computer, and the doctor essentially an engineer rather than a healer. Indeed, what healing actually is, the power of life to restore health, is unknown and uninvestigated in modern medicine. The values of modern medicine are based on this. For example, the major value of therapeutics is the 'Magic Bullet'. This is the concept of a medicine that goes like a bullet to its target, and knocks out some invading organism or fixes a step in a process that has gone wrong. Examples such as penicillin, cortisone and beta-blockers are given as the top successes of modern medicine. But with a different philosophy, such as that of the Chinese traditional medicine, such drugs are seen as too powerful, toxic, and not very helpful. They are called the 'servants' and used only when all else has failed. Instead, the most valuable remedies in Oriental medicine are the mild, restorative and preventive remedies like licorice, called the 'kingly' remedies, which are valuable because they integrate with self-care in order to prevent disease. This is of course the opposite to modern medicine which would consider such remedies worthless.

The problem is that diseases, especially the diseases of today, are often caused by lifestyle. Chronic diseases grow because of, and through an individual biography. To try and attack such diseases with a Magic Bullet approach is childish. More than that it can lead to suffering and waste. For example, consider cancer research. It has been calculated that about 1 million, million dollars has been spent on cancer research in the last 20 years, since the 'war on cancer' was launched. The result has been a virtual failure, and a terrible waste of resources. The US President's Cancer Panel was recently told that the war has been lost, and indeed the statistics show that in the last 20 years the 5-year cure rate has hardly improved. I believe that it is because the wrong model is used. A more vitalistic, lifestyle-oriented, biographical approach would succeed in understanding and treating such diseases much more successfully.

3. THE DEFINITIONS OF HEALTH

3.1. The Conventional Perspectives on Health

Within conventional medicine there are two separate kinds of health definition: the practical/medical and the utopian. The practical/medical is that found in medical texts and that which medical students learn in university. A healthy person is a symptom-free person. Symptoms are defined as abnormalities which are recognised by professionals. They are not necessarily connected to the patient's subjective experiences of illness. For example a patient with non-specific discomfort, malaise, reduced function, withdrawal, insomnia and feeling off colour may not be recognised by the doctor as being ill, and may therefore remain untreated. Even if the doctor wishes to help he may find that extremely limited treatment options available within modern medicine for problems outside its definitions of sickness. However the patient may indeed feel subjectively very ill. A specific example of this is the so-called post viral syndrome (M.E. or Myalgic Encephalomyelitis). Patients were not recognised as having a legitimate disease until they mounted an international public campaign to have their disease named and recognised. Only then could it be taken seriously in treatment and research. There has been a long critique of this biomedical description of health, beginning with Illich[13]. The main points of this critique are the lack of acknowledgement of the wider influences on health, including social forces, psychological disposition, stress, etc, and the pathological bias that sees in health a simple negative, an absence of disease, which is essentially a statement that there is no such thing as health.

There are those within modern medicine, especially paramedics such as art therapists, occupational therapists, speech therapists, physiotherapists, etc., who do recognise the need to go beyond this restrictive view of health, and who attempt to expand it in practice to include function and social relations. However with only a few minutes per patient, the conventional model dominates primary medical encounters, and it certainly is the conceptual basis for secondary (i.e. hospital) care.

The utopian model goes beyond this. It is typified by the definition arrived at by the World Health Organisation in its 1977 conference at Alma Ata: "A state of complete physical, mental, and social well-being, and not merely the absence of disease or infirmity". It is an ultimate definition of health, determining it as the fulfillment of the human potential. However it is of limited usefulness, because modern medicine does not have the tools nor the ability to serve the human potential. There are no well-being medicines in the pharmacopoeia. This definition is unrealistic even in western countries where there is a cultural awareness of human performance issues. However in Third World countries, where most of the World Health Organization's efforts are directed, their definition is even more inappropriate as a basis for action. Modern medicine is hard pressed to contain life-threatening diseases such as malaria, AIDS, tuberculosis, bilharzia, etc. In Third World countries, well-being is often identified as having enough to eat, and easy access to clean water. Thus there are obvious limitations with both the above definitions of health.

3.2. The Folk Depiction of Health

A further perspective is that based on common sense and life experience, and it is embedded in folk culture and ordinary language. Here health is one of those terms which run right across cultural and individual consciousness like 'goodness', 'life', or 'rightness'. Everyone will define it in a different way. For example someone who has chronic pain will define health as a pain-free period. In the modern fitness movement, health will be defined as physical performance. However in traditional village cultures, health is often defined as peaceful undisturbed existence. In rural France, the average lifespan of women is several years longer than those in the US or northern Europe. They smoke, drink wine, and eat meat, and view the essence of health as being close to the family, and being able to rest and do little. They regard the fitness obsessions in the Anglo Saxon countries as driven and therefore fundamentally unhealthy.[14]

Folk views of health are set by role models and cultural norms. In some cultures fatness is a sign of health, in some thinness, and European culture has had both at various times in its history. Ancient Greece viewed health as perfect bodily proportions. The American Indians viewed it as endurance to physical strain. In China health is regarded as equivalent to longevity, by definition. In India health is regarded as the luster of the skin and the shine in the eyes. All of these perspectives are partial signs of health, none is all inclusive. Such common-sense views of health change all the time and are a reflection of present-time experience. Since they are changing, and they reflect human function and well-being rather than symptoms, the folk views are closer to those of alternative rather than conventional medicine.

3.3. Biopsychosocial Descriptions of Health

There has been a movement on the fringe of medical science to break previous disciplinary boundaries and introduce social, ecological and cultural/political issues into the considerations on health and disease.[15] The introduction of new fields such as medical sociology, social epidemiology, and community medicine, imply a wider theoretical basis to disease causation and the nature of health than that described above. It has become clear that most of the major advances in health during the last part of the last century and the early part of this century were the result of hygienic social measures and not of medicine or the biomedical model.[16] Today, at a time when the main causes of morbidity and loss of life are degenerative chronic diseases which are strongly influenced by the habits of the modern world, such as smoking and chemical pollution of the biosphere, the emphasis is returning once more to the social sphere. Many studies have demonstrated that diseases such as heart disease are much higher in groups such as poor urban blacks in the USA, which are under social pressure and stress.[17] There is a complex relationship between risk factors such as obesity or smoking and social pressures, such that it is quite clear that we cannot consider individual health behaviors in isolation from the social and cultural environment. Consideration of these aspects has led to a number of descriptions or definitions of health which are based on the way humans negotiate through the sea of environmental, social and informational influences.

These concepts are often based on systems theory, since they imply a complex hierarchy of relationships. Health is a process of alignment to norms of human function and well-being that are defined by society and by professionals. Illness implies deviancy from these norms. These norms will vary from time to time and social group to social group. Health is a process, not a fixed defined condition of the organism as in the biomedical model. The process is one of dynamic balance where the capacity of the organism to self-repair, self-support and renewal is not overwhelmed by the interactions with the world within and outside the organism. The organism, in this case, is not only the individual, but also the community, and the entire socioecological system, so that one can talk about the ill-health of a group as well as of a person.[18] Health is determined by two aspects, namely the degree of distortion or stress imposed by the environment, which can be described as the 'health balance' and the extent to which the organism is able to cope, that is the resources of the individual or group available for coping, which can be described as 'health potential'. This kind of systems thinking has been very useful in developing the field of health promotion. For example the improvement in health potential of a community will include promoting nutritional status, physical fitness, immunological capacity, healthy lifestyle and health knowledge. Improving the health balance will imply actions to reduce stress in the workplace, pollution from chemicals, radiation (VDU's) and noise, and social integration.[19]

Perhaps the best known proponent of the biopsychosocial concept of health is Antonovsky. He has very little sympathy with the biomedical position, not because it is mechanistic since Antonovsky's position is also mechanistic, but because it is concerned with the individual. According to Antonovsky, concepts of disease as an external influence ignore the ability of the individual or group to cope and adapt, and concern with individual behavior, i.e. risk factors for disease, is blaming the victim. The biomedical definitions are a kind of medical imperialism in which medicine defines health according to its interests, and for the sake of its continuing power.[20] Instead he proposes that there is a spectrum from ease (salutogenesis) to dis-ease (breakdown) according to the individual's ability to cope with stresses. The stresses are generally social, and include war, lack of food and clean water, family discord, unemployment, and discrimination. Salutogenic influences would include social stability, rewarding occupations, and social acceptance and integration. Antonovsky continued by considering that the way these influences act on the individual in his world is by information. Input that is conflicting and chaotic (e.g. anxiety, persecution) or excessive (e.g. classical stress) or lacking (e.g. isolation, loneliness) will reduce the sense of coherence of the individual in his society and create conditions of stress and dis-ease.[21] This view is the other extreme from that in which the individual is responsible for his fate. Indeed Antonovsky has some harsh words for the well-being movement, described below, which "prevents understanding the social burdens that pressure people to behave in pathogenic fashion and that block them from behaving salutogenically. From a moral point of view the focus on the 'health within' is at the very least a passive and unconscious approval of the social status quo...the world of technomedicine is not better. This comforts me not at all."[21]

The dialogue between health as an individual issue and health as a social phenomenon is interesting and deep. The philosophy of health as an individual practical and philosophical direction is well expressed in Larry Dossey's book: Space, Time and Medicine.[22] Such a personal view of the nature of health have arisen from considering those people who know themselves to be healthy, and indeed arrive at health by themselves even in the face of devastating illness or social conditions. Studies of cancer patients who cured themselves or lived far longer than expected found that they seemed to get better by individualistic health promoting attitudes and practices, rather than by therapy. Psychological treatments and psychotherapy which started with the question: "what is wrong with this person, and how did she or he get this way?" were less effective than asking: "what is right with this person, what are her or his unique strengths, enthusiasm, and zest for life?"[23] In addition, there is some research on those who cope well, who are healthy and appear to live long. The qualities which appear to be expressed by such people are those of activity, exercise and energy, involvement in life and commitment to some engaging life activity, a conviction that life has meaning, and good social connections.[24,5] The late Norman Cousins discussed this approach in his personal journey to health, and described health itself as such a personal odyssey, a process of fulfillment, of becoming what you truly are.[26] To which Antonovsky has commented that it is all very well if you can afford it. Despite its limitations on the level of biography and the individual, the social and systems approach to health is a step towards the more open descriptions implied by alternative medicine, because it stresses health as a balance, an ecological process, within and without the individual.

3.4. The Alternative Perspective on Health

Health within alternative medicine is drawn from a tradition which is far more ancient than that of conventional medicine. Because of that it does not need to formulate its perspectives on health with biomedicine in mind, either in accord with it or in opposition to it. Thus it would view the symptom-based definition of health of biomedicine, the function-based definition of the WHO, the fitness-based definition of the well-being movement, the autonomy-based definitions of the psychological movement, and the systems theories of the salutogenic movement as partial and limited. It would draw on all of these as needed, but transcends them all. It is much more vitalistic and life oriented, without the need to constrain itself into western mechanistic, reductionist models of how the human being functions. The definitions of health arise from the main themes of alternative medicine described earlier. Considering these themes will demonstrate the inclusive yet pragmatic definitions of health implied within them. For example, health in Oriental medicine involves a harmonious relationship with all the energies and influences within which man is immersed. These include but are not limited to material, natural, environmental and social influences. But it also implies having a good constitutional and genetic basis, termed 'inherited energy', and corresponding perhaps to the systems theory's 'health potential'. In addition, it is recognised that health is a mind-body-heart issue, without acknowledging any boundaries between them. Oriental medicine and the major alternative medical systems never passed a Cartesian phase, so there is no need to postulate or evoke concepts like psychosomatic, or even autonomy. These are qualities observed naturally within the mind-body-heart continuum, expressed as the total energetic body of man.

In alternative medicine both patient and practitioner share an alternative world view based on the themes described above. It is a more open, contextual definition. Health is viewed, in the clinic and in the home, as a process without a beginning and end. There is no absolutely healthy person, and no absolutely diseased person. There is, rather, a journey in which greater and greater health is achieved through a combination of life wisdom and health practices. For example, alternative medicine has a major role to play in the area between complete symptom-free well-being, and actual disease. This 'third state' or pre-disease state, is viewed as a lack of health needing attention. "Why dig a well after you are thirsty?" states the Oriental medical classic: Yellow Emperor's Book of Internal Medicine. Thus some of the work of alternative medicine may be with essentially symptom-free people. Alternative medicine deals not only with disease, but also with vulnerabilities, concerning which it has a huge knowledge and a large catalogue of remedies.[27] The skills of prevention and health maintenance rely on a concept of health in which all the many subtle levels of susceptibility and risk are regarded as states of health requiring assistance.

Since conventional medicine has drawn a line at the symptom boundary it has few tools for health maintenance (apart from public health). In consequence it also lacks an operational language to assess health itself. The various states of health are describable only in common language (e.g. 'well-adjusted', 'vital', 'energetic', 'glowing') even by doctors. Whereas in alternative medicine the practitioner is able to call on a rich language to describe subtle states of health, and to differentiate these different states of health by diagnostic signs. For example in Oriental medicine the practitioner determines the flow of energy and materials through organs and the functioning of the main organ systems. He may ascertain that an organ is too 'full' or 'empty', 'hot' or 'cool', etc. The quality of the 'chi' or vitality of the organ systems may be stuck or scattered, rising or falling, or balanced, spread in the right way around the body, full, resistant, etc. In other alternative systems this subtle health measurement is made using different conceptual frameworks and languages. Thus the balance between elements or fluids is tested in herbalism and naturopathy. Health is a dynamic state where these qualities are well-adjusted, and heat, moisture, etc. are balanced and in their right proportion and location relative to the constitution of the individual. In addition there are measures of the quality of organs (skin elasticity, brightness of eyes, clarity of skin, warmth, firmness of body, lack of waterlogging, etc.) which are similar to folk descriptions and to some signs used by good conventional doctors.

Although health is relativistic, alternative medicine does have some standards. Very often these standards are set by the therapist himself who is required and assumed to be healthy. In Oriental medicine the speed of pulse and breathing of the patient can be measured against that of the therapist himself. In therapy further measures of pulse are taken which reveal the individuality and characteristics of the patient, and build a constitutional picture which would then act as a reference point. For example a patient may naturally have a very deep and slow pulse. This is his state when healthy. The therapist will attempt to raise and lighten the pulse a little by means of acupuncture so as to adjust the patient towards even better health and reduce his natural vulnerability to cold diseases. If the patients pulse became too light, erratic or fast the therapist would know a disease was brewing and be able to take corrective action.

The definitions of health employed in alternative medicine have as their philosophical basis a view of nature different from that underpinning the modern world. There is more respect for what is 'natural', meaning unfabricated and less manipulated according to transient human intentions. There is a greater sense of co-ordination with global flows of energy. There is a strong vitalism. There is a lack of urgency to construct explanatory models – an empiricism incorporating a greater sense of the unknown and respect that goes with it. This leads directly to the lack of strong interventions that characterizes alternative medicine compared to conventional medicine, and the greater trust of self-healing capacities even if they cannot be understood.

3.5. Some Characteristics of Health Arising from the Vitalistic World View of Alternative Medicine

It is not easy to give one formula describing health within the various alternative medical systems. However some features of health that are common can be listed. They give the flavour of the definitions of health implicit within the alternative techniques and traditions:

1. Reducing input of physical and mental toxins.
2. Being sensitive to deep signs of function and dysfunction.
3. Understanding your constitution and its patterns and needs.
4. Tending towards a state of harmony and balance between internal and external worlds, e.g. seasons, environment, social relations.
5. Respect for the unknown and changeability.
6. Knowing health as a journey, a process.
7. Discourse with therapeutic activities: knowing when to use what remedies or professional help.
8. Vitalistic, life-affirming attitude: the will to be well.
9. Longevity
10. Energy
11. Subjective sense of well-being
12. Total accommodation to life – and death.
13. Purpose

14. Shining in the eyes
15. Movement towards harmony between the internal and external world

Consider, for example, the case of a young man of 40, Dennis, who had childhood diabetes. At the age of 25 he went blind as a result of diabetic pathology. Doctors noticed that his kidneys also began to fail as a result of his diabetes. Besides attempting to keep his keep his blood glucose levels as constant as possible, there was very little the doctors could do to protect his kidneys. His doctors could only helplessly record the deterioration towards an inevitable kidney transplantation. In this case, the definition of health of Dennis was exclusively that of his diabetic symptoms, and not at all of the total energetic functioning of all his organ systems and life processes. However one day he went to a traditional acupuncturist. The acupuncturist, using a different model of health, asked how each of the organ systems were functioning and adapting to the problem of erratic glucose supplies. The acupuncturist essentially worked with the processes of adaptation within the body to strengthen them and the co-ordination of all the organ functions and life processes. He began to feel psychologically better, less threatened and depressed about his future, and more fit and energetic. His general health indices, including the stability of his blood glucose, improved. Then his kidneys began to improve, something that is inconceivable within conventional medical explanations. Dennis's kidneys are now fine, and he is well through the use of acupuncture and herbal and dietary treatment, although he must also take insulin. Years later, Dennis is still in good health and leading a constructive life despite his blindness and diabetes. It seems that only the subtle methods of complementary medicine can help Dennis to adjust to his disturbed glands. Conventional medicine (insulin) is keeping him alive. Complementary medicine is helping him live.

4. IMPLICATIONS OF ALTERNATIVE DESCRIPTIONS OF HEALTH ON HEALTH CARE

We are not getting more healthy today despite a massive and crippling investment in health services. In the UK, in the last 20 years, there has been a 50% increase in the percentage of the Gross Domestic Product spent on health. Yet there has been a third increase, to 34% of the population, in those suffering from long term illness, and a 64% increase in incapacity, or days of certified sickness.[28] Though life expectancy has risen, largely because of the drop in infant mortality, US statistics show a steady rise in ill-health: from an average of 0.82 episodes of disabling illness a year in 1920 to 2.12 in 1988.[29] This is the age of "the vertically ill".

Evidently something has gone wrong. All of the many people who have offered critiques of the biomedical approach have provided partial answers to what is wrong, and there is no doubt that the biomedical model is in a slow crisis and in process of change towards a more holistic paradigm, in the Kuhnian sense. Here I propose that the limitations of the biomedical model are inherent in its definition of health. This definition, though largely implicit and unconscious, nevertheless informs everything that is carried on in the name of medicine. More than that, the definition of health produces the kind of results in the population that are inherent in that definition. Thus the definition of health as freedom from obvious symptoms has created a generation which are free of obvious symptoms, but are not fully well either. The clarity with which acute sickness can be defined as ill-health compared to chronic and early stage degenerative conditions has resulted in huge numbers of chronic cases, which, as the death certificates show, are largely untreated and untreatable.

Scientists and health professionals could learn a great deal from alternative medicine about defining and measuring health. Health care today would benefit from an awareness of more meaningful ways of describing and measuring health offered by alternative medicine, for example by including constitution and the quality of vitality. It could also provide a rich source of ideas and criteria for assessment to efforts at prevention, some of which (including lower fat, raised fiber, certain vitamins to lower cancer risk, the health risks of food additives) have already been acknowledged by modern medicine. That new criteria for assessment are urgently required is illustrated by the example of blood cholesterol. The question whether lowering it contributes to overall health has been tested using billions of dollars of research money and many years of investigation, yet the question is still not finally decided, and official advice on this question remains ambiguous. If more appropriate assessments of health were used it would be much easier to ascertain what is healthy and what is not, for each person.

Alternative medicine is rapidly increasing in popularity. As its conceptual basis is prior to, and in many respects opposite to, conventional medicine, it provides a challenge to it which has not yet been accepted. Alternative medicine's concentration on healing the healthy as well as the sick, and its familiarity with the origins of disease on the ground of human life, give it special skills at understanding states of health. It sees health as a process which is intricately related to the way a constitution and personality develops through time within the matrix of all the influences and relationships that are experienced.

But it is also important to realise that there is a competing world of views within society, and that this creates both problems and opportunities. For example, typically, a patient may arrive at an Oriental therapist after seeing his doctor, which is the case with most patients of alternative medicine. He enters with the medical diagnosis, talking about gastric ulcer, and as his dialogue with the therapist continues, he or she begins to think in the language and concepts of 'Qi', energies and 'yin/yang'. This can create confusions about the validity of each model and how treatment should proceed. When the symptoms decline he will be cured from the medical model, but his oriental treatment may just be beginning, for it may be intended to build a renewed resistance and energetic structure so that the problem will not return. He has to choose his view of health. If the doctors, he will not wish to pay for further Oriental treatments that do not relieve any obvious symptoms. But then he may soon have to return to the doctor if the symptoms return. Or he can choose the deeper model of health implicit in the Oriental treatment, carry on until the end, and feel committed to continue with this preventive approach.

Not only the patient may have to sort out competing models. In the UK and other countries there are now medical centers and clinics in which alternative medicine is offered along with conventional medicine. These clinics are usually structured so that the conventional medical practitioners are the gatekeepers and retain overall control of the patients, the financing, and the clinics themselves. But this creates problems of the dominance of the medical model. The alternative practitioners and their models of health don't fit well into such arrangements. For example, rarely can the medical and the alternative medical practitioners discuss common patients with a common language, and agree, for example, when treatment should finish, or how many treatments are necessary, or what systems of the body have gone wrong. Many times, alternative practitioners in such arrangements complain that doctors dump on them their failed cases in which much damage has already been done by both not treating the disease with alternative methods before, and by medical side effects. In such cases of deep seated damage, and long-term chronic disease, alternative treatment is performing at its worst, and requires a lengthy treatment which the doctors will not pay for. The doctors demand a rapid throughput and turn-around of large number of patients, and the alternative practitioners cannot fulfill these expectations.

These problems can be solved by an understanding of alternative views of health, that should be spread throughout the healthcare system. It is simply good medicine to treat the human being as human, as whole (the word health comes from the word 'hale' which is the same as 'complete'), and as a unique individual. The alternative paradigms have a role to play in teaching conventional practitioners not just a few techniques, such as electro-acupuncture, but actually the basis of alternative medicine—a more vitalistic view of health.

REFERENCES

1. Rosenberg, C.E. (1977) The therapeutic revolution: medicine, meaning and social change in the nineteenth century. *Perspectives in Biology and Medicine,* 20: 485–506.
2. Pietroni, P. (1988) Alternative Medicine, *R. Soc. Arts. J.,* 136:791–801. Pietroni, P. (1990) *The Greening of Medicine,* London: Gollancz.
3. British Medical Association (1993) *New approaches to good practice,* Oxford. Oxford University Press.
4. Cameron-Blackie, G. (1993). *Complementary therapies in the NHS.* Birmingham. National Association of Health Authorities.
5. Fisher, P. and Ward, A. (1994). Complementary medicine in Europe. *British Medical Journal,* 309: 107–11.
6. Eisenberg, D.M., Kessler, R.C., Foster, C., Norlock, F.E., Calkins, D.R. and Delbanco, T.L. (1993). Unconventional medicine in the United States. *New Eng. J. Med.,* 328: 246–252.
7. Reilly, D. (1983) Young doctors views on alternative medicine, *Br Med. J.,* 287: 337–340.
8. Perkin, M.R., Pearcy, R.M. and Fraser, J.S. (1994). A comparison of the attitudes shown by General Practitioners, hospital doctors and medical students towards alternative medicine. *Journal of the Royal Society of Medicine,* 87: 523–5
9. Budd, C., Fisher, B., Parrinder, D. and Price, L. (1990). A model of co-operation between complementary and allopathic medicine in a primary care setting. *British Journal of General Practice,* 40: 376–8. Richardson, J. (1995). Complementary therapies on the NHS: the experience of a new service. *Complementary Therapies in Medicine,* 3: 153–7
10. Himmel, W., Schulte, M. and Kochen, M.M. (1993). Complementary medicine: are patients' expectations being met by their general practitioners? *British Journal of General Practice,* 43: 232–5.
11. Bensoussan, A. (1991) *The Vital Meridian: a Modern Exploration of Acupuncture.* Churchill Livingstone, Edinburgh.
12. Fulder, S. (1996) *The Handbook of Complementary Medicine,* Oxford. Oxford University Press.
13. Illich, I. (1976). *Limits to Medicine,* Harmondsworth. Penguin
14. Siegel, B. (1992) *Peace, Love and Healing* Harper Collins, New York.
15. Engel, G.L. (1977). The need for a new medical model: a challenge for biomedicine. *Science,* 196:129–36.
16. McKewan, T. (1976). *The role of medicine,* London. Nuffield Provincial Hospital Trust.
17. Pappas, G., Queen, S., Hadden, W. and Fisher, G. (1993) The increasing disparity in mortality between socioeconomic groups in the United States, 1960 and 1986. *new Eng. J. Med.* 329: 103–9.
18. Capra, F. (1983). *The turning point,* London. Fontana.
19. Gopel, E. (1993) Human health and philosophies of life In: Lafaille, R. and Fulder S. (eds.) *Towards a New Science of Health* London. Routledge.
20. Antonovsky, A. (1981). *Health, stress and coping,* San Francisco. Jossey-Bass.
21. Antonovsky, A. (1994) A social critique of the 'well-being' movement. *Advances,* 10: 6–12.
22. Dossey, L. (1982). *Space, time and medicine,* Boulder. Shambala.
23. LeShan, L. (1979) *Cancer as Turning Point,* New York. Dutton.
24. Sheehy, G. (1985) *Pathfinders,* New York. Bantam.
25. Rijke, R. (1993) Health in medical science In: Lafaille, R. and Fulder, S.J. *Towards a new science of health* London. Routledge.
26. Cousins, N. (1979) *The anatomy of an illness as perceived by the patient,* New York. Norton.
27. Fulder, S. (1993) *The Book of Ginseng and Other Herbs for Vitality,* Healing Arts Press, Rochester, Vermont.
28. Department of Health (1992). *Compendium of health statistics* London. Her Majesty's Stationery Office. Office of Population Census and Surveys. (1995). *Social Trends.* London. Her Majesty's Stationery Office.
29. Barsky, A. (1988). The paradox of health. *New England Journal of Medicine,* February 18, 414–8.

6

TOWARDS A NEW EPISTEMOLOGY FOR HEALTH AND HEALING

Willis Harman and Marilyn Schlitz

Institute of Noetic Sciences
475 Gate Five Road #300
Sausalito, California 94965

Beginning in the 1970s a series of innovative approaches to health care emerged in the west as viable adjuncts to traditional allopathic healthcare. These included: wellness, holism, behavioral medicine, and alternative and complementary therapies. These developments represent more than an expanded range of available approaches and treatment options, they reflect the emergence of deeper and persistent cultural forces that lead us toward a fundamental reconsideration and reconfiguration of our existing constructs of health and illness.

Although there has been considerable dialogue regarding the merit of one approach or another, this dialogue has failed to address the broader issues of change, specifically, the underlying assumptions of the dominant medical model. In the absence of an open dialogue on these issues, it is possible that our efforts at change may prove to be ineffective as innovative initiatives are subtly and invisibly reshaped to fit the constructs of the existing healthcare system rather than serving as a nidus for change. This effect can already been seen in the reshaping of the traditional values and practices of the alternative and complementary therapies as they are mainstreamed into the institutional structures of the existing health care system.

One important factor in this situation is the form of science that has shaped modern medicine; a form that has evolved with a certain passion and world view about the nature of reality. Indeed, modern Western science fundamentally entails three important metaphysical assumptions that limit it's ability to integrate alternative medicines.

These are:

1. *Realism* (ontological—leads to epistemological conclusion). There is a real world which is, in essence, physically measurable (positivism). We are embedded in that world, follow its laws, and have evolved from an ancient origin. Mind or consciousness evolved within that world; the world pre-existed before its appearance, and continues to exist and persist independent of consciousness.
2. *Objectivism* (epistemological and ontological). That real world exists independently of mind, and can be studied as object. That is, it is accessible to sense perception and can be intersubjectively observed and validated.

Potentiating Health and the Crisis of the Immune System
edited by Mizrahi *et al.* Plenum Press, New York, 1997

3. *Reductionism* (epistemological). That real world is described by the laws of physics, which apply everywhere. The essence of the scientific endeavor is to provide explanations for complex phenomena in terms of the characteristics of, and interactions among, their component parts.

These assumptions, and the methods they require, currently dominate biomedical science. The present situation in medicine involves a deepening understanding of such factors as the role of DNA in determining the nature of the organism, an expanding reliance of advanced and expensive technology, and a growing faith in the power of modern biomedical theory. To some extent the metaphysical premises have been reevaluated through an understanding of the relativistic views of science recently explicated by works in the history, sociology, and philosophy of science. More central to our concern here, however, are the new findings in the areas of alternative and complementary medicine that have profound implications for science and its application in the biological and medical domains. In particular, many alternative epistemologies involve a world view in which human experience, including thoughts, feelings, and intentions, are believed to interact in causal ways with subtle forms of "energies," "forces," or "spirits" to create a healing response. Such beliefs currently have no place in the Western scientific paradigm.

A fundamental difficulty appears to be that Western science continues to be caught in a basic dualistic trap—that of considering the subject doing the mapping as separate from the map. Getting a more accurate map (more based on modern physics, more "holistic", more "systems") will not solve this problem. Rather, it may be useful to reflect on the possibility that thoughts are not merely a reflection on reality, but are also a movement of that reality itself. The map maker, the self, the thinking and knowing subject, may actually be a product and a performance of that which it seeks to know and represent (1–2). The critical epistemological issue is whether we humans have basically *one* way of contacting Reality (namely, through the physical senses) or *two* (the second being the deep intuition) (3).

The importance of the issue shows up in a central ontological question: is consciousness caused (by physiological processes in the brain, which in turn are consequences of the long evolutionary process) or is it *causal* (in the sense that consciousness is not only a causal factor in present phenomena, but also a causal factor throughout the entire evolutionary process)? Western scientific method urges toward the former choice, whereas some of the phenomena of alternative and complementary medicines suggest the latter choice.

There is in the medical world much faith that explanations of some of the claimed results in alternative and complementary medicine (as well as debunking of some other claimed results) will be forthcoming from this strengthening biomedical science. At the same time, much of complementary and alternative medicine does not fit even with the accepted new views of science. It seems to be true that, taken together, these diagnostic, therapeutic, and health-promoting practices pose a fundamental challenge to the metaphysical foundations of Western science. We wish to explore the extent and implications of that challenge.

INCOMPATIBILITIES BETWEEN ALTERNATIVE MEDICINE AND SCIENTIFIC WORLD VIEW

There are areas of alternative and complementary medicine where not only do we lack scientific models that would help explain the "mechanisms" of healing, but the models that do arise from the various complementary medical systems do not seem compatible with the Western scientific world view. Examples include the areas of homeopathy; acupuncture, qigong, and traditional Chinese medicine; ayurvedic (Indian) medicine; Tibetan medicine; and various practices utilized by First World peoples. Even herbal therapies, nutritional supplements, aromatherapy, meditation and biofeedback, guided imagery, body work, etc. may be of questionable compatibility.

Intentionality

The role of the mind and intentionality represents one of the key features of most alternative medical systems and is an important challenge to the Western scientific epistemology (4). Broadly defined, intentionality involves the projection of awareness, with purpose and efficacy, toward some object or outcome. It includes ways in which people are able to interact with their own bodies, such as in self-healing; ways in which people's intentions can influence others through direct or indirect communication, such as in placebo and "noxebo" effects; and more difficult to reconcile with our current scientific world view, ways in which intentions may influence others through some "non-local" means.

While psychological approaches that assume intentionality, including imagery techniques, have been used by alternative practitioners for eons in order to help mobilize the healing process, such concepts as psychosomatic illness, stress disorders, placebo effect, dissociation, and mind/body medicine, have met with considerable resistance. This is due, in large part, to the fact that consciousness as a causal factor is excluded from the scientific world view. Today, guided imagery is widely used for relief of chronic pain and other symptoms, and for accelerating healing and minimizing discomfort from injuries and illness symptoms (see 5–10). Imagery has been used to bring about healing from serious illness; spontaneous patient imagery has been used to better understand the meaning of symptoms or to access inner resources. But not only is intentionality implicit in all of this; a connection between imaginary image and real physiological effects is assumed.

As of now, we have no clear understanding of how this works within the context of our bio-medical models. The influence of a healers' or patients' intentions (including expectations) on the physical state of the patient puzzles or disturbs some medical professionals, and is a troubling artifact for many researchers. But it can also be viewed as an untapped resource in healing. Although the typical view of placebos is that they should be controlled or eliminated, they may in fact turn out to be powerful agents in linking intention, belief, expectations, and bodily responses. The challenge of delineating all significant variables is considerable and may be one reason that so little has been done to integrate placebos into clinical practice. More research could be done to analyze nonspecific factors including rapport, anticipation, and hope, in a way that begins to clarify their roles in healing. At the same time, we must develop reliable holistic methods and approaches that allow us to understand the healing relationship in other than reductionistic terms.

More difficult conceptually are the claims made by healers that they can use intentionality to access some form of "transpersonal" consciousness—consciousness that seemingly originates from a "higher source," passes through one person, the healer, to another, the patient—at a distance. The idea of transpersonal or "nonlocal" healing has widespread support in most cultures. Further, it is widely believed that people can obtain information about the world around them without any direct sensory contact. So we find, for examples, that in the Spanish folk medicine of *Curanderismo*, or in the *Obeah* practices of the southern Caribbean, or among the Kaluli peoples of the Papua New Guinea rain forest, healers believe that are able to physically affect other people at a distance through some kind of direct mental or spiritual interaction. What's more, they say they can heal this way without other people necessarily knowing about the effort–presumably eliminating any direct placebo effect. On the other hand, our Western scientific model holds these as observations that are impossible, despite increasing data to support the claims (11–15). But there is currently no explaining such results with accepted scientific concepts, even those that purport to involve "subtle energies."

Subtle Energies

Many alternative healing practitioners associate healing with what are commonly referred to as "subtle energies," or "fields." This concept may include electromagnetic and other energy fields throughout, and in the space surrounding the body, but it may also involve factors which are, at least so far, non-physically measurable (16). The admitted ambiguity allows a co-convening of those who believe all phenomena will eventually be understood in terms of energy fields which fit within the known models of science, together with those who find psychological and spiritual phenomena to involve aspects of reality not representable in terms of measurable fields.

There are many cultural instances of such "subtle fields." In the eastern cultures, we find such concepts as *prana, chi*, or *ki*, concepts that find no easy place in our scientific lexicon. Traditional Chinese Medicine, for example, is a comprehensive professional discipline, based on a complete system of thought. Within this epistemology, the human body is seen as a reflection of the natural world—a whole within a larger whole. Energy and fluids in the body are spoken of as flowing like channels and rivers. A medical diagnosis describes the body in terms of the elements—wind, heat, cold, dryness, dampness–concepts that have no place in Western diagnostic categories.

The complementary terms *yin* and *yang* are used by the TCM practitioner to describe the various opposing physical conditions of the body. *Yin* refers to the tissue of the organ, while *yang* refers to its activity. TCM also introduces a major component of the body, *qi*, that Western medicine does not acknowledge. This "vital life energy" flows through the body following pathways called meridians, which move along the surface of the body, and through the internal organs. According to this view, organs can be accessed for treatment through their specific meridians, and illness can occur when there is a blockage of *qi* in these channels. TCM incorporates a wide range of methods of treatment, including herbal medicine, acupuncture, dietary therapy, and massage. These have all become more or less acceptable in Western medicine; however, the conceptual model including the central *qi* concept is by no means compatible with Western science.

For another example of subtle energies within the framework of alternative medicine, we turn to homeopathy, which was founded in the late 18th century by German physician Samuel Hahnemann (17). It remains reasonably acceptable in northern Europe, where in one sense it is more or less integrated with conventional medicine. Within the American medical model, however, the underlying principles comprise a challenge to conventional medicine's concepts of illness and healing. These basic underlying principles are:

1. Like cures like; the same substance that in large doses creates symptoms like those of the disease, in minute dosage can be used to cure it.
2. Dilution increases potency; potency is greatest after dilution has reduced the amount below chemical detection.

3. Illness is specific to the individual; an illness generally defined will be treated in homeopathy only after finding the symptom pattern unique to the patient.

Like TCM, homeopathy is a complete self-contained alternative system of medicine that can supposedly reportedly have a therapeutic effect on almost any disease or health condition. On the other hand, the implied causality in homeopathy does not fit those of Western science.

POSSIBLE ALTERNATIVE ONTOLOGICAL ASSUMPTIONS

At a fundamental level, these alternative approaches imply pictures of reality that are not in accord with the Western scientific world view. The pose both an epistemological and an ontological challenge. It might seem more reasonable to take up the challenges one by one—the possibilities of subtle energies, the role of the mind in healing, the puzzle of intentionality, the mystery of remote diagnosis, etc. Science has often progressed by focusing on the simplest and most tractable case first, and later proceeding to the more complex.

However there is an alternate strategy which also has precedent in the history of science. Consider the origin of the evolutionary hypothesis. In the mid-nineteenth century there was much to be learned from studying separately the great variety of microorganisms, plants, and animals with which the planet is populated. But Charles Darwin boldly turned his attention to the synthesizing question: How can we understand *all of these together?* The result was the concept of evolution, around which practically all of biology is now organized.

There would seem to be an analogous situation in the multifaceted challenge posed by alternative and complementary medicine. We appear to need some sort of conceptual framework within which to understand a broad range of phenomena and experiences. What sorts of conceptual frameworks and organizing metaphors could be used to help us understand the many facets and dimensions of Western medicine and complementary medicine *all considered together?*

A step toward resolving this long-standing impasse may be the recognition that it is, in a sense, a historical accident that physics was taken to be the root science. That led naturally enough to such ideas as seeking objectivity through separating observer and observed; taking reality to be essentially that which can be physically measured; and seeking explanations of the whole in terms of understanding the parts.

But what if the study of living systems had been taken to be the root science, rather than physics (18)? Had this been the case, science would undoubtedly have taken a more holistic turn. It would have recognized that wholes are self-evidently more than the sum of their parts, and would have adopted an epistemology more congenial to living organisms. It might well have adopted a different ontological stance in viewing reality.

Such an alternative ontological stance was proposed by American philosopher Ken Wilber (1–2), that of considering reality as composed of "holons," each of which is a whole and simultaneously a part of some other whole—"holons within holons." (For example, atom-molecule-organelle-cell-tissue-organ-organism-society-biosphere.) Holons at the same time display agency, the capacity to maintain their own wholeness, even as they are also parts of other wholes. A holon can break up into other holons. But every holon also has the tendency to come together with others in the emergence of creative and novel holons (as in evolution). The drive to self-transcendence appears to be built into the very fabric of the universe. The self-transcending drive produces life out of matter, and consciousness out of life.

Holons relate "holarchically." (This term seems advisable because "hierarchy" has a bad name, mainly because people confuse natural hierarchy [inescapable] with dominator hierarchy [pathological].) Thus cell-holons are parts of organ-holons, which in turn are parts of organism-holons, which are parts of community-holons. For any particular holon, *functions and purposes* come from levels higher in the holarchy; *capabilities* depend upon lower levels.

In the holarchic picture of reality, the scientist-holon seeking to understand consciousness is in an intermediate position. Looking downward in the holarchy (or to the same level, in the social sciences), and exploring in a scientific spirit of inquiry, it is obvious that the appropriate epistemology is a participative one. Looking upward in the holarchy, it is apparent that the appropriate epistemology involves a holistic view in which the parts are understood through the whole. This epistemology will recognize the importance of subjective and cultural meanings in all human experience, including some religious or interpersonal experiences—that seem particularly rich in meaning even though they may be ineffable. In a holistic view, such meaningful experiences will not be explained away by reducing them to combinations of simpler experiences or to physiological or biochemical events. Rather, in a holistic approach, the meanings of experiences may be understood by discovering their interconnections with other meaningful experiences.

If this ontological stance is accepted, a good many seemingly opposing views in Western thought become reconciled. From the level of the human-holon, the scientist looks mainly downward in the holarchy; the mystic looks mainly upward. Science and religion are potentially two complementary but entirely congenial views; each needs the other for more completeness. In Western philosophy there have been three main ontological positions: the materialist-realist, the dualist, and the idealist. Again, the materialist looks downward, the idealist upward, and the dualist tries to reconcile fragments of the two—all represent but partial glimpses of the holarchic whole.

This new ontological stance takes some living with to fully appreciate how successfully it helps resolve many of the time-honored puzzles of Western philosophy—the mind-body problem, for example, and free will versus determinism. Since everything is part of the one holarchy, if consciousness or purpose is found anywhere (such as at the level of the scientist-holon), it is by that fact characteristic of the whole. It can neither be ruled out at the level of the microorganism, nor the level of the Earth, or Gaia. Nor need we be nonplused by evidence of alternative and complementary medicine about experiences that don't fit with a materialist, reductionist ontology.

IMPLICATIONS FOR SCIENTIFIC EPISTEMOLOGICAL ASSUMPTIONS

As within the presently dominant concept of medical science, the epistemology implied by this ontological stance, and to some extent defensible even without it (19), will insist on *open inquiry* and *public (intersubjective) validation* of knowledge; at the same time, it will recognize that these goals may, at any given time, be met only incompletely. Taking into account how both individual and collective perceptions are affected by unconsciously held beliefs and expectations, the limitations of intersubjective agreement are apparent.

This epistemology will be "*radically empirical*" in the sense urged by William James (19) in that it will be *phenomenological* or experiential in a broad sense. In other words, it will include subjective experience as primary data, rather than being essentially limited to physical-sense data. Further, it will address the totality of human experience—no reported phenomena will be written off because they "violate known scientific laws"). Thus, consciousness will not be conceptualized as a "thing" to be studied by an observer who is somehow apart from it; research on consciousness involves the interaction of the observer and the observed, or more accurately, the *experience* of observing.

This adequate epistemology will be, above all else, humble. It will recognize that science deals with *models and metaphors representing certain aspects of experienced reality*, and that any model or metaphor may be permissible if it is useful in helping to order knowledge, even though it may seem to conflict with another model which is also useful. (The classic example is the history of wave and particle models in physics.) This includes, specifically, the metaphor of consciousness. That may sound strange.

Indeed, it is a peculiarity of modern science that it allows some kinds of metaphors and disallows others. For example, it is perfectly acceptable to use metaphors which derive directly from our experience of the physical world (such as "fundamental particles," acoustic waves), as well as metaphors representing what can be measured only in terms of its effects (such as gravitational, electromagnetic, or quantum fields). It has further become acceptable in science to use more holistic and non-quantifiable metaphors such as organism, personality, ecological community, Gaia, universe. It is, however, taboo to use non-sensory "metaphors of mind"—metaphors that tap into images and experiences familiar from our own inner awareness. We are not allowed to say (scientifically) that some aspects of our experience of reality are reminiscent of our experience of our own minds—to observe, for example, that some aspects of animal behavior appear as though they were tapping into some supra-individual nonphysical mind, or as though there were in instinctual behavior and in evolution something like our experience in our own minds of *purpose*.

The epistemology we seek will recognize *the partial nature of all scientific concepts of causality* (see 20–21). (For example, the "upward causation" of physiomotor action resulting from a brain state does not necessarily invalidate the "downward causation" implied in the subjective feeling of volition.) In other words, it will implicitly question the assumption that a nomothetic science—one characterized by inviolable "scientific laws"—can in the end adequately deal with causality. In some ultimate sense, there really is no causality—only a Whole evolving.

It will also recognize that prediction and control are not the only criteria by which to judge knowledge as scientific. As the French poet Antoine Saint Exupéry put it, "Truth is not that which is demonstrable. Truth is that which is ineluctable." Here we find that the unquestioned authority of the objective and detached observer is challenged. In particular, the double-blind controlled experiment, considered the gold standard of clinical research, is thrown deeply into question if the consciousness of the experimenter or the clinician is causal (see 22). An engaged epistemology will involve recognition of the inescapable role of *the personal characteristics of the observer*, including the processes and contents of the unconscious mind. The corollary follows, that to be a competent investigator, the researcher must be *willing to risk being profoundly changed* through the process of exploration. Because of this potential transformation of observers, an epistemology which is acceptable now to the scientific community, may in time have to be replaced by another, more satisfactory new criteria, for which it has laid the intellectual and experiential foundations.

BROADER IMPLICATIONS

Science and society exist in a dialectical relationship. The findings of science have a profound effect on society; none of us have any doubts about that. But science is also a product of society, very much shaped by the cultural milieu within which it developed. Western science and medical science have the forms they do because science developed within a culture placing unusual value on the ability to predict and control.

Research on perception, hypnosis, dissociation, repression, selective attention, mental imagery, sleep and dreams, memory and memory retrieval, all suggests that the influence of the unconscious on how we experience ourselves and our environment may be far greater than is typically taken into account. Science itself has never been thoroughly re-assessed in the light of this recently discovered pervasive influence of the unconscious mind of the scientist or the healing practitioner. The contents and processes of the unconscious influence (individually and collectively) perceptions, "rational thinking," openness to challenging evidence, ability to contemplate alternative conceptual frameworks and metaphors, scientific interests and disinterests, scientific judgment—all to an indeterminate extent. What is implied is that we must accept the presence of unconscious processes and contents, not as a minor perturbation, but as a potentially major factor in the construction of any society's particular form of science. Again, we may have to reevaluate the role of the experimenter effect in outcome studies, as well as our firm reliance on double-blind control studies, and other assumptions about objectivity, materialism, and reductionism.

The implications of research in these areas go even further. They suggest interconnection at a level that has yet to be fully recognized by Western medical science. The ontological stance of the universe as holarchy appears to have great promise as the basis for an extended science in which consciousness-related phenomena are no longer anomalies, but keys to a deeper understanding; a science of medicine that transcends and includes the science we have. But the most important thing is not to accept a particular answer, but to open the dialogue about the metaphysical foundations of Western science and their relationship to understanding mind-body-spirit health and healing.

What assumptions underlie the attempt to marry alternative and complementary medicine to the U.S. allopathic, science-based medicine? On the one hand, these approaches encourage openness to whatever has seemed to work in the past; diversity of approaches for a diversity of persons; empowerment of the individual to choose and hence be more highly motivated. On the other hand, if alternative medicine in the U.S. is to be fitted into the fee, power, HMO, and assumption structure, it is likely to be subtly shaped by that structure so that its effectiveness may not be the same as in its original cultural context.

Besides the choice to ignore or adapt to the existing structure, there is a third choice—whole-system change (23). We need to look at the forces that might make this plausible. How might society move toward a really integral system of healing?

REFERENCES

1. Wilber, K. Sex, Ecology, Spirituality. The Spirit of Evolution. Boston: Shambhala, 1995.
2. Wilber K. A Brief History of Everything. Boston: Shambhala, 1996.
3. Harman, W. and DiQuincy, C. *The Scientific Exploration of Consciousness: Toward an Adequate Epistemology*. Research Report, Sausalito: Institute of Noetic Sciences, 1994.
4. Schlitz M. Intentionality in healing: mapping the integration of body, mind, and spirit. *Alternative Therapies* 1995;1(5):120–119.

5. Achterberg, J. *Imagery and healing: Shamanism and Modern Medicine*. Boston: New Science Library, 1985.
6. Achterberg, J. and Lawlis, F. *Imagery and Disease*. Champaign Ill.: Institute for Personality and Ability Testing, 1984.
7. Benson, H. *The Relaxation Response*. New York: Morrow, 1975.
8. Dacher, E. *Psychoneuroimmunology. The New Mind/Body Healing Program*. New York: Paragon House, 1993.
9. Dienstfrey, H. *Where the Mind Meets the Body*. New York: Norton, 1991.
10. Locke, S. and Colligen, D. *The Healer Within*. New York: E.P. Dutton, 1986.
11. Benor DJ. *Healing Research* (in four volumes). Munich: Helix Editions, 1993.
12. Braud W, Schlitz M. A methodology for the objective study of transpersonal imagery. *Journal of Scientific Exploration* 1989;3(1):43–63.
13. Braud WG, Schlitz MJ. Consciousness interactions with remote biological systems: anomalous intentionality effects. *Subtle Energies* 1991;2(1):1–46.
14. Schlitz MJ. Intentionality and intuition and their clinical implications: a challenge for science and medicine. *Advances: Journal of Mind-Body Health* 1996;12(2):58–66.
15. Solfvin J. Mental Healing. In: Krippner S, ed. *Advances in Parapsychological Research*. Jefferson, NC: McFarland and Co., 1984.
16. Schlitz, M.J. Subtle Energies and Consciousness: An Overview of Research. In *Where Eastern Philosophy and Western Sciences Meet: The Smithsonian Lectures*. New York: Random House, in press.
17. Jonas, W. and Jacobs. Healing with Homeopathy. The Natural Way to Promote Recovery and Restore Health. NY: Warner Books, 1996.
18. Harman, W. and Satoris, E. xxx
19. James W. *Essays in Radical Empiricism*. New York: Longmans, Green and Co., 1912.
20. Harman W. The scientific exploration of consciousness: towards an adequate epistemology. *Journal of Consciousness Studies* 1994;1(1):140–148.
21. Harman, W. and Clarke, J. (Eds.), *New Metaphysical Foundations of Modern Science*. Sausalito: Institute of Noetic Sciences.
22. Wiseman, R. and Schlitz, M. Experimenter Effects and the Remote Detection of Staring. Annual Proceedings of the Parapsychological Convention, San Diego, CA., 1996.
23. Harman, W *Global Mind Change. The Promise of the Last Years of the Twentieth Century*. Sausalito: Institute of Noetic Sciences.

THE WOUNDED BODY

Towards a Phenomenology of Health and Illness

Guillermo Michel

Universidad Autónoma Metropolitana
Xochimilco, Mexico

With the drawing of this love
and the voice of this Calling
We shall not cease from exploration
And the end of all our exploring
Will be to arrive where we started
And know the place for the first time.
—*T. S. Eliot*

THE BODY AS A PRESENCE IN THE WORLD

I have reached the stage of being, in the world, in this body. There is nothing closer to perception itself, and possibly nothing as obscure and amazing as this awakening and renewal of the experience of being-in-the-world, thanks to which I have learnt that it is only *"with the attraction of this Love,"* living flame, that all art, including the art of healing, is possible.

I am here, now, thanks to my body. Perceptible, audible, visible, tangible to all... In the impenetrable silence, I can state, without a shadow of doubt, "I am my body." It is true that, by saying this, I would seem to be saying, "I *have* a body," as though a consciousness outside me perceived me as "someone who has a body." However, by declaring that "**I am** my body," at that moment I recognize that consciousness fully occupies the body that utters this phrase. In other words, I am a body inherent and implied in a consciousness, or if you wish, an embodied consciousness, that is, incarnated in a body, whose presence, far better than any discourse, reveals the situation in which it is immersed to those who perceive it, in the same way that other bodies, by their presence alone, express their own situation: the very one we are currently experiencing, and in this situation we can be aware that consciousness shows itself as the anonymous, pre-personal life of our flesh.

Potentiating Health and the Crisis of the Immune System
edited by Mizrahi *et al.* Plenum Press, New York, 1997

However, despite knowing that I am my body, I cannot say that I understand it. Knowing and understanding, in this case, as in others, are not synonymous. Knowing that I am a body means that for an indefinite number of years, I am this being-in-the-world, indecipherable, alive, obscure, opaque, present here, now. Tomorrow, when I die, I shall cease to be a body and become a corpse, dust and ashes. I know that I am **my** body, simply undergoing the experience of being alive, like a "knot of relationships" in the multicolored fabric of Ex-istence. Indeed, my body determines whether I perceive or understand my world superficially or penetrate the density of Being. [Cf.Merleau-Ponty, Maurice. **Le visible et l'invisible**, p. 24.]

Therefore, by stating *"I am my body"* I discover that I have been shaped by my personal and social history, the people who have taken part in it, the culture in which I was born and grew up, my language, my own environment, my personal and familiar world, the particular way I have developed and existed..., projecting myself onto a life-world (*Lebenswelt*) that emerges as I approach it and in which I can never cease perceiving my own subjectivity as inherent. In this respect, saying that "I am my body" is equivalent to saying that *I am my world.* Not a representation nor an idea of both world and body, but the lively experience, pre-objective, of being-in-the-world, as an active, dynamic and concerned being.

In other words, my body is not like all the other beings in the world but the thread, the hinge, the tissue, the center of all my perceptions, the space across which specific smells (gunpowder, gasoline, roses) waft at the same time, together with the taste of the moment experienced and the view of trees, people or houses, while I feel on my skin the rough scab of the world and the rotten smell, that Hamlet noticed in Denmark, spreading throughout our world.

Once again, I undergo the experience, through all my senses, of being in the world as an embodied consciousness as if my senses slid over each other to weave a single perceptual fabric, a single, intercorporeal, existential plot, a single world, a unique **system,** in which the **phenomenological** body is a heart which animates and nourishes from within, just as our hearts give our bodies life.

The world/body relationship possesses a wealth of meanings, that are metaphorically expressed in the heart/body symbiosis, since we are all aware of the heart's symbolism. Not only as the center or engine of Ex-istence,* but as the seat of love, the erotic energy which imbues our lives with hope and passion.

From a purely biological point of view, we know that the heart animates the body, at the same time as it receives blood and oxygen, as a result of the multiple interweavings that constitute our bodies, and shape a system, whose parts depend one upon the other. Likewise, our phenomenological body form a sort of fleshy fabric out of its world: it animates and nourishes it from within, through its ceaseless, reciprocal interactions. Consequently, it is also possible to state, with absolute certainty, that the world is in our bodies, like the heart which gives it life, animates and nourishes it in many ways, throughout its experience of being-in-the-world: using it to create an eco-system, a living organism, a single multiple being, which Merleau-Ponty calls, with every justification, ***la Chair,*** *the Flesh.* [See Merleau-Ponty's **Le visible et l'invisible**, 199 ff.] And it is for this very reason that the body itself expresses total Ex-istence, since it is here that the latter is fully re-

* Here I make a distinction between "existence" and "Ex-istence". The former refers to any being, any thing, plant or animal. The second concerns the human way of being-in-the-world, intentionally aimed at the realization of one's life project, even if unconsciously or involuntarily. For Felipe Boburg, *"Ex-istence designates not only a subject's being, but movement in and out of the Flesh."* See, **Encarnación y Fenómeno**, pp. 95, 137 ff.

alized, so that distinguishing between body and spirit, or between sign and signified, or between visible and invisible is an abstraction. Indeed, what is the spirit if not the invisible 'side' of the body, and what is the latter if not the visible 'face' of the soul? And what are both if not living corporeality, an open being, a permanent questioning that opens itself to the world?

Thus, the phenomenology of the body that has been experienced implies a theory of perception and the latter, as noted by Jerry H. Gill, [Gill, Jerry H. **Merleau-Ponty and Metaphor**, Humanities Press, New Jersey-London, 1991.] should start from the phenomenological body, if it wishes to prove fruitful.

I think that the above reflections allow one to state that without delving into this experience of being a body, i.e. a **phenomenological body** (neither object nor subject), it is impossible to overcome the belief that we are an *object* of sciences (whether physiological or psychological). Therefore, we must rebel against this supposed *scientific objectivity* because reducing the body, or the psyche, to a mere object among objects, an empirical datum that can be clearly classified in the light of some scientific theory, or weighed, measured and turned into an index card, as a patient, means depriving it of its soul, which one could call undoubtedly *the light of the spirit*, which shines in the density of being-a-body -of being Ex-istence,- in other words, the possibility or project of being anonymous, living among living beings. [See Boburg, **op. cit.**, pp 91–106.] Perhaps for this reason, Merleau-Ponty warns, from the first pages of his **Phénoménologie de la perception,** that any scientific explanation starts from our perceptual experience of the life-world, without which *"the symbols of science would mean nothing,"* [**Phénoménologie de la perception**, Gallimard, p. ii] since the latter is only a fairly complex abstraction of what is real.

Therefore, I think there is a pressing need to overcome the dualistic view that has separated mind from body, the psyche from our being-in-the-world, and turned "my body," all human bodies, into yet another object among the innumerable objects present in our perceived world. It is therefore necessary for us to undergo the experience of being a body simultaneously with the experience of the world that has been lived through, since world and human bodies co-exist in communion, like a *chiasm* or *interlace*, [**Le visible et l'invisible**, 268. For further explanation see John O'Neill's **The Communicative Body: Studies in Communicative Philosophy, Politics and Sociology**, Northwestern University Press, Evanston, IL, p. 104 ff.] like a single duality, or a dual unity, with all the force of the interweaving that takes place in our flesh.

Indeed, I depend on the world to be able to understand and perceive that "I-am-**in**-the-world," to discover that "I am my body," mysteriously shaped by a time and space which I did not choose but which I am forced to *inhabit*, subjected to unknown forces whose presence I perceive when I am hurt, wounded, sick, anguished, or cast out of the world in a hospital bed or even my own bed.

THE BODY: VULNERABLE BEING

Our bodies rarely reveal themselves to us more mercilessly than when we feel "something is wrong" or "something hurts." It is then that we discover the experience of being a body as something which should be in harmony with its world and itself. Something threatens our lives: a fairly serious illness, an accident that leaves us badly hurt, damaged, lacking energy. "Why me?", we ask. "Why did I have to be affected by this disease or why did I have to break an arm or be hit by a stray bullet?"

Perhaps out of fear, or unawareness, we hardly ever face the protean, parasitic evil that emerges like a stinking sore from any part of our bodies, or everything that we regard as Evil, which disturbs our existential balance. I myself, victim of my own self-deception, have felt that Someone, outside, treacherously, as with the suffering Job, *"multiplies my wounds without cause, (and) will not let me get by breath, but fills me with bitterness"* (9:17–18).

My body resounds with the words of Job, not the only symbol of the "man of sorrow." It was a Man whose name is Jesus who received this name. I shall not attempt, like Carl Jung, in his **Response to Job**, to "give expression to a voice that speaks for many who feel the same," but rather, "give voice to the trembling" produced by realization that the Evil present in the world, throughout our history, reproduces itself absurdly, like a many-headed hydra. Consequently, it appears as an imperishable, destructive phenomenon, not only of bodies, lives, entire populations, but of hope itself. This, perhaps, is still locked up in the famous *Pandora's Box*, while all kinds of evil, present in our world, threaten to undermine and shrivel the seeds of faith and hope, when they discover, like Job himself, *"divine ferocity and lack of consideration."* †

In the density of the present, I attempt to converse with a Job who is always contemporary, since we can all be linked together in some way, in this present, and particularly in this part of the world. Job's presence, always so vital, does not only exist because I happen to recall it now, but because, by reading his words, I draw him out of his poetic world to communicate with him, in this very same present, in my presence. And by making him **present,** it is the very pain of **his** body, that speaks with **my** voice, and **our** words:

> Terrors are turned upon me;
> my honor is pursued as by the wind,
> and my prosperity has passed away like a cloud.
> And now my soul is poured out within me,
> days of affliction have taken hold of me.
> The night racks my bones,
> and the pain that knows me takes no rest...,
> But when I looked for good, evil came;
> And when I waited for light, darkness came.
>
> —*Job, 30:15–26*

The roots of my self-justification lie in the awareness of being **innocent.** In times of affliction and sadness, my body rebels, since it hopes for happiness and is met by unhappiness. My body, wracked with pain, glimpses a ray of hope in the friendly gesture of that Other who comes to meet him. However, I fail to find sufficient reasons to warrant the Good that will bring me joy or the Evil that burns my innermost parts: remorse, anguish, wounds in the flesh, a thousand fractures that make me aware of the vulnerability of this body I am and have been, and whose fears often paralyze it, since experiencing life makes the certainty of its own death even more distressful.

The wounded body, my own body, grows in this fertile, fragile earth. It is true that, as Existence, we are vulnerable; open to all winds, all diseases, all suffering. To our own

† Unlike Jung, who focuses his analysis of the **Book of Job** on the divine personality, in *"the image of a God without restraint in his emotions,* I shall try to emphasize my own emotions as a being-in-the-world, as a Man subject to pain, inclined to Evil, submerged and inherent in a world, where, as in a bloody ocean, human life elapses, without ceasing its search for happiness, well being and joy.

Self and the Other, who, like my own body, is open, and as such, vulnerable. Yet, as Levinas points out, this is not vulnerability as it is usually understood, since, *"vulnerability is obsession for the Other or the Other's approach: an approach that is not... the representation of the Other, nor the awareness of proximity... All love or hatred towards one's neighbor, like an attitude, reflects, assumes this prior vulnerability; pity, a stirring of one's innermost parts" (Jeremiah, 31: 20).* [Levinas, Emmanuel. **Humanismo del otro hombre**, pp. 122–125. Levinas points out that he thinks *"of the Biblical term* ***Rakhamin****, which is translated as pity, but contains references to the word* ***Rekhem****, or womb: it is a kind of pity like a mother's innermost emotion."*]

Thus, I am vulnerable not only because my body is exposed to pain and suffering, but primarily because, as regards pity, I see the wounded Face of the Other, his sick flesh, his flaws, and, since he is vulnerable, I can do nothing other than be-for-the-other, look into his naked Face and become responsible for his pain, his loss. Thus assuming responsibility means responding to a cry, nearly always mute, but which, in today's times, has risen to a shout, not only in the Lacandona jungle of Chiapas, in Mexico, but in numerous places throughout the world.

As far as Mexico is concerned, the Indians from Chiapas have denounced the way in which the **system** in which we live, known throughout the world as neo-liberalism, has caused them to be *"dying of hunger and curable diseases, while those in power are unconcerned that we have nothing, absolutely nothing, not even a proper roof over our heads, or work, health, food or education..., and no peace or justice for ourselves or our children."* [*"Declaración de la Selva Lacandona"*. January 2, 1994, in **EZLN: Documentos y comunicados**, Era, México, 1994, p. 33.]

From this point of view, the body itself is obviously not the body of someone who reveals himself as an anonymous Self but also, and far more deeply, the body of the poor, suffering people, hounded and excluded, whose vulnerability makes them "cannon fodder," exposing them, day after day, to "dying from hunger and curable diseases." Their faces, although concealed under frayed balaclavas, have revealed themselves in all their nudity and have made us dramatically aware that it is our Flesh, our social body, that is sick: it contains the cancer of corruption, the AIDS of indifference and selfishness, the hypertrophy of bureaucratization and the Alzheimer's of the forgetfulness of Being. As a macro-system and **eco-system,** we also undergo the experience in communion with jointly shared vulnerability. And for this very reason, as Levinas points out, *"despite myself, the Other concerns me...(and) the Ego which has returned to itself (is) responsible for the Other, in other words, he is a hostage to all..."* [Levinas, **op. cit.**, pp 110–111.]

These words may seem obscure to many of us since, once again, we conceive of the Other as a thing, not as a Co-existence, a Presence, a Face, an Openness. Like Kafka's **K**, accused by the "lord of the Castle," we have learnt to experience what Karl Jaspers called "anonymous responsibility" which is ultimately irresponsibility. We have even forgotten the ancient wisdom that arose from the fertile terrain of the Middle East, such as that of the **Talmud of Babylon,** whose anonymous author asks, *"If I do not answer to myself, then who will answer to me? But if I only answer to myself, am I still me?"* [Quoted by Levinas, **op. cit.**, p. 112. According to Dr. Shulamit Gunders from the Bar-Ilan University (Ramat-Gan, Israel), the author of the **Talmud-Mishna** is Hillel Hazaken ("the Elder"). She friendly made me know this during the 3rd. Dead Sea Conference. I appreciate her remarks on this subject. She also suggests that a better translation of this quotation could be: *"If I am not for myself who will be for me?* Thanks for her kindness.]

THE MYSTERY OF BEING-BODY

In the web of Ex-istence that is woven as we wander through the world, joining up with other Ex-istences, the phenomenon of illness is revealed to us like an unfathomable enigma. It would seem that Gaia, our living Mother Earth, has been mortally wounded, and is swinging crazily on her axis, dragging all the peoples of the world with her in a dizzying descent towards self-destruction. In this gigantic technological machine, that some call *Western civilization*, for whatever reason, we have all become "patients." Like Job, we have failed to discover where the hand that wounds us is hidden. When we most long for happiness, we are met by unhappiness, the closer we look at the light, the more we are swamped by darkness. Millions of innocent beings are thrown on to the streets of big cities every day to live and die, drugged, hungry, or victims of curable illnesses. Others are mutilated as the result of terrorist attacks. Thousands more die every day or are wounded in absurd, incomprehensible wars which reveal the bloated face of Power.

Our sick body, which is not only vulnerable but wounded and lacerated, as a vital, everyday experience, must be discussed in view of the global crisis of mankind, which shows the general imbalance, disharmony and discordance present in Gaia's body, in whose lap, as in Michelangelo's *Pietà*, lies the crucified Man. Because Man, as a totality, is not only each anonymous individual, each supposed Ego, but that being-in-the-world, "my body," part of the perceptual phenomenon from which the epiphany of being emerges: the *Flesh*, that invisible flimsy "tissue", or fabric, in which and through which we are all interlinked, and in whose weave we are one in everything, since everything is one, as Heraclitus discovered many centuries ago. [See **Le visible et l'invisible**, pp. 192–3.]

From this perspective, the Other reveals itself to us as a bottomless well, *"broader than the ocean and its islands"* in the words of the Chilean poet, Pablo Neruda. However, at the same time, it perceives in its own body the general vulnerability, and when it falls sick, it seems that it rebels in some way, and refuses to continue sharing its Ex-istence with other Ex-istences. The sick man is one who isolates himself, who narrows his perceptual field considerably and reduces the vast density of his Ex-istence and diminishes his vitality. As an experience, from a personal point of view, like Job, he feels that someone outside is beating and wounding him. However, when he talks to his own incarnate consciousness, his illness, the lacerated parts of his own flesh, the world he has experienced, or, in one word, his Ex-istence, he may be able to discover the roots of his illness far more accurately than from a discussion with a physician or the supposedly "objective" results of magnetic resonance imaging, an encephalogram or an X-ray. [To a certain extent, this is the method proposed by Thorwald Dethlefsen and Rüdiger Dahlke in **La enfermedad como camino** ("Sickness as Path"), Plaza y Janés, Barcelona, 1996. However, I think their proposal is based on a dualistic, Cartesian assumption, since it regards consciousness [the mind] as something transparent to itself, which can give matter [the body] orders and effect its cure.]

Yet the body, "my own body," is not that network of nerves, muscles, bones, cells and organs that compose it, but "embodied consciousness," living corporeality, which cannot be thought of separately. As Ex-istence, my own body is neither subject nor object, nor body-and-soul, nor an isolated individual or world, but a being-in-the-world, flesh of *Flesh*, being in the density of Being, unfathomable mystery. Hans-Georg Gadamer describes this situation of dual unity inherent to human beings when he states that *"the body, in any case, is what is alive; the soul is what animates this body. At bottom, each is so reflected in the other that any attempt at objectivization by one, without the other, leads, in some way, to the absurd"*: [Gadamer, Hans-Georg. **El estado oculto de la salud** (The Hidden State of Health), Gedisa, Barcelona, 1996, p. 113.] a true **chiasm**, a dual unity.

Nevertheless, it is not enough to indicate and reiterate the way the entire soul is in the body, nor the way all consciousness is in-flesh, since like "being in the world," like Ex-istence, each one is also "the heart" which animates the spectacle of the world from within, thereby forming **a system.** For this reason, the phenomenon of the sick human being cannot be understood if one does not also understand the mundaneness inherent in the body and the phenomenon of humanity inherent in the world: its cruelty, its mysterious uncertainty, its impenetrable opacity. Consequently, this situation places us in an ambiguous perceptual field, since no embodied consciousness will ever be able to become transparent to itself or to the Other, rooted as it is in this anonymous being, individual and collective unconscious, which inhabits the body itself, as though it were too an Other. The illness perceived in my body, as an ailment, as pain, as a wound, as weakness, as cancer, or as simple discomfort, appears, on the one hand, as vulnerability, but on the other hand, this same vulnerability is, in a paradoxical way, my perceiving body. In my own body, I undergo the experience of discomfort and the being-perception of discomfort, since *"to perceive is the same as being perceived."*[Boburg, Felipe, **Op. cit.**, p. 146.] Nonetheless, no one can tell precisely who is perceived and who is perceiving, as when I stare at the eyes of anybody else gazing at me: who is then the perceiver?

Therefore, in the same way, it is impossible to establish precise limits between well-being and discomfort. I think that we can only hope to see the health-sickness link from the experience of Ex-istence, in which the body is immersed; in other words, from the shadows where it dwells because of its human condition. While one is on harmony with oneself, **in** oneself, and with one's world, one will experience balance and harmony in the flesh; that invisible state of health, which may be expressed as being *"one in everything,"* according to Heraclitus. Insofar as discomfort, disharmony, uncertainty, insecurity, the feeling of being unprotected, occupies *everything* (our world) or *one* (our individual body) as part of everything, the experience of illness will make itself felt in our bodies, in Co-existence.

This situation may become clearer if we compare the state of health to the experience of listening to a symphony, in which all the instruments play the score perfectly, with no discordance or wrong notes. It takes just one instrument to be out of tune or come in at the wrong moment to destroy the harmony and cause the perturbation and discomfort of both the players and the audience.

This analogy may help us understand why Plato compared rhetoric to the *art of healing* and why he suggests that it is impossible to cure the body without knowing the soul and understanding the nature of everything. Recalling this passage from **Phaedrus**, Gadamer considers that *everything*, according to Plato, is what *"shapes the movements of the stars, the climate, the composition of the water, the nature of the fields of crops and woods, which surround the general state of man and the risks to which he is exposed. Medicine seems to be a genuine universal social science, especially if one adds the sum of our social world to this."* [Gadamer, **op. cit.**, p. 130.]

Thus for at least the past two thousand five hundred years, a holistic art of healing has been advocated that is not merely somatic, psychic, or even psycho-somatic but Ex-istential. An art of healing that includes man as Co-existence, as a living body among living bodies, rooted in a world, focused on it, projected to it, and linked in some way that is neither visible, nor conscious, but intentional, with everyone in it. Just how far away we are from achieving this is shown by the multiple specializations and sub-specializations through which not only the body as a whole but each of its parts and functions are increasingly objectivized. And, since this is the reality of our situation, it should come as no surprise that all over the world, diseases are proliferating whose visible expression is to be found in the crisis of the immunological system. How then, are we to fulfill ourselves as responsibility, and how are we to ensure that our own vulnerability is effectively shown as pity, i.e. *a stirring of the innermost emotions*?

It is my hope that we may be able to clarify some things by this means, and, I believe, the hope of many, although, at the moment, to keep this hope alive, we shall have to follow the now-forgotten motto of the great medieval physician, Paracelsus, regarding the art of healing, *"The most necessary of all things is pity... (since) where there is no love, there is no art."* And this brings us back to our starting point with T.S. Eliot:

Not farewell,
But fare forward, voyagers.

—*Zichron Yaakov, Israel, October 1996*

REFERENCES

Anónimo. *Libro de Job* [Book of Job], en **La nueva Biblia latinoamericana** [The New Latin American Bible], Ediciones Paulinas, Madrid, 1972.

Boburg, Felipe. **Encarnación y fenómeno** [Incarnation and Phenomenon], Universidad Iberoamericana, México, 1996.

Dethlefsen, Thorwald y Rüdiger Dahlke. **La enfermedad como camino** [Sickness as Path], Plaza y Janés, Barcelona, 1996. (Original: **Krankheit als weg**, Bertelsmann Verlag, München, 1983).

Eliot, T.S. **Four Quartets**, Bilingual Edition and Translation by Esteban Pujals-Gesali, Red Editorial Iberoamericana (REI), Mexico, 1991.

EZLN. **Documentos y comunicados (I)** [Documents and communiqués], Prólogo by Antonio García de León, Ediciones Era, México, 1994.

Gadamer, Hans-Georg. **El estado oculto de la salud** [The Hidden State of Health], Gedisa, Barcelona, 1996. (Original: **Über die Verborgenheit der Gesundheit**, Suhrkamp Verlag, Franfurt am Main, 1993).

Gill, Jerry H. **Merleau-Ponty and Metaphor**, Humanities Press, New Jersey-London, 1991.

Jung, Carl G. **Paracélsica**, Kairós, Barcelona, 1988. (Original: **Paracelsica**, Rascher Verlag, Zurich, 1942).

———. **Respuesta a Job**, Fondo de Cultura Económica, México, 1992. (Original: **Antwort auf Hiob**, Rascher Verlag, Zurich, 1952).

Levinas, Emmanuel. **Humanismo del otro hombre**, Siglo XXI, México, 1993. (Original: **Humanisme de l'autre homme**, fata morgana, Montpellier, 1972).

Merleau-Ponty, Maurice. **Phénoménologie de la perception**, Gallimard, Paris, 1945. (**Phenomenology of Perception**, Transl. by Colin Smith, Atlantic Highlands, N.J.: Humanities Press International, 1963).

———. **Le visible et l'invisible** (Texte établi par Claude Lefort), Gallimard, Paris, 1964. (**The Visible and the Invisible**, edited by Claude Lefort, transl. by Alphonso Lingis, Northwestern University Press, Evanston IL., 1968).

Neruda, Pablo. **Canto General** [General Chant], Losada, Buenos Aires, Argentina, 1955.

O'Neill, John. **The Communicative Body: Studies in Communicative Philosophy, Politics and Sociology**, Northwestern University Press, Evanston, IL, 1989.

8

FOOD AS MEDICINE FOR BODY AND SOUL

Geraldine Mitton

Medical Director
Cleto Saporetti Foundation
High Rustenberg Hydro
P. O. Box 2325
Stellenbosch, 7601, South Africa

Let Food be thy Medicine and thy Medicine be thy Food.
—*Hippocrates 460–377 B.C.*

ABSTRACT

There is a great body of evidence which clearly shows that nutritional factors including phyto chemicals are beneficial in prevention of cancer, heart disease, chronic degenerative disease, and premature aging. Anti oxidants, bioflavonoids and lycopenes as well as specific plants and foods with healing properties, are discussed. A modified fruit fast acts as a catalyst to promote inner healing and enhanced immunity. A mainly vegetarian regime would seem to be the most beneficial dietary program to promote optimal health.

INTRODUCTION

Today it is generally recognised that we are, in fact, what we eat, and that if we do not eat correctly, we invite illness, chronic degenerative disease, and premature ageing. Modern science has shown that certain plants are rich in substances which may be as effective as any pharmacological medication prescribed by doctors. These substances are known as phyto chemicals or nutri ceuticals, and the foods are sometimes known as functional foods. Recent research has shown that many common foods for example, carrots, onions, broccoli, cabbage, beetroots and artichokes, to mention just a few, each contains ingredients which are beneficial in treatment or prevention of illness. In general the trend today is towards eating a wide variety of plant foods and there could well be a major revolution in the therapeutic communities within the next decade.

Fasting has been a therapeutic remedy for many centuries, and has been used in many countries throughout the world. There are many types of fasting including the pure

Potentiating Health and the Crisis of the Immune System
edited by Mizrahi *et al.* Plenum Press, New York, 1997

water fast, herbal teas, the Mayr cure, or the more popular mono fruit fast. This latter diet has been shown to be an effective form of cleansing and detoxification. Not only is fruit palatable, but it has a low glycaemic index, and the person undertaking the fast does not suffer from hypoglycaemia. If this is followed after a period of several days by a mainly raw food vegetarian diet, the overall effect is one of enhanced energy and wellbeing. It has been our experience that while patients are undergoing an elimination and detoxification diet in this fashion, they are susceptible to lifestyle modification including smoking cessation, increased exercise and general health related practices.

Research has shown that diet is a significant factor in treatment of heart disease. The recent lifestyle heart trial conducted by Ornish in the US, has shown that it is possible to reverse coronary heart disease by prescribing a ten percent fat vegetarian diet together with relaxation exercises and physical activity, for a period of at least a year. Certain foods have been found to be cardio-protective. These include garlic, onions, avocado, walnuts, legumes, olive oil, red wine and certain herbs for example, rosemary. Garlic has also been shown to be effective over a period of time in helping to lower total serum cholesterol in hypercholesterolaemic individuals. The cardio-protective diet has gained popularity in the form of the Mediterranean diet. The so-called 'French paradox' spotlighted red wine which has significant anti-oxidant properties. Although red wine possesses about six times more phenols than white wine recent studies have shown that white wine may also be effective. Fruit and vegetables have also been shown to be protective against stroke.

Vitamin E is the major anti-oxidant in benefiting heart disease. The Cambridge Heart antioxidant study examined 2000 high risk patients and found that Vitamin E reduced cardiovascular deaths by 36% and non-fatal Myocardial infarction by 66%.

Bioflavonoids have been prominent in the scientific press, including such foods as citrus fruit, tomatoes, green peppers, pawpaw, broccoli, grapes, cherries and dark red berries, red wine and Chinese green tea. Dietary flavonoids can protect against cardiovascular disease by scavenging free radicals and protecting LDL against oxidation.

It is said that 35% of cancer cases are due to incorrect diet, and once again certain foods have found to be protective, particularly against cancer of the head and neck, lung, and digestive tract. Phyto chemicals are found in broccoli, brussels sprouts, cabbage, carrots, yellow fruit and vegetables, to mention a few. The emphasis is on consuming a diet which is high in vegetables and fruit. These phyto chemicals have been shown to inhibit the growth of specific cancers. Curcumin, found in the rhizomes of the plant Curcuma Longa, has attracted interest because of it's possible chemo preventive effects. A high fibre diet including both soluble and insoluble fibre may reduce the incidence of colon cancer, and the spotlight here is on the beneficial effect of legumes.

Probiotics are useful for colonic health, to combat yeast overgrowth in the gut and to enhance the immune system. Probiotics such as the lacto-bacilli and bifid bacteria in yoghurt and fermented foods, whey and certain plants are protective against colon cancer. Licorice root contains over four hundred medicinally active phyto chemicals. It is a patent adaptogen and is useful in enhancing immunity. Soya products are an excellent source of vegetable protein. They also have lipid lowering effects and are a rich source of phyto-oestrogens.

Juicing therapy has been a popular mode of treatment for many years, and has been used by naturopaths and many health practitioners. The benefits of using fresh vegetable juices aids in cleansing and detoxification, and also provides many plant enzymes in the concentrated form. Juices containing celery, beetroot, parsley and carrot juice are excellent for general cleansing and liver support. Cranberry juice is beneficial in preventing urinary tract infection. Potato juice is useful in prevention of gout. Various green juices, for example wheat grass, parsley, spinach, beetroot, greens, water cress, dandelion, barley grass and alfalfa have been used to help boost immunity and for those people undergoing chemotherapy and radiation. Chlorophyll rich juices have a cleansing effect on the bowel and lymphatic system.

Lycopenes have gained attention as being a most potent source of carotenoid, thus tomatoes and red grapefruit are valuable foods to be included in one's diet.

The glycaemic index has been useful in categorising appropriate foods for diabetics. Foods with a high index such as sugar will stimulate the pancreas and encourage a rise in blood sugar, whereas food with a low glycaemic index such fruit, vegetables, legumes and whole grains will create a stable normal blood sugar. This of use not only for diabetics, but for people who are requiring weight control, and also for those suffering from hypoglycaemia.

Whereas there is a relationship between diet and psychological symptoms, that is, food and mood, the link between nutrition and mental illness remains somewhat controversial. It is true that the food we eat and the nutrients therein, do have an effect on the brain, on the nervous system and on immunity. Thus, nutritional advice should be available for all patients who have psychological and mental problems. Another important factor is the problem of food sensitivities whereby certain foods such as wheat, dairy, citrus fruits and chocolate as well as additives and preservatives may cause symptoms. Thus food sensitivities should be ruled out in psychological problems including depression.

The general concensus is that we should include more plant foods in our diet. A study by Knekt et al in Helsinki measured the diet of 5000 Finnish men and women, and followed them for fourteen years. They found a 34% reduction in the risk of fatal heart attack in the third eating the most fruit and vegetables compared with the third eating the least. Manson et al. in Boston, measured the diet of over 87,000 American nurses, following them up for eight years, and found a 26% reduction in the risk of stroke in those eating the most vegetables. It has been found that vegetarians have an estimated 25% lower incidence of hypertension, stroke, ischaemic heart disease, diabetes, constipation, gallstones, and arthritis. The International bodies have recommended a mean consumption of 400 gm or five portions of fruit and vegetables per day which would appear to be in reach of the general population. Hertog et al from the National Institute of Public health, in Bilthoven found that the risk of heart attack was 49% lower in men eating 110 gm or more of apples a day. Thus, it may be that an apple a day really does keep the doctor away.

Modern man has strayed a long way from our ancestors' 'Hunter-Gatherer' diet, venturing into the realms of gastronomic pornography. Eating should be a pleasure. Meals should be unhurried. We need to chew our food thoroughly, mixing it with our salivary juices before it enters the stomach. When we consider that during our lifetime we consume approximately twenty five tons of food, it is important that we do all we can to aid and facilitate digestion and assimilation of the nutrients and phyto chemicals by ensuring that we have healthy livers and a healthy gut. If we wish food to be our medicine, we need to know that it is adequately absorbed.

SUMMARY

Together with convincing evidence that we need to increase our intake of fresh plant foods there is exciting new research in the field of phyto chemicals encouraging us to add variety to our daily diet and to use specific plants for treatment and prevention of disease.

REFERENCES

Stephens et al. Cambridge Heart Antioxidant Study. (CHAOS) Lancet 1996: 34; 781–78.

Hertog et al. Flavonoid intake and long term risk of Coronary heart disease and cancer in Seven Countries Study. Arch Intern Med 1995; 155: 381–386.

Knekt P. et al. Am. J. Epidemial 1994; 139 : 180–9.

'Diet, nutrition and the prevention of chronic diseases' Report of WHO Study Group. 1990.

Hertog et al. ibid; 1993; 342, 1007–11.

Effective Nutritional Medicine. Position Paper by BSAEM and British Society of Nutritional Medicine. 1995.

Dwyer J.T. Health aspects of vegetarian diets. Am. J. Clin Nutri 1988: 48: 712–38.

Beecher C. Cancer preventive properties of Brassica. Am. J. Clin Nutri 1994; 59: 11665.

Ornish D. Lifestyle Heart Trial. Lancet 1990: Vol 336; July 21: 1299–133.

McNutt K. Medicinals in Food Nutrition Today 1995; Vol 30: 5: 218–22.

THE MAGIC OF THE IMMUNE SYSTEM

Shin-ichiro Terayama

Japan Holistic Medical Society
4–28–14–210 Asagaya-kita
Suginami-ku, Tokyo 166, Japan

ABSTRACT

Disease, like cancer, changes human beings' attitude toward their existence and during the healing process, people have an opportunity to change their life style magically. When healing in the body happens through positive action, in a harmonious way, the body comes back into balance naturally.

There are many miraculous things about the immune system that become clear during recovery from cancer. Contemporary medicine seems to forget the words "natural healing power of the body" which Hippocrates wrote about more than two thousand years ago.

This process happens beyond scientific comprehension. It transcends mind, body and spirit, to reach the love of God.

I was a physicist with cancer eleven years ago. I transformed, and have been free of metastasized kidney cancer for seven years. I tell the story of my recovery from cancer with cello playing, confessing how I loved my cancer instead of fighting it. I changed to a vegetarian "macrobiotic" diet, selected good mineral water, and most importantly, watched the sun rise in the morning. It was in front of the morning sun that I made an exciting discovery. I found I was becoming very positive, very relaxed and healing energy was entering me first through my heart, and then all seven chakras. I began to play cello again after a long absence. These things were practiced harmoniously by intuition and not by instruction.

Now is the time for modern medicine to recognize the ancient wisdom of human beings to bring about their own recovery from cancer. The healing power within will save many people with serious and chronic diseases.

1. INTRODUCTION

A long time ago, in ancient times, people were born naturally and died naturally. During their lives people survived by the natural healing power of the body and ended

Potentiating Health and the Crisis of the Immune System
edited by Mizrahi *et al.* Plenum Press, New York, 1997

their lives. In every moment of lives healing in the body happened by immune system of the body and balanced automatically and kept the health. Everything was closely connected with nature itself.

AT the beginning of Western medicine, Hippocrates wrote: "Disease can be healed by nature" more than two thousand years ago. But now year by year the progress of modern medicine seems to be far from Hippocrates. It is more scientific than before but it loses the intuition and the heart of medicine and more over some medicines have strong side effect. Sometimes it kills the immune system of the body and consequently weaken the power of natural healing of the body and apt to lead the people to death. So it is the era of the crisis of the immune system.

2. WISDOM OF MEDICINE: HEALING SYSTEM

Originally human body has the healing system in the body. For instance when the person cut the surface of the skin there occurs the pain and then the wound will be naturally healed after some time as the immune system is good condition. But when the immune system is weak so sometime have a pus and it takes more times than he is in good health.

Being good health means that the immune system is strong and it is the key to recover quickly. Every body have its own healing system in the body even if it is strong or weak and always heal the body naturally. But nowadays most people forget the importance of the healing power of the body and apt to rely on to take medicine instead of increasing the healing power of the body.

When the medicine has the strong effect it has also the strong side effect and sometimes it affects the other organ of the body and consequently induce another disease of the body.

Actually today modern medicine has been successful to treat the disease in many cases especially in the treatment of trauma, bacterial infections and surgical emergencies. These treatment became dominant in the field of medicine and medicine became allopathic medicine. Most allopathic medicine use more refined medicine and became more dangerous and more expensive so side effect of the medicine results to weaken the healing power of the body and in addition reduce the immune system of the body.

Patients are apt to rely on the doctors and doctors treat patients with allopathic medicine and the importance of the natural healing power of the body is forgotten.

Now it is called the time of the crisis of the immune system and medicine induce many cases of tragedy which do not recover from the disease. For instance the major treatments for cancer patient are as follows:

1. Operation to remove tumor
2. Chemotherapy
3. Radiation therapy

and these treatments kill the immune system of the body eventually.

In Japan cancer is the number one death rate for the causes of death from 1981 and still the rate is increasing year by year. Almost 30% of the people die by cancer and it is believed to be most incurable disease by the doctors so ordinary people are apt to believe cancer as the most incurable disease and they are afraid of it.

But is it true? I met many people who healed the cancer themselves during these ten years of my recovery and medical practitioners do not believe it due to a little document of medical record.

Dr. Andrew Weil, MD., a graduate of Harvard Medical School, who is currently Associate Director of the Division of Social Perspectives in Medicine and Director of the Program in Integrative Medicine at the University of Arizona in Tucson U.S.A., wrote about it in his book "Spontaneous Healing" that the stories like spontaneous healing are just stories, not taken seriously, not studied, not looked to as sources of information about the body's potential to repair itself by the medical doctors.

3. MY CASE: RECOVERY FROM CANCER

I was the cancer patients of right kidney in 1984 and had major treatments of the cancer mentioned before and eventually I faced with the near death experience. Maybe I was in the midst of the crisis of the immune system at that time.

3.1. How It Was in the Beginning

I shall tell my story how it was and how I healed myself.

It was the Autumn of 1983 I felt something wrong in my body when I had a very high fever lasting a month. Until that time I was very healthy person and very strong and could sleep deeply even in a short time. So for me to be a patient was beyond my thought. I could not stand or walk at that time and went to several hospitals to check but all medical tests were normal.

As I was a solid state physicist and formerly engaged with the development of medical test equipment I understood the low accuracy of medical test. If patients feel not good but it still can not measure by the machine. This is the level of modern medicine.

But even though I did not believe the accuracy of the medical test machine I had a complete faith in doctors and hospitals. A few months later on a cold February morning after my walk, I found the urine red in the toilet and I felt something thrill in my mind. It was very beautiful red and was the first time to see such a red urine in my life. I went to the hospital to consult the doctor but nothing found out by the medical test and after that I was very tired.

At that time I was completely a workaholic. As I worked without rest for 18–19 hours a day and even midnight and after I talked with people.

3.2. A Friend Gave Me a Chance to Change

Fortunately, I have a wonderful friend, Koshiro Otsuka, who is a holistic medical counselor. At the beginning of February, he came to my office. We talked, and he looked at my face and said, "Your kidney is a little bit wrong". Only by observation he said this. He said, "Your diet is completely wrong. You should avoid eating meat, chicken, eggs, chocolate, sweets, and don't drink coffee." At that time, I ate meat a lot, and drank ten, maybe twenty cups of coffee a day. He said, "You need to change your diet, eat brown rice, organic vegetables and fruits." He said this six months before I was diagnosed by my doctor!

But I couldn't understand or believe in my diet, or what he was talking about. I continued to eat meat. When I got tired, I ate even more meat, for energy! I thought it was good for me.

My red urine continued every month, like a lady's menstruation. Four times, every 28, 29 days. Gradually, I felt very tired so I couldn't work. So I went to consult the doctor, but he couldn't find anything.

Six months passed. My health got worse and worse. After many tests, my doctor said that my right kidney was twice as large as ordinary, and he recommended an operation.

He told me it was a tumor, but he did not mention cancer, and I confess I had no medical understanding about my disease. Previously, I was so healthy! It's a funny thing, but I always called myself a doctor of companies!

I asked my doctor, why did I suffer from my kidney?

He said, You're so nervous, don't worry.

He didn't answer my question. As a physicist [Terayama's major study at university and subsequent early career], my question is always "why".

Finally, the doctor said that there is no why in the Japanese medical textbooks. The system of education always says when there is something wrong to have an operation first. Of course, some prominent doctors are very sensitive to their patients' questions – it depends on the individual. But mostly, they never answer why, and that is important, I think. In Japan, the diagnosis of cancer is still routinely with held from patients lest it depress them unduly. This leads inevitably to subterfuge.

Anyway, on December 4, 1984. I had the operation to remove my kidney.

After that, the doctor said injections are necessary for prevention and it has a strong side effect. The first day, the second day, I didn't feel anything. On the third day, I felt something bad, so I didn't eat anymore. On the fourth day, when I saw my face in the mirror, I had a big surprise. My beard had turned white. Overnight!

I asked my doctor to stop this treatment, but he said it was very important. He strongly insisted I finish. Actually this treatment was cisplatinum and later I knew by myself. My hair began to fall down and soon after I was almost bald.

Maybe three or four weeks after, the doctor said it's better for me to have emission several times. I thought it would be like artificial sunlight, because at that time one of my friends used infrared light for better health. Actually, it was high-energy radiation. X-rays. I had it 30 times, 180 rad each time.

My condition became worse. My red blood and white blood counts decreased day by day. December, January, February, March, gradually my vitality went down and down.

After my operation, Mr. Otsuka brought me a macrobiotic book on the life of Georges Ohsawa. Inside, he wrote, "Without food, there is no life." That impressed me very much. Sometimes I asked my wife to bring brown rice onigiri[rice balls]. But still I didn't understand the relationship of diet to health.

3.3. Near Death Dream

Then around 5:00 am in the morning of March 4, I had a dream. A very strange dream. My body is lying in a coffin and I am watching from the ceiling. Many people are coming to see me. Just before the lid of the coffin is to be closed, I yell, "I'M LIVING!" This woke me up.

I told my wife about that dream. She cried and said, "That's a good dream. She knew that my life wasn't supposed to be longer than a few months. There was another tumor in my rectum; also, a shadow in my lung. I was almost like bones.

About that time, my taste in music changed. Previously, I loved Bach, Beethoven, Brahms, strong classical music. But after this strange dream, I was afraid of listening. This kind of powerful music that reached the inner depths of my heart was too much for me. But one melody continued to play in my ear. I found out what it was after I got out of the hospital and attended my friend's funeral. They played it there—it was funeral music! "Ave verum" by Mozart.

3.4. Sense of Smell Changed

Suddenly, my sense of smell dramatically increased. I couldn't stand to stay in the hospital bedroom because of the smell of antiseptics and other people's bodies. I could even tell the nurses apart just by their smells. I was like a wolf, a wild animal. I couldn't eat the hospital food anymore-it didn't smell good. I tried to find where there was no smell. My hospital had six floors and a roof, and my bed was on the second floor. I finally found that the only place with no smell was on the roof top, so 9:00 pm after the lights were shut off, I went up there. When the next check came at midnight, the nurses came to my bed and I wasn't there. They looked all over for me. Finally, about 12:30, they found me on the roof in a blanket and sweater. They insisted I come back to bed, and five nurses carried me, like luggage. Next morning, my doctor came to me and said, "You caused so much distress last night, the nurses were worried, they thought you had jumped off the rooftop and committed suicide. If you want to stay in the hospital, that's OK. But if you want to return back home, that's also OK—it's your choice".

"Of course, I want to go home," I replied.

If I had committed suicide, maybe it would have done some damage to the hospital.

So I was saved.

After I returned home, Mr. Otsuka came to my home and once again, he told me everything about how to change my diet. This time, I did. Very gradually, I got stronger. I could move, I could walk.

3.5. The Sun Is the God

One very wonderful thing also happened at this time. Every morning, I thought to myself, "Oh, I am still living, that's wonderful." I wanted to see evidence of the morning. I live in an apartment, and from the roof I can see the sunrise. So every morning, I would use the elevator, go to the roof and pray to the sun as it was coming up. As I watched the sunrise, I would meditate, concentrating on healing my seven chakras and chanting "Namu Amida Butu"[a Buddhist chant meaning "I sincerely believe in Buddha]. Meditation is really good to increase the body's immune system. Any kind of meditation.

I thought, this is wonderful, I can see a sunrise. And praise the sun, thank the sun. It's so simple. And I'm still continuing it.

On one occasion I had a most profound experience. The sun poured right into my heart when I opened my hand when I was playing. I burst into tears. When I returned to my family, I was surprised to see that everyone had auras. I'm sure they had them before, but I had never been gifted with the ability to see them before. Everyone was like Jesus Christ.

I also noticed when I saw the sunrise every morning that the birds were already singing. I tried to find out when they begin to sing. I got up earlier and earlier. Finally, I checked the time they started; it was about 40 minutes before sunrise and changes according to season. I'm a physicist, so I asked why, what happens?

One morning, it was still dark. I was sitting in the woods near my house and looking at the birds on a tree, and I noticed something coming from the back side of the leaves. Oh, this is oxygen, I thought. So I did a little experiment. I bought a small gas cylinder and waited until night. My family kept small birds. At about 11:00 p.m., I sprayed a little bit of oxygen to the cage. The bird began to sing suddenly. I found out! The birds know, they understand the wonderful gift from nature, the sun's energy. Most humans in modern society have lost such intuition.

Rubber trees also produce their juice just before sunrise, and never during the day—a special coincidence.

From that point, I became very anxious about the environment of our planet earth.

Around June, I suddenly said to my wife, "I'm sure to cure my cancer." She said to me, "You knew it!" But I didn't know I was suffering from cancer. Yet some part of me did. When I realized that, it was enough. My condition was getting better and better.

3.6. Encounter with Holistic Medicine

From August 1985, I started going to Hotaka Yojoen [a natural food and health retreat in the mountains]. During my stays there, I had strong exercise; yoga, jogging, hiking, and deep breathing with meditation.

Around that time, I still had difficulty of walking. But after Yojoen, I changed.

3.7. Cello Is a Healer

I started cello again. When I was a schoolboy, I was trained as a professional player. But when I joined Toshiba [right after university], I stopped playing. During my recovery, I sometimes didn't like to listen to music, but I began playing the cello again. I think it is a wonderful exercise: using muscles, using the heart, and using energy. Maybe there's something about the vibration, too, that helps the body.

I had much time to reflect. I learned so much from my cancer. I was changed so much by my cancer.

3.8. I Loved My Cancer

The most important thing I learned is that cancer is my body. It's not an enemy; it's still my body. And I create my cancer. It's like my child. So I said to my body, "I made a mistake. Oh, I'm very, very sorry. But are here. So I love you." Many image methods see cancer as a enemy to defeat and to fight. But for me, no.

I thank my cancer always because I changed so much because of it.

3.9. Love of Findhorn Was Strong

In February of 1988 I received an invitation from Findhorn Foundation as a speaker of the conference in October. At that time still I had a small shadow in my right lung. I talked at first to my wife. She said it was early to go abroad for my body. I asked many friend about that but there no reply to go.

Finally I asked a friend of mine, who is the excellent translator and had a channelling ability, and she said by phone that by channelling I was to go Findhorn and my spirit was changed to go there. I was delighted to hear that so I decided to accept the invitation and sent FAX to Findhorn to go.

Actually I came to Findhorn and stayed there for about 40 days. It was my first experience to me to hug. Usually Japanese are very shy to do that but it was a wonderful moment every day and after I returned back from Findhorn my shadows of the right lung disappeared completely. There was the true love in Findhorn and by accepting the love from people I could healed by myself.

It took me almost 3.5 years to be healed and love action had a strong power to heal. After that I recommended Findhorn for many cancer people of good will who feel the power of healing. To love and to be loved is the best medicine for all the people and there are no side effect at all.

4. DEFINITION OF HOLISTIC MEDICINE BY JHMS

During my recovery from cancer the Japan Holistic medical society (JHMS) was established in 1987 and I joined as one of the founder of the Society. Through our activity I felt the meaning of "Holistic" and it related to "whole" "relationship" "connection" "harmony" and "balance". This concept comes from the original philosophy of the Orient.

The definition of Holistic Medicine by JHMS is as follows:

1. Holistic Medicine is based on the concept of holistic health.
2. In Holistic Medicine, one's Natural Healing Power is the main key to recovery.
3. In Holistic Medicine, the patient is the true healer, while the medical agent is the helper.
4. In Holistic Medicine, various types of treatments are implemented in a harmonious way.
5. In Holistic Medicine, illness is seen as a great opportunity toward self-awareness. This can lead to a deeper self-realization of oneself.

These definitions are still not perfect. The reason is that 4 is not totally reach the true wisdom but most important items for the patient are 3 and 5. Especially 5 is the most important for the patient. Illness is the kind of "Warning" to the patient and stepping stone toward self-awareness and self-discovery and it is also to feel it as a great opportunity to redirect our lives in a more meaningful way. That means that through self-awareness, a higher self-growth and self-realization can emerge.

5. "BEYOND" THEORY

I am a management consultant and also consulted many cancer patients until now. Every time I ask at first patients how they make their cancer themselves. Most people are very surprised at the beginning but gradually they begin to feel deeply what the meaning of my question and apt to tell their stories of making cancer by themselves. This method is almost the same as my management consulting and the patient to understand that he/she is the president of his/her body and they made their cancer by themselves.

The patient who recovered and healed themselves are the type of achieving higher-self during their healing process and they are mostly the president of the company and the artists. The most difficult type of patients are as it is called "well-educated people" like professors of university, medical doctors, lawyers, members of the Diet and journalists and they are not so skillful to forget. But there are some exceptional people above mentioned. They are so much willing to give and have a deep mind of gratitude and always smiling. This means that these people are beyond the level of science.

By education we learn to remember things but on the contrary we loose the sense of feeling. To think is between to remember and to feel. Fig. 1 shows the relation between that.

Another aspect of the medicine is the relation between body and mind. It is well known fact that mind affects body and body affects mind. This is shown in Fig.2.

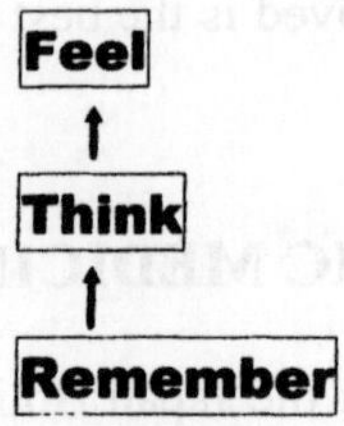

Figure 1.

Figure 2.

In addition to the relation to the body and mind there is another expression of body, mind and spirit. You can see it in many medical articles and places, for instance, YMCA and YWCA. This relation is showed in Fig.3.

These relations did not satisfied me and during my recovery from cancer I experienced that as it is called the transformation of my consciousness and I noticed that science was the bottom level of understanding. As I am a scientist I know this fact and I realized the level of understanding in the deep sense.

The level of science is the body. Beyond the body there is the level of religion and this is mind. Between mind and body there are many kinds of emotion and feeling and science can not measure it by machine. Beyond mind there is the level of tao and spirit and when the patient reach this level healing occurs in the deep sense of the body and heals totally from the inside. This is called "Spiritual"in the broad sense of meaning.

Beyond spirit there is the level of light and soul. In this level we can communicate without word. Maybe most of the people have experienced in daily life when they encountered the moment to communicate or to talk with nature like trees and flowers etc. Beyond this light level I cannot confirm it by now but I think maybe it is the level of love and God and farther more there will be the level of nothing and universe.

My proposal is shown in Fig. 4. This is vertical phenomenon so if someone cannot feel it you will still remain in the level of science or religion.

As I mentioned in the beginning of introduction, in ancient people were born naturally and died naturally but nowadays people are born unnaturally and die unnaturally. This is the typical phenomenon of modern medicine. Most doctors are well educated but on the contrary lose the intuition to feel which is already equipped in the deepest heart of

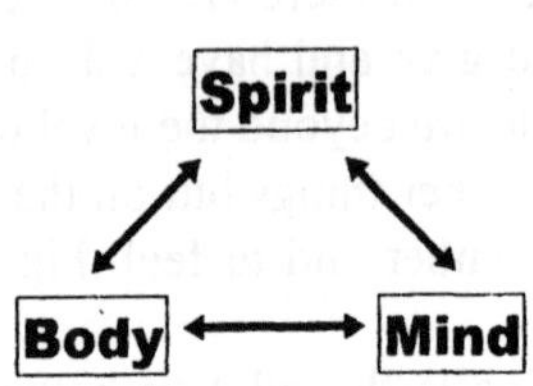

Figure 3.

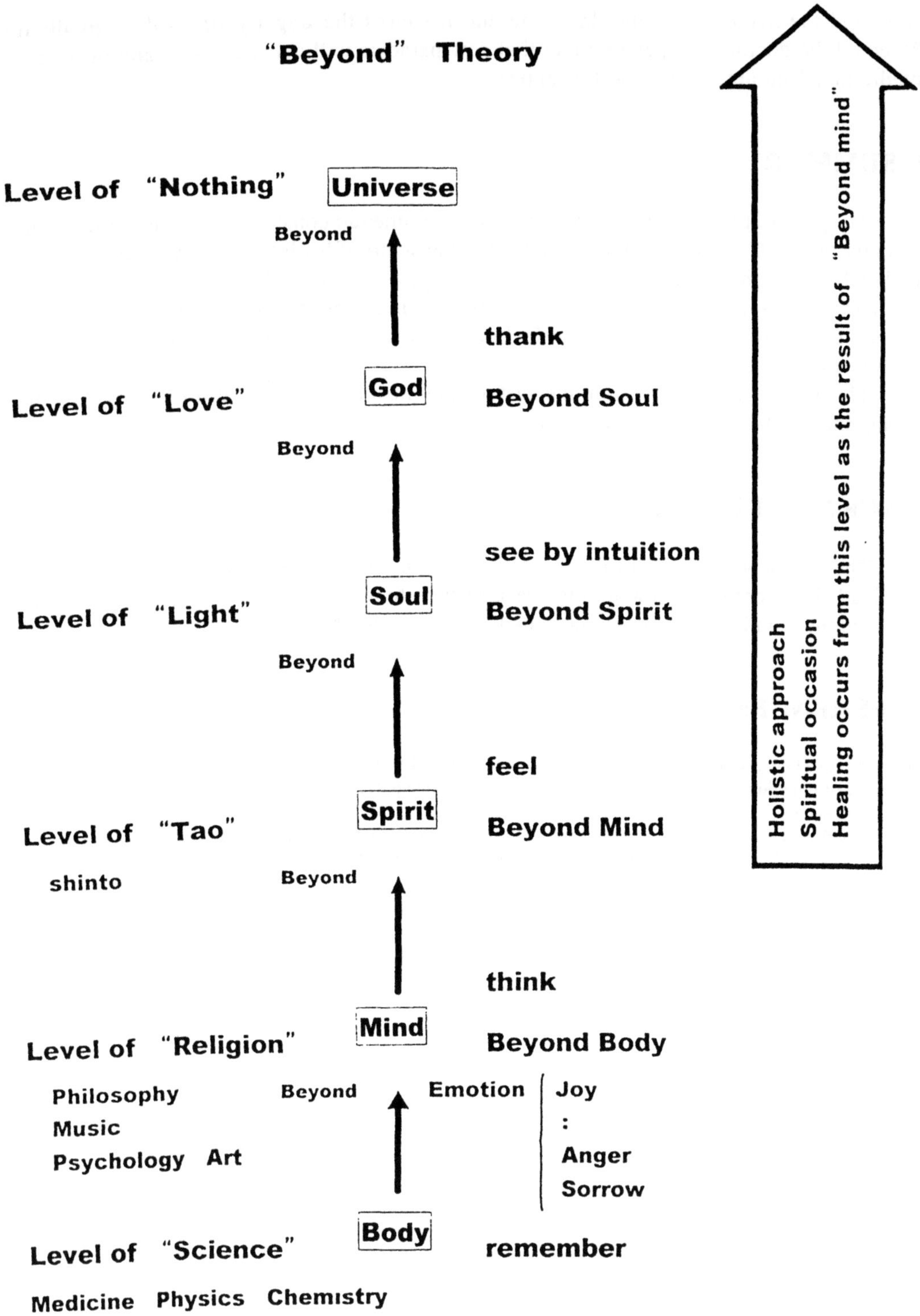
"Beyond" Theory
Level of "Nothing" Universe
Beyond
thank
Level of "Love" God Beyond Soul
Beyond
see by intuition
Level of "Light" Soul Beyond Spirit
Beyond
feel
Level of "Tao" Spirit Beyond Mind
shinto
Beyond
think
Level of "Religion" Mind Beyond Body
Philosophy
Music
Psychology Art
Beyond
Emotion
Joy
:
Anger
Sorrow
Level of "Science" Body remember
Medicine Physics Chemistry
Holistic approach
Spiritual occasion
Healing occurs from this level as the result of "Beyond mind"

Figure 4.

mind and the neglect it in the daily life and it causes the tragedy of modern medicine. When all the doctors and patients feel that new paradise will come chronic and deficiency syndrome of immune system will disappear.

6. SUMMARY

My proposal is the vertical existence of phenomenon consist from body, mind, spirit soul and maybe God. Spiritual level is the another word of healing and when patient reach the level of spirit most people begin to notice the power of natural healing with deep gratitude and healing occurs in the body very strongly and heal themselves but all the people are sure to die. So for the patient is to feel the importance of "Just now" and enjoy the life of "Just now".

This is the magic of the immune system and most scientists should feel that they are staying the bottom level of consciousness.

7. ACKNOWLEDGMENT

The author would like to thank deeply to Prof. Avshalom Mizrahi the president of the Dead Sea Conference for giving me a chance to write this article. I felt something magic connection between Israel and Japan in the near future.

REFERENCES

A reference of a book: Andrew Weil. Spontaneous Healing. Alfled A. Knoph, N.Y.,1995.

A reference from a journal: Iris Ansell. (1991 Fall) Insights. Tokyo. Here comes the sun, An interview with Shin Terayama 8–13.

A reference from a newspaper: Lucy Birmingham Fujii. Mainichi Daily News Sept. 3, 1996 One man's remarkable battle against cancer: Oct.1, 1996 Holistic medicine: Redefining the West's approach.

10

SETTING UP THE IMMUNE SYSTEM

Perinatal Influences

Annemarie Colbin

Affiliation: The Natural Gourmet Institute for Food and Health
Home mailing address: 365 West End Avenue
New York, New York 10024
Phone: 01–212–580–7121; Fax: 01–212–721–3336

1. ABSTRACT

The major factors influencing the development of the immune system in the perinatal period include: the parents' health; the health, diet, drug and alcohol use of the mother during pregnancy; the type of infant feeding at breast or bottle; the child's diet; and the use of antibiotics and vaccines. This paper presents a brief overview of the role of each of these influences, with some suggestions for maximizing immune system health.

2. INTRODUCTION

There are two major definitions of the immune system and its function. The best known states that its major role is in distinguishing self from non-self. It is a *function*, a system of communication between cells and organs. Immunologist Steven B. Mizel, in his book *In Self Defense*, states "Immune cells engage in something very much like conversation: an elaborate network of announcements, commands, and counter-commands. The placement, the timing, and the nature of every signal is exquisitely precise; the effect of a single command extremely powerful."[1] This conversation happens throughout the body, but most particularly between the lymphatic system with its lymph nodes; the blood stream with its white blood cells (lymphocytes), B and T cells, and antibodies; the spleen, the thymus, the mucosal lining of the respiratory system and the intestines, and possibly even other organs made of lymphatic tissue such as the tonsils, the adenoids, and the appendix.

Matzinger and Fuchs delineate a more inclusive function of the immune system: its purpose is to protect the individual from harm or danger. Their definition of danger is "anything that causes cell stress or necrotic cell death," as opposed to normal, programmed cell death. The danger is sensed by the tissues themselves and communicated to the immune system cells, which become activated as necessary.[2]

Potentiating Health and the Crisis of the Immune System
edited by Mizrahi *et al.* Plenum Press, New York, 1997

Human beings are born with a digestive system, respiratory system, excretory system, and sensory apparatus that are already working from birth on. These systems cannot be forced to work; they either do or they don't. They can be helped along with drugs, herbs, exercise, glasses, hearing aids, and the like; but essentially the function of each organ systems is inborn. It is the same with the immune system:

- It is there from birth.
- It cannot be forced to exist where it doesn't.
- It can be helped along and allowed to flourish.
- It can be disturbed, confused, inappropriately challenged, or not given what it needs to develop.

A number of factors influence the immune system both before and after birth. These factors are: the parent's health; the mother's nutrition and lifestyle during pregnancy (including the use of alcohol and pharmacological or recreational drugs); breast-feeding or the lack of it; the child's diet; and the use of antibiotics and vaccines.

3. TEXT

Healthy parents provide the genetic material to make a healthy child. During pregnancy, the developing fetus can be affected by the pregnant woman's dietary and other health habits. Alcohol use during pregnancy, which is usually associated with poor nutritional status, may lead to the condition known as *fetal alcohol syndrome*. This condition is characterized by growth deficiencies, central nervous system disfunction, poor coordination, hyperactivity, learning difficulties, and developmental delays.[3]

Both medical and recreational drugs can affect the developing child's health. The most dramatic example was Thalidomide, an anti-nausea drug given to pregnant women in the Sixties; babies born to these women had deformed, undeveloped, flipper-like arms. During the 40's and 50's, women who appeared at risk for miscarriage were given DES (Diethylstilbestrol), now no longer prescribed. The children of those women, known as DES babies, developed cancer and other illnesses of their reproductive organs in their adolescence and young adulthood; others had a higher prevalence of autoimmune diseases such as rheumatoid arthritis and pernicious anemia,[4] there were higher rates of unfavorable pregnancy outcomes and psychiatric illnesses in DES daughters, and more testicular cancer, infertility, and less heterosexual experiences in sons.[5,6]

The use of prescription drugs during pregnancy has been studied extensively. As drugs invariably cross the placenta,[7] they will affect the baby one way or another; however, there is little general agreement about the seriousness or extent of the damage, or even if any occurs. Some studies have found an association between prescription and other drug use by pregnant women and certain problems in their babies, such as birth defects and musculoskeletal changes,[8] toxic effects on the heart,[9] and adverse effects on the brain and central nervous system.[10] Approximately 10% of birth defects are associated with exposure to environmental substances including drugs.[11] See chart 1 for a list of medications that could cause birth defects.

3.1. Breast Feeding

The most important of all perinatal variables for a healthy immune system is breast feeding. In the first few days, "before the milk comes in," the breasts secrete the highly

- Androgens (male hormones),
- some antibiotics (tetracycline, streptomycin, gentamicin, kanamycin, erythromycin estolate, nalidixic acid, nitrofurantoin);
- some anticoagulant drugs (dicumarol, warfarin);
- some antiepilepsy drugs (Dilantin);
- benzodiazepines (Valium, Librium);
- some thyroid drugs (propylthiouracil, iodide, methimazole);
- tolbutamide;
- trimethadione (Tridione) and paramethadione (Paradione);
- valproic acid (Depakene);
- corticosteroids (including cortisone cream or ointments);
- diethylstilbestrol (DES);
- isotretinoin (Accutane);
- lindane (Kwell);
- lithium;
- meprobamate (Miltown);
- podophyllin;
- thalidomide;
- alcohol,
- vitamin A,
- warfarin,
- hypoglycemics (*i.e., insulin*),
- non-steroidal anti-inflammatory drugs (NSAIDs) (including aspirin and ibuprofen).
- some cancer chemotherapy drugs (methotrexate, aminopterin);
- chlorpropamide;
- estrogens,
- organic mercury compounds,
- inhalational anesthetics,
- live vaccines (such as polio),
- penicillamine,
- progestogens,
- progesterone (high doses),
- radiographic contrast materials (*e.g., barium*)

Chart 1. Medications that can cause birth defects.

nutritious liquid known as colostrum. This substance clears the mucus and meconium out of the baby's digestive system and sets up the appropriate intestinal flora which protects the child against digestive disorders. Colostrum contains the mother's immunity factors, including immunoglobulins IgA and IgM,[12] which will protect the child against many varieties of infections.

As the milk comes in, it brings immunity factors that are specifically formulated for this mother's infant. As she is exposed to the various toxins and pathogens around her infant, as she holds, kisses, and nuzzles him, she breathes in the air and particles around him; both her milk-producing glands and her immune system then create the specific antibodies needed by the baby, and these go through the milk into the child.[13] Stored breast milk, however, is not as effective, as studies have shown that its cellular components destabilize after heating, and cannot tolerate either boiling or freezing.[14]

Lawrence found that breast milk is protective against the following virus and bacteria: poliovirus, *E. coli*, staphylococci, *Enterobacteriaceae, Candida albicans, Salmonella*, herpes virus, leukemia virus, influenza, and many others. In addition, the breastfed infant's intestinal flora is composed almost entirely of *Lactobacillus bifidus*, which inhibits the growth of pathogenic bacteria. The bottle-fed infant's colon one tenth of the bifidobacteria, and more enterobacteria which is associated with more illness.[15]

Human milk has never been shown to cause allergy, whereas cow's milk allergy is common. Among the symptoms associated with these allergies we can count digestive distress (colic, diarrhea, spitting up and vomiting, colitis), skin rashes, runny noses, chronic cough and mucus, slow growth, and sudden infant death. Sudden infant death is extremely rare among breastfed infants.[16] Margaret K. Davis at the National Institute of Child Health and Human Development, of the National Institutes of Health in Bethesda, MD, found that infants bottle-fed or breast-fed for less than six months had an increased risk for lymphoma and other cancers. Her evidence suggested that breast fed infants are "less severely affected by infections in infancy than are artificially-fed infants. The type of infant feeding may affect later obesity, cancer, and diabetes mellitus." The study established that "human milk has substantial antimicrobial effects;" therefore, "if human milk increases resistance to infections in infancy, artificial feeding, which provides no immunological benefits, may alter the child's responses to early infection."[17]

Essentially, mother's milk does two things besides providing perfect nourishment: it provides immune factors, *and* it doesn't make the child sick. Formula allows for sickness also through two aspects: it provides no immunities and it has the potential for actually causing illness and allergies.

Breast feeding is best instituted immediately after delivery. In order to maintain an abundance of milk, no supplementary bottles or solids should be given to the baby for 6 months. Once solid foods are introduced, breast feeding can continue for as long as mother and child want, in a gradually decreasing manner.

3.2. Diet

Diet is an important variables for children's health. Giving them natural, unprocessed foods is an essential element for strengthening their immune system. Natural food always has nutrients not yet discovered. In factory-processed foods, the law allows a certain amount of rodent hairs, insect fragments, and assorted other impurities considered harmless in most canned or frozen foods.[18]

For good nutrition and a strong immune system, children need:

- Fresh, natural foods
- Vegetables (cooked and raw)
- Whole grains and breads
- Protein foods: beans, fresh fish, naturally raised fowl or meats
- Fruits, nuts and seeds
- No milk products if the child is prone to infections.[19]

3.3. Chemicals in the Food

Chemicals in the food include pesticides, herbicides, colorings, preservatives, and other additives. Practically all the man-made pesticides used on our food supply are known toxins, and have been associated with cancers,[20] birth defects, liver toxicity, peripheral and central neurological problems, and reproductive problems in both animals and humans.regulations.[21] Because they eat more food relative to their body size, children are in greater danger from pesticides; the 1993 report of the National Academy of Sciences, *Pesticides in the Diets of Infants and Children*, states that low but steady exposure to pesticides early in life can lead to a greater risk of cancer, neurological impairment, and immune dysfunction.

Fruits and vegetables are most commonly contaminated, but so are grains and meats, fish that have been caught in polluted areas, as well as imported unregulated produce. The NAS further states that exposure to common levels of pesticides in available foods might be high enough for some children to cause symptoms of acute pesticide poisoning.[22] The best way to avoid chemicals in the children's food is to purchase only organic produce, grains, beans, nuts, and to buy products from animals allowed to graze or run, fed natural chemical-free food, and not given any antibiotics.

■ Allergic reactions:
- changes in the skin,
- contact dermatitis,
- "hairy" or "black" tongue,
- development of fungi in the mucous membranes,
- gastrointestinal disturbances;
- inflammation of the stomach and pharynx;
- urticaria,
- itching,
- herpes simplex,
- rosaceaform dermatitis,
- purpura;
- acute fungus infections of the skin;
- edema with swellings of lips, eyes, mouth, tongue, or epiglottis;
- anaphylactic shock
- sudden death.

■ Specific organ reactions:
- Most frequent allergic manifestations: Asthma and hay fever
- pulmonary symptoms,
- changes of the blood, liver, veins, arteries;
- increased tendency of the blood to clot faster than usual;
- **narrowing of the coronary arteries with subsequent heart damage** (author's emphasis);
- damage to the kidneys;
- toxic reactions in the central nervous system such as convulsions and coma;
- restlessness, hallucinations, psychosis;
- endocarditis,
- infections of the kidneys,
- pneumonia;
- infections of the genital organs;
- staphylococcal gastro-enterocolitis (a complication of tetracycline therapy), which manifests as fever, nausea, vomiting, flatulence, diarrhea, and collapse.

Chart 2. Unexpected reactions to antibiotics.

3.4. Antibiotics

Antibiotics, discovered by Alexander Fleming in 1928, have been in widespread use since the mid-1940's. In 1957, Leo Schindel, MD, published a review of the medical literature on the effects of thirteen different categories of antibiotics, including penicillin, streptomycin, tetracyclines, erythromycin, bacitracin, and polymyxin. He found a large number of "unexpected reactions" duly described in the journals (see chart 2). Schindel also points out that "most of the resistant strains of staphylococci are to be found in hospitals, especially in surgical and pediatric wards".[23]

This review study is credible because it dealt with studies made in the early days of antibiotic use, shortly after World War II, and the adverse events could be clearly noted in populations exposed to the drugs as compared to those not exposed. Current studies would be more difficult to evaluate, as antibiotic use is widespread throughout the world and it is nearly impossible to find non-exposed individuals as controls.

Antibiotics eradicate beneficial bacteria in the intestines, including the bacteria that help synthesize nutrients such as vitamin K, biotin, riboflavin, pyridoxine, and folic acid. For that reason, frequent antibiotic use results in vitamin deficiencies because of the inability of the intestines to synthesize or absorb these nutrients. Amoxicyllin in particular reduces drastically the numbers of the helpful bacteria, such as the important lactobacillus types.[24]

In chronic infections where they have been used repeatedly with minimal success, the use of antibiotics is questionable, if not damaging.[25]

- Encephalopathy (acute and chronic)	- optic neuritis
- aseptic meningitis	- sensorineural deafness
- subacute sclerosing panencephalitis	- sterility via orchitis
- residual (chronic) seizure disorder	- anaphylaxis (cessation of breathing)
- Guillain-Barré syndrome (GBS)	- arthritis
- transverse myelitis	- erythema multiforme
- paralytic poliomyelitis (in recipient or contact)	- insulin-dependent diabetes mellitus
	- early susceptibility to Hib disease
- myelitis	- thrombocytopenia
- neuropathy	- death

Chart 3. List of vaccine adverse events, Institute of Medicine, USA.

3.5. Vaccines

The immune system can also be damaged by vaccinations, especially live and attenuated viral vaccines.

The Institute of Medicine in a 1994 publication, recognized 20 different possible adverse effects of commonly administered vaccines; 11 of these are neurological damage. Also recognized was the fact that there are not enough studies in many cases to either establish or dismiss the vaccines as being the cause of these common adverse effects[26] (See chart 3).

Live or attenuated viral vaccines can be particularly dangerous, according to Professor Richard DeLong, a retired microbiologist formerly of the University of Toledo and Del Mar College, and author of *Live Viral Vaccines: Biological Pollution*. (Live or attenuated viral vaccines include the oral [Sabin] polio, measles, mumps, rubella, varicella [chicken pox], influenza, and hepatitis A.) These vaccines can:

- cause mutations
- cause chromosomal aberrations

1) Gene recombination occurring among two or more different viruses infecting the same cell;

2) Contaminating viruses in the live vaccine from the cells that were used to produce the vaccine

3) Vaccine viruses changing their genes during reproduction in the *in vitro* phase of vaccine production

4) Vaccine viruses changing their genes during in vivo reproduction in the vaccine recipient.

Chart 4. How live viral vaccine viruses effect genetic changes.

- cause birth defects
- cause cancer
- cause new diseases
- revert to virulence
- be contaminated with other viruses and other microbes.[27]

Professor DeLong theorizes that live viral vaccines originate new diseases in four different ways (see chart 4).

Other recognized adverse effects of vaccines that relate to the immune system include swelling, redness, induration, high fevers, inflammation, and immediate and delayed-type hypersensitivities.[28,29]

Clearly, we need much more information on the real effect of vaccines and antibiotics on the immune system of children.

3.6. Natural Remedies

The children's immune system should be allowed to do exercise by recuperating from minor illnesses on its own, with the help of soups, herbs, and natural remedies, instead of being pushed and pulled by pharmaceutical drugs, all of which have adverse effects. These drugs are best left for serious illness and life or death matters where the benefits outweigh the risks.

The following are essential to avoid the occurrence of infections:

- Good nutrition
- Enough rest
- Fresh air and exercise
- No milk products

To manage minor illnesses, the following can be used:[30]

- Natural home remedies (garlic, soups, teas, lemon, fasting)
- Herbs
- Homeopathy, naturopathy
- Massage, chiropractic

Considering that chronic immune and autoimmune diseases are a modern phenomenon, it is the author's suggestion that more study is needed to ascertain their connection with iatrogenic factors and bottle feeding.

4. SUMMARY

The basic needs for a healthy immune system are: Healthy parents; nourishing food, no alcohol or drugs during pregnancy; breast-feeding for at least six months. The immune benefits of breast-feeding include: Colostrum colonizes the intestinal flora for life; the mother's own immunities are passed on; the mother continuously creates new immune factors based on the environment; mother's milk does not make children sick.

For good nutrition and a strong immune system, children need fresh, natural foods: vegetables (cooked and raw), whole grains and breads, protein foods (beans, fresh fish, naturally raised fowl or meats), fruits, nuts and seeds, and no milk products if the child is prone to infections.

Among the influences that damage the immune system are a) Chemicals in the food; 2) Antibiotics; 3) Vaccinations. Chemicals in the food can cause cancers, birth defects, liver toxicity, peripheral and central neurological problems, reproductive problems in both animals and humans.regulations, and immune malfunction. Antibiotics damage the immune system because they damage the intestinal flora, destroy beneficial bacteria as well as harmful ones, and create resistance in pathogenic bacteria.

Vaccines also damage the immune system. There are numerous different possible adverse effects of commonly administered vaccines, including anaphylaxis, inflammation, immediate and delayed hypersensitivities, urticaria, seizure disorder, brain damage, and death. There are not enough studies in many cases to either establish or dismiss the vaccines as being the cause of these common adverse effects. Live viral vaccines can cause mutations, chromosomal aberrations, birth defects, cause cancer, and new diseases. They can also revert to virulence, and be contaminated with other viruses and other microbes.

More study is needed to ascertain the role of iatrogenicity in immune disorders.

5. REFERENCES

1. Mizel, Steven B., and Jaret, Peter, *In Self Defense: The Human Immune System- the new frontier in medicine*, p.7. Harcourt Brace Jovanovich, New York: 1985.
2. Matzinger, P; Fuchs, EJ, "Beyond Self and Non-Self: Immunity is a conversation, not a war." *The Journal of NIH Research*, July 1, 1996.
3. Lawrence, *Nutrition in Pregnancy and Lactation*, p. 179.
4. Kenneth L. Noller, MD, et al, "Increased occurrence of autoimmune disease among women exposed in utero to diethylstilbestrol," *Fertility and Sterility*, Vol. 49, No. 6, June 1988, pp. 1080–82.
5. Meara, J.; Fairweather, DV, "A randomized double-blind controlled trial of the value of diethylstilboestrol therapy in pregnancy: 35-year follow-up of mothers and their offspring," *British Journal of Medicine*, May 1989, 620–22.
6. Gerald G. Briggs, Roger K. Freeman, Sumner J. Yaffe, *Drugs in Pregnancy and Lactation*, Williams and Wilkins, Baltimore: 1986, p. 140/d.
7. Niebyl, Jennifer R., "Drugs and related areas in pregnancy, "University of Iowa, College of Medicine, Department of Obstetrics and Gynecology. *Zent.bl. Gynakol.* 113(1991) 375–388.
8. Lawson, J.P., "Drug-induced lesions of the musculoskeletal system." Yale University School of Medicine, New Haven, CT. Radiologic Clinics of North America, March 1990; 28(2):233–46.
9. Balazs, T., "Cardiotoxicity mechanisms from the point of view of preclinical or premarketing safety evaluation." Archives of Toxicology - Supplement, 1986; 9:171–7.
10. Redmond, G.P., "Physiological changes during pregnancy and their implications for pharmacological treatment." Clinical Investigations in Medicine, 1985; 8(4):317–22.
11. Cope, Ian, "Medicines in Pregnancy," (letter). Chairman, Working Party on the Categorisation of Drugs in Pregnancy, Commonwealth of Australia. *Medical Journal of Australia*, Vol. 155 August 19, 1991. pp. 214–14.
12. Lawrence, Ruth A., MD, *Breast-feeding: A Guide for the Medical Profession*. The C.V. Mosby Company, St. Louis: 1985, p. 127.
13. Marano, Hara, "Breast or Bottle: New Evidence in an Old Debate," *New York Magazine*, 6/29/1979..
14. Lawrence, p. 125.
15. Lawrence, pp. 133–137.
16. Lawrence, ibid., p. 138.
17. Davis, M.K., Savitz, D.A., Graubard, B.I., "Infant feeding and childhood cancer." *The Lancet*, 1988 August 13. 2(8607):365–368.
18. *The Food Defect Action Levels: Current Levels for Natural or Unavoidable Defects for Human Use that Present No Health Hazard*, Department of Health and Human Services, Public Health Service, FDA, Center for Food Safety and Applied Nutrition, Washington, DC 20204: current through January 1989.
19. Colbin, Annemarie, *Food and Healing*. Ballantine Books, NY: 1996.
20. *Health and Environment Digest*, "Agricultural Health: Pesticides and Cancer." A publication of the Freshwater Foundation, Volume 6, No. 5, September 1992.

21. Gray Davidson, Osha, "Pesticides: The Killing Fields." *Woman's Day*, September 20, 1994.
22. Lefferts, Lisa Y., "A Commonsense Approach to Pesticides," *Nutrition Action Healthletter*, September 1993. Center for Science in the Public Interest, Washington, DC.
23. Schindel, p. 83.
24. LappJ, p. 52.
25. Schmidt, Michael A., MD, Smith, Lendon H., MD, Sehnert, Keith W., MD, *Beyond Antibiotics: More Than 50 Ways to Boost Immunity and Avoid Antibiotics*, p. 55. North Atlantic Books, Berkeley, CA: 1994.
26. *Adverse Events Associated with Childhood Vaccines: Evidence Bearing on Causality*, National Academy Press, Washington, DC. 1994. Edited by Kathleen R. Stratton, Cynthia J. Howoe, and Richard B. Johnston - sponsored by the Vaccine Safety Committee of the Division of Health Promotion and Disease Prevention, Institute of Medicine.
27. DeLong, Richard, *Live Viral Vaccines: Biological Pollution*, p. 9. Carlton Press, NY: 1996.
28. "Diphtheria, Tetanus, and Pertussis: Guidelines for Vaccine Prophylaxis and Other Preventive Measures," a Recommendation of the Immunization Practices Advisory Committee (ACIP). MMWR, Volume 34, No. 27, July 12, 1985, p. 411.
29. Physician's Desk Reference, 1995 edition.
30. Zand, Janet; Walton, Rachel; Rountree, Bob, *Smart Medicine for a Healthier Child*. Avery Publishing Group, New York, 1994.

DISCOVERY OF THE HUMAN BEING'S SUPERSTRESS ADAPTATION BIOLOGICAL ORGAN

Arkadii G. Mogilevskii*

Anti-Stress Center
Marx Avenue 10, Apt. 94
Omsk-10, 644010, Russia

THE ABSTRACT

The loosing strategy used by the nature and medicine in the struggle with the stress is observed. The wide synthesis of Pavlov's, Bekhterev's, Asratyan's, Vvedensky's, Ukhtomsky's nervist schools ideas at the macro and micro levels (particle - wave) united with the cibernetics, synergetics let us discover the very best strategy of human being's adaptation.

The new medical paradigm is offered: not to heal the disease, but to teach the person how to adapt to the stress, having discovered his or her antistress self-adaptation organs. Every patient can eliminate disadaptation diseases and to prevent them with the help of these organs according to the sanoreflexogenesis and reflexosynthesis laws. The discovery of the natural antistress biological reactions (reserves) and their new self-arrangement into the new system-organ let us provide human rationally-emotional development and the Mind control over the distress.

INTRODUCTION

Stress is a multiform biological phenomenon that accompanies us all through our life from the very first till the very last breath. This is stress, energizer of Mind, that civilization should be obliged to for the achieved power, beauty, and splendor. When we say that labor created human being we mean the constructive aspect of stress, great human efforts. However, uncontrolled stress, excessive stress, distress according to G.Selier(12) is a great destroyer of health and killer of mankind. Stress diseases are the diseases of misin-

* Tel. 00–7–3812–314877; Tel. in Israel: 06–545–937 (Igor Mogilevskii); e-mail: as@opsb.omsk.su.

formation, disadaptation, disregulation, disintegration, civilization. These diseases which are predominantly neuroses and psychosomatic disorders have killed more people than all the wars on our planet. Cruel and ever growing stress—emotional psychotraumatic epidemic swept the whole of the world. Stress-"virus", stress-"bullet", stress-"bomb" of biological character are targeted at everybody. The principle shooting marks are our heart, brain, spirit, and quality of our life.

In the course of this battle with diseases of civilization the human mind is, evidently, loosing. Stress escalation and emotional disadaptation of population, regardless of ever growing might of health service and development of science, assumes a character of pandemic. According to the WHO data the rate of neuroses has been increasing 24 times for the previous 65 years. From 10% to 15% of the population in highly developed industrial countries suffer from initial hypertension.

Crisis of medicine reflects scientific way of thinking crisis and today's human being pathogenic emotional style of living manner.

We have found ourselves unprotected at the sphere concerning our spirit. Let us comprehend where do the sources of emotional fragility and bioemotional imperfection of a person lie? Applying to the common sense, logic, and the results of clinical investigations, let us try to see the typical traits of a person anti-stress psychoprotection behavior under the critical and chronicle stress conditions.

DEFICIENCY OF HUMAN BEING'S ANTI-STRESS EMOTIONAL BIOPROTECTION

The civilization deprived the primitive man of his natural superstress self-protection with moving. We are speaking about the aggressive defending motor dominant (to beat or to kill an enemy or to run away, saving one's life). Excessive emotional energy was feeding the strength of muscles. All the dangerous hormones of the stress were burnt down in the muscle "boiler" of the motor behavior reactions and its "evil spirit" did not destruct the body. Physiology of human being was protected from superstress by the explosion of muscles. The reality of a civilized life turned out to be cruel for a human being. The avalanche of stress-superstress, being deprived of the natural barrier-motion protection-acquired autoaggressive destructive properties. As an evil demon it began to destroy and "break" health of the own organs and systems. This pathological process resulted in the anti-stress deficiency of the brain and the organism state and self-regulation breach. Acceleration of social life tempo, more complicated nature of modern personality, technical expansion and urbanization have led to the increase of "atmospheric pressure" of stress on the organism and explosion of disadaptation diseases.

Clinical experience manifests that the excessive stress, the superstress has a destructive effect on the organism. As a rule, this is a long-run, exhausting negative emotions or critical shock stress having no relief. In what way does the nature protect us from superstress?

The clot of aggressive psychic energy of superstress presents serious threat to the life of the whole organism. Apparently, to survive it is more expedient to decrease dangerous tension, to direct the bigger part of this energy to one or another organ. Energy of stress often "feeds" some reactions: hypertension, pain, for example, etc. Pain causing anti-stress protection works in favor of our health. Pain plays here the life rescuing role of the anti-stress defense first line. If pain has occurred and has disappeared—it is wonderful. It means the defense completed its function. If stress has materialized, stuck in the organ and does not come out, neurotic sick dispensary starts its own life, autonomous one, separate from the rest of the organism, accumulating stress "gun powder".

Approaching avalanche of newcoming stress, mixing up with the excitement of painful center, becomes dangerous and breaks all the limits. In this case brain, to protect to life, switches on alarm program - anti-energetic mechanism of anti-stress. The energy of brain and organism is switched off by general tonic "lever". The program, let us call it half-dream, is switched on. It helps to extinguish the fire at the periphery, in any organ.

The patient now complains of general weakness, asthenia, somnolence, unwillingness to move, heavy head, feeling of unclear, not fresh, slow-witted mind. Physiological self-protective limited inhibition mechanism is the organism's anti-stress bioprotection second line.

The next line of the defense is switched on by the consciousness. This is the mechanism of reflection or subjective evaluation of the situation. Inability to mobilize oneself by means of determined efforts to carry out active and, at times, hard activities, serious subjective perception of the disease are taken by the personality as dramatic circumstances. He pins labels of appalling estimation—auto-suggestions on his disease and himself—"I am seriously ill. It may be even cancer. I am doomed, I am defective and soon go mad... etc."

The following chain of self-protection reactions may be named as the secondary stress having its roots in the patient's personality, in his wrong, panic evaluation of his health state. Anxiety about the course of disease results in the search of the way out from unpleasant situation, activating this searching activity. However, the stress energy favors the growth and strengthening of the disease on the whole.

Thus, we observe here the process consisting of two indissoluble united parts of physiological reactions. From the one hand, anti-stress biological strategy of the organism is wise and purposeful (expedient) because it defends life. From the other hand, mighty stereotype of chain pathogenic reactions disturbing self- regulation to even bigger extent and, thus, aggravating the disease is formed. We obtain, in this way, the inert mechanism of unconscious protection from stress which has the reverse side—even more serious damage, adaptation disease.

You may well ask, where is Mind, powerful human being's Mind that created the civilization? What does it oppose to this bodily–unconscious element of self-destruction, the name of which is emotional madness?

We can only be surprised here. The small island of materialized stress, neurotic painful dispensary of a pea size for years and decades tortures and destroys a person. And Mind is of no help in this situation. Since the age of stone ax of a primitive man he has been seeking for the anti-stress medicine. A great number of drugs and methods of healing, tranquilizers included, were discovered. Unfortunately, they all provide only temporary interruption of the disease, touching no bases of structure and system of self-organization.

What is the striking power of stability and steadiness of the stress in the disease lie in? As the experiments reveal, this force is conditionally-reflex and has vertical manner of self-organization of psychic substance. Applying the adequate approach, according to the laws of stressogenesis, the problem of the sound stress self-control may be successfully settled.

What is the essence of anti-stress medicine crisis? For the period of more than 2500 years medicine has been proceeding along the "path of diseases". That is high way of the medicine. Its scientific-and-practical credo means affecting the substance, material substratum, as long as the most significant principle of life-metabolism is in disorder. Such kind of notion is correct but is non sufficient and primitive. Stress diseases are accompanied by reflex system exchange: substances-energies – information-self-regulation. The major forces of the medicine are directed at the search of more effective remedies and methods or "psychotechnology". This is a blind alley of development that leads away from its aim—self-restoration of any "breach" in emotional regulation. The traditional methods of exterior stress control having been existing for a long time are inefficient and non sufficient. They accomplish momentary, accident adaptation, which means interruption of the disease attack. This is partial and non perfect adaptation.

The described above paradigm of healing realizes the protective strategy of Mind. It is doomed to failure. The disease attacks, grows stronger, and the organism defends itself, tries to get rid of the disease, "runs" away from the disorder, suffering and getting weaker. The result of this strategy is paradoxical: more pills, more doctors, more hospitals, more expenditures for health programs but... more and more patients !!

The way out from the crisis is in the development of new ways of inner self-regulation of stress based on full and complete adaptation. We need qualitatively new paradigm of the civilization diseases healing -offensive, preventive strategy of Mind. The unity of inner and outer self-control of stess, both switching (switching off) the disease and transformation of the disease system into the anti-system of non-suffering (health) form the basis of this strategy. If the target of the influence under outer control are the stress disease forces, in the second case there should be both forces of the disease and anti-stress, healthy forces of the organism.

CRISIS OF THE SCIENTIFIC WAY OF THINKING AND NEW HEALING STRATEGY

The existing healing paradigm is based on the pathology of organs primacy. It is based on the supposition that the primary cause of the disease is the damage, the "breach", the disorder of the structure and the function. This notion is wrong. The problem is put upside down. Both damage and "breach" of self-regulation is evident, but it is not the cause, it is effect. The cause is underdevelopment of reasonably-emotional anti-stress system - biological specialized organ of adaptation to superstress.

Nature, civilization, and culture did not make a human being perfect, they did not develop him as a creature with reasonable emotions. The civilization diseases are at the first rate the disease of under-civilization, i.e. misinformation meaning shortage of necessary knowledge and skills to control stress by the Mind, and only then diseases of disadaptation-disintegration-disregulation.

Emotionally civilized person (reasonably-emotional one) should have developed anti-stress analyzer (AA) absolutely protecting him from superstress. This is the best kind of remedy which is called healthy self-regulation.

Presence of anti-stress analyzer demands the developed consciousness. In the moment when superstress switches on sensitive to pain structures of the organ, it automatically switches on memory mechanisms as well. The process of materialization, self-regulation of the disease takes place. Informed Mind immediately mobilizes anti-stress reactions of suppression and eliminate the "breach" in embryo.

In case when stress has penetrated deeply into the organism, the Mind armed with knowledge and method of fighting against it, drives the stress out quickly and reliably. To achieve this we are to carry out sano-reflexo-synthesis: to take deficient reaction and combine it with the stronger health reaction, according to Pavlov-Asratyan-Vvedensky. In this case the stress tears itself off the deficient reaction and implants itself in the reaction of anti-system (non-suffering). Presence of anti-stress analyzer or health-emotional homeostat makes it possible to realize the dream of the mankind—to live according to G. Selier, namely: "stress without distress", which means – get excited but not get sick.

On the basis of everything said above the behest of great Russian scientist S.P.Botkin – "there should be found such natural reactions in the organism which are able to interrupt and abolish any disease" is practically realized.

Let us make a conclusion. Human being's Mind should oppose smart strategy of self-rescue and stress self-control to physiological unconscious stress madness. Modern human being develops its emotional sphere. As the history of medicine and this analysis prompts, anti-stress physiological "equipment" came into contradiction with the modern mode of life. Person is always seeking freedom from distress and historically ascends to the acquiring of anti-stress analyzer that can give him this cherished freedom - consciously flexible self-regulation and self-control over crazy behavior in distress.

Three circumstances promote the breakthrough of human being to the discovery of anti-stress analyzer—this psychic space of the brain.

The first: a wider view on the cause of the health "breach" at macro- and micro- life levels and discovery of more powerful levers of Mind self-influence on the organism.

The second: the discovery of the organism anti-stress reactions system and their new self-organization in healing health super dominant.

The third: the discovery of the "secret" of converting the of disease "copper" into the of non-suffering (health) "gold" on the basis of physiological law of distress disease reactions transformation into the anti-stress reactions.

If we have a look upon the history of the anti-stress human self-development as upon a long way of ascending the emotional common sense and still higher quality of psychological protection from stress, we may come to an interesting idea.

The idea is as follows: apparently, all available historically proved anti-stress methods of treatment is just the development of natural anti-stress "sub-organs" that a human being possesses. And the evolution "vertical line" of this process is their synthesis and integration into more developed forms of self-organization, the peak of which is the anti-stress analyzer or biological organ of adaptation and distress self-regulation.

Let us look at ourselves in a new way. Which anti-stress "sub-organs" or natural anti-stress reactions do our organism have? Where are our natural interrupters of stress reactions situated?

Let us make a psycho-biological inventory of our organism.

The most important of these organs is our Mind, its creative thought, and verbal reaction aimed against the disease. The "thinking" word, the word that sounds, ordering signal of brain, at all times was the most powerful healing remedy for psychic disorders. The energy of thought or information and the energy of the word being pronounced are health-creating. This is the golden "pill" of psychotherapy. To become a "brilliant" one, the word-thought of a person should be magic which means—efficient. The medicine now is to settle a very important task: to discover the mechanisms amplifying the healing power of word thousands of times, the same way as it is done in technical sphere. In this case word-thought obtains new healing quality: does not only heal psychic but to recover it.

The principle methods of treatment the distress disorders are created on the basis of the verbal " sub-organ" of anti-stress. They are: hypnosis, suggestion, rational psychotherapy, autogenic training.

The second "sub-organ" is the eye with its inner mental vision, thoughts and images of any phenomenon that are a powerful force of psychical behavior control of the distraction mechanism. Let us include healing color optical reactions of organ of sight. It is known from psychology and our experience proves it that "warm" colors, orange, yellow, red, and, especially, pink, possess mighty properties of anti-stress and anti-pain psychic protection.

These words belong to great Bekhterev: "The day comes when mankind will be healed by color".

On the basis of anti-stress visual analyzer such methods of self-regulation as meditation, contemplation of the beautiful, and healing with the help of color appeared.

The third "sub-organ" of anti-stress self-regulation is "caressing hand" possessing great spiritual properties of manual distress control.

This is laying of hands, healing passes, methods of non-contact massage and psychic healing.

The forth "sub-organ" of anti-stress self-regulation is breathing analyzer. The breathing that can heal is realized in yoga breathing, in the methods created by Buteiko, Strelnikova, Groff (pneumocatharsis) and others.

The fifth "sub-organ" of anti-stress is a certain specific "geometry" of throat muscles, actualization of a joyful anti-stress outline of mimic muscles as of natural sloppier and interrupter of pain and modulator of joy and comfort feeling.

The sixth "sub-organ" of anti-stress is of systematic and integrative character. It is a center of positive emotions of emotional analyzer bearing psychic energy of any healing anti-stress reflexes. All the conditioned reflexes are emotional. For our investigations the statement written by E.A. Asratyan, representative of Pavlov's school, is of great significance. He wrote, "Tonic conditioned reflexes (read – emotional ones) open certain nervous paths and close the other ones, defining the destiny of any nervous reaction".

The seventh "sub-organ" of anti-stress are skin and muscles that bear tactical feeling of physical skin and muscles joy-comfort.

On the basis of the property of this "sub-organ" arise bodily "remedies" against stress-massage and self-massage, rhythmical soft strokes and rubbing of a sensitive point, all the methods of chemical and physical influence on the skin.

DISCOVERY OF THE ANTI-STRESS ANALYZER ELEMENTS – STRESS SELF-ADAPTATION BIOEMOTIONAL ORGAN

The enumerated above seven anti-stress reactions of organism form the nuclei of the person's outer anti-stress psychic protection. We have called them "sub-organs" underlying their functional insufficiency and elementary nature.

Organ, according to A.A.Ukhtomsky, is combination of functional elements (reactions) for realization of one or another activity. To become anti-stress organ of person's adaptation to uncontrolled superstress, i.e. to render harmless pathogenic stress, the enumerated seven organs should be integrated into functional system of anti-stress analyzer.

Ordinary stress is not dangerous for an organism. Directed at realization of vital purposes it is easily neutralized by mechanisms of self-regulation.

The other thing is superstress—the stress exceeding the limit of healthy self-regulation and, thus, spoiling the health.

The Golden Rule of emotional health may be expressed as the following: organism is to oppose by more powerful super-anti-stress action to every superstress action. Here is the secret and fundamental problem and drama of human being's life – elementary and imperfect psychobrain self-organization of organism. The seven anti-stress reactions of an organism named above are not strong enough and, thus, not able to develop super-anti-stress opposition and to become super-anti-stress dominant. Super-stress of life easily sweeps off each of them.

What is the way out? The salvation is in integration, unity of all the anti-stress forces of the reactions mentioned above till the level of system functioning is achieved. The designated tendency of uniting all the forces against stress is seen from the analysis of clinical experience taken in its evolutionary aspect.

Thus, speech therapy is united with sight and respiratory system. Non traditional methods of healing, yoga, Tibet, Chinese schools of healing unite reasonable anti-stress reactions with motion, visual images, breathing and skin massage.

The highest level of anti-stress integration is autogenic training according to I. Schulz. All the anti-stress reactions are represented here, but they do not achieve the target of adaptation due to the reason of methods' insufficiency.

CRISIS OF THE ANTI-STRESS MEDICINE METHODS

One of the main reasons of insolvency of all the treatment techniques of stressogenic disorders (neuroses, psychosomatic and others) is "linear" or consecutive character of dominant self-organization of psychic protection reactions.

Anti-stress reactions mentioned above act in collaboration, successively joining into the fight with superstress of diseases.

Stressogenic disorders have resonance-vortex (wave) conditioned reflex principle of self-organization. That is why non-systematic anti-stress reactions of psychic protection are ineffective. Anti-stress psychogenerator of their dominant can not achieve the "super" quality, which is so necessary for self-regulation.

The matter is that anti-stress dominants provide only temporary interruption of the disease and do not touch the essence of their system self-organization.

The problem is that the Mind should impose its victorious, offensive strategy to the disease, the strategy that leads to the liquidation of the disease and restoration of healthy emotional self-regulation of life—the essence of valuable, fool-blooded adaptation.

The basis of this victorious strategy should form systematic method of psychic protection reactions anti-stress integration.

ASCENT OF THE HUMAN BEING'S FLEXIBLE REASONABLY-EMOTIONAL ANALYZER

Let us set up the problem. It was admitted, that biological (brain's) crisis of the human being is the reason of the civilization diseases. Modern person, leading a highly emotional style of life in the superstress environment, does not have bioemotional "equipment" necessary and efficient for this life. Challenge of time is superstability under the stress. The damage of the person's life-emotional self-regulation is the result of this bioemotional disharmony. The most vulnerable parts under the stress are the brain and the heart, that have brought dramatically increased level of cardio-cerebro diseases.

Modern life situation is the following—the volume of stress overcomes the brain's antistress protective capabilities. The conclusion can be made: the human being should be developed, his antistress protection should be erected.

The following question appears – what do the person and the body need to get the antistress neuro-psychic immunity? The answer is evident. Specialized antistress biological organ of the self-regulation is required, the organ which could work according to the laws of the analyzing functional system and conscious flexible and unconscious automatic self-regulation.

The analyzer should include three levels as well as anyone of this kind. Its main feature should be high reliability. The main goal of its work is to keep emotionally healthy stability of inner state, in other words to change disease causing stress into the neutral stress, which would not be a threat for the health.

It is clear from presented above, that the idea of the antistress analyzer is based on the laws of the reliable bioemotional automate or emotional health psychohomeostat, working according to the words of G.Seleir "stress without distress".

We believe, that antistress analyzer functional system or stress adaptation biological organ, providing neuroemotional immunity of the body should carry three functions, determining its activity goal.

Let us name them.

1. To be responsible for the distress defense in the case of the accident using instant conscious flexible switch off the stress by the outer inhibition mechanism,
2. To suppress stress-neurotic activity in any part of the body by changing the stress into the health antisystem with the help of the inner inhibition mechanism,
3. To perform the role of the rational stress "teacher" and the stress trainer: do not switch on the disease reaction.

The problematic situation includes a whole complex of the very difficult problems of theoretical and methodical character, and those which have no solution by now as well.

Nowadays human being does not have universal biological organ of adaptation (BOA) to the superstress, as specialized analyzing functional system (FS). It's necessary to erect this building-logic system using presented constructive materials.

Six antistress suborgans mentioned above are the constructive blocks for the future building. Everyone of them is just a little block of the future adaptation to the emotional accident–disease, because no one of them is capable to achieve the supergoal by itself.

Let us define our goals. The main one is determined by the essence of the adaptation. To have the specialized organ of adaptation to superstress—it means that the person is protected from the stress diseases.

But there are three kinds of superstress. The first one brings the threat of the future health disorder, if it would not be neutralized in time. The stress that was not inhibited in time will have all the qualities of the grounded disease seed. It is clear from the presented above that we are talking about the problem of the primary psychoprophylactic-protection from the future biocatastrophe.

The second kind of the disease creating stress take place when the person suffers from disadaptation disease. The essence of the adaptation will be totally different in this case. Conscious person should get the program concerning the restoring of the healthy emotional behavior self-regulation to avoid stress disease in this case. Here we meet the realization of the secondary psychoprophylactic principles with the help of BOA.

The superstress of the third kind could get the might of the flow and to destroy erected constructions of the health-restored stress psychoprotection. In this case the activity of the tertiary psychoprophylactic has the goal of the preventative defense and strengthening of the antistress healthy forces of non-suffering.

Let us make a conclusion. We have fulfilled the operation of the goal structuring. The supertask is not to be sick because of the stress, to get antistress neuroimmunity, that means presence of three linked goals and three programs of the psychoprophylactic, that should be realized by the sick person with the help of the BOA.

Let us make a resume. Biological organ of adaptation to the superstress, fulfilling the task not to be sick because of the stress disease, should be rational or in other words conscious (informed) automatic system stabilizing antistress reactions of the health (non-suffering).

THE HEURISTIC MODEL OF THE SEARCH FOR THE BIOLOGICAL ORGAN OF ADAPTATION

The supposed discovery of the BOA and the search for it can be presented as the ascent of the mental mountain, to the top of it – the goal. The desired key of the mystery is situated at the top of the synthesis. We are standing now at the bottom of this mountain – "Mont Blanc" of the knowledge. Every stair up is the process of the brain ascent onwards new knowledge–superknowledge, which means capability. If we would present every stair of knowledge as a step of the spiral, getting more narrow towards the top, than we would get the dialectic model of the learning. The brain ascent would be the ascent from the phenomenon to the essence, from the concrete phenomenon to the common one, that means to the law.

For getting the goal – discovery of the BOA, our euristic model of the mental mountain should have twin peaks and two ways. The first mountain is theoretical, here the goals, programs and hypotheses are born and set up. The second one – methodical. The set up goals could be realized practically with its help.

Let us begin the ascent of our goal – brain storm of the problems. Let us do the first mental jump.

MENTAL STEP 1 (SYNTHESIS OF THE CIBERNETICS AND NERVISM IDEAS)

The bases of work for the antistress and technical automates have common control laws, that are based on the feedback mechanism. The synthesis of the nervistic schools ideas and technical cibernetics and synergetic ideas should be heuristic.

The goals of the automate, autopilot of the aircraft, for example, and BOA are equal: to avoid the accident, to stabilize the flight, to get the aircraft to the arrival point. The human being's antistress automate has the goal to avoid biological accident in case of the superstress shock for the organs and systems and to fly through the life being emotionally healthy (stress without distress).

Evidently, the the models of alive and technical automate should have a lot in common. As far as the work of technical automates has developed control theory, we will compare the principles of the BOA activity with the technical one's.

Let us admit the most important parts of activity and synergetic.(8)

Both automates should have two control levels: peripheral or inner contour of self-regulation, which supports the system's work main regime and central one, that compares the activity of the unit and the etalon system and makes the corrections. The etalon model sets up the goal for the system of contours and the comparing unit controls it to minimize the difference between the etalon and the inner contour reactions.

High reliability of the technical automate is achieved due to the contour of self-regulation, synthesis of autoregulated systems, etc. We should admit the law of the adaptation and avoiding of accident, which is known in regulation theory as declaration for adaptation: since the error is located in the work of the automate it should be immediately corrected, because it carries the threat of the accident.

Let us make an ascent of the superknowledge.

MENTAL STEP 2 (SYNTHESIS OF THE SYNERGETICS AND NERVISM IDEAS)

This synthesis will help us to realize deeper the essence of the psychophysiology on the conditioned reflex ultrastructure level. The synergetic model of the conditioned reflex, idea of the psychological solitone provides the great increasing of the behavior science level. It also presents in the clear and simple form the model of the goal achievement. Synthesis of the synergetics and nervism ideas is to our mind the synthesis of the particle-wave in the psychology.

To realize how we can escape from the psychotrauma, the pathological stable part in our subconscious, we should address to the synergetics, the new physics department, researching the processes of self-organization in the complex systems.

From the synergetic point of view neurogenic reactions are bioenergetic whirlwinds of mental, emotional, bodily and spiritual particles, the mental solitones, which have a lot in common with the whirlwind of the air and dust.(13) The ciclones and anticiclones have the same nature.

It is known in physics that one of the most important condition for the appearing of the stable whirlwind is the presence of highly intensive flows of energy. The analogy of this flows in the mentality is superstress. Taking part in the healthy situation the stress is utilized by the required behavior and does not reach the nonlinear level usual for the intensive emotions. The superstress reaching the pathological whirlwindgenic level is getting to its mental ciclon the flow of the stress mental particles and mental particles of the body' preceptors of pain, hypertension etc.

This mental ciclon of the unhealth, whirlwind of the neurotic stress and suffering particles comes out of the conscious control, becomes an uncontrolled irrational unit of the behavior and shows up as the disease symptom.

This flow of the neurogenic mental particles begins to live its own life, getting any energy of the life stress for itself. Its goal is to save its own essence, to defense against the owner, keeping the stress around the pain receptors, increasing the number of solitones, growth and strengthening of neurotic part.

From this point of view we could declare that the neuroses and psychic reactions are stressogenic energoinformational new formations of behavior.

Taking into consideration the ideas presented above development of the disease is the including new and new mental receptors into the stress mental ciclon, formation of new mental solitons - whirlwind's microunits of the disease conditioned reflex.

Let us make the ascent of the presentation, widening the conscious level.

MENTAL STEP 3

Every appearing of the neurotic and psycho-somatic reaction is strengthening of the conditioned reflex, in other words the strengthening of the disease. This phenomenon could be understood as the formation of the new disease unit, in other words the new psychosolitone, and this is one more block for the stressogenic building of the disease. Every pathological conditioned reflex is the mental centrifuge roating the psychounits of the stress and suffering.

Let us make one more ascent towards the superknowledge.

MENTAL STEP 4 (SYNERGETIC HEALING KEY FOR THE STRESS DISEASES)

The basic question is – which direction are the psycho-whirlwinds of the "psychocentrifuge", stress diseases conditioned reflexes rotating? The answer – sure, the direction, contrary to the whirlwinds of the norm and healthy reflexes. All the nature is rotating together with the Earth around the Sun, according to the common sense, following the clock hand. It means, that stressogenic psycho-whirlwinds are rotating following the contrary direction. Thus, the new practical principle of treatment of the stressogenic diseases can be declared.

To switch off the disease symptom-reflex, it should be put into the more powerful "psycho-centrifuge" - contrwhirlwind of the contrreflex, rotating to the contrary direction. Just this principle is used as the base for the flexible personal self-regulation with the help of the 7 reflexes of the BOA.

Let us make one more ascent of the superknowledge.

MENTAL STEP 5 (WIDENING OF THE EMOTIONAL SELF-CONCIOUSNESS)

If the stress switched on the pain symptom-reflex and the Mind is indifferent to this fact - it means that the Mind sleeps because of the low knowledge level. Even more, this indifferent position is the fact of the biological immorality towards ourselves.

Let us try to explain. The fact of letting the stressogenic pain into our life without immediate elimination of it is, from the point of view of physiologist, the fact of strengthening–increasing of the conditioned reflex. Stress is materialized in new pain receptors and the disease has grown till the new level. The biological accident takes place, the contradiction between the brain's and body's behavior.

The moral biological imperative "Don't kill... Don't damage yourself with the stress..!" is coming from the deep comprehension of the mental genesis as the disease development on the base of the psychological culture and the physiological knowledge.

The knowing of the disease inner picture means getting of the mental vision, view of the Mind and awareness of the emergency.

Let us make a conclusion: the person having the civilized mentality is neutralizing the complaints with the help of the BOA while the emotionogenic symptoms appear. The moral biological imperative "Don't kill... Don't damage yourself with the stress..!" is forbidding the damage of our own life because of the distress.

Let us make a new ascent of the mystery.

MENTAL STEP 6

Let us present BOA with the image of the alive person protecting us from the terror of the stress.

Four important parts are working for the salvation:

1. "the head" should take the decision, define the goal and choose the strategy of the behavior;
2. "the body" is performing the role of the functional transformer in the automates, it breaks the disease and transform it into the contrary thing;
3. "the arms" are fulfilling the commands of the Mind, they act as the mental channel of the communication between the conscious and the body of the BOA with the anti-disease reaction;
4. "the legs" are going towards the health, one inhibition – one step from the disease towards the kingdom of non-suffering.

The first part is the "head". It corresponds to the rational (conscious), informed part of the specialized antistress organ-analyzer. The "head" is making the analysis-syntesis of the stress situation—accident and making the decision—goal. It should choose the behavior, forbidding the activity of the disease. The system should be informed enough to act in this way and reach the goal. Information is considered to be complete, if the system knows how to answer 5 questions: what to do-how-when-why (the law) - what for (the reason).

Let us ask ourselves the question. Does the Mind know the goal of the self-restoring the health while the disease manifestation? The answer looks evident. To escape from the suffering here and now by any means. The first part of the question is logical. The second part is determining the disease destiny: to live and to torture the person by years and decades or disappear. The answer for it is contained in the question: what to do-when-why-what for? The answer looks like the following.

The act of switch off-inhibiting the disease reaction should be made not by any way, because in this case it will be just the act of switching without the elimination of the disease program.

If we would inhibit the disease reactions only with the help of the BOA, in other words using the controlling feedback, then we will get the process of the changing of the disease reactions to the healthy reactions of non-suffering. According to the psychologist's point of view we will teach the stress not to enter the disease with the command of the Mind, or we will humanize its animal's nature.

Let us now answer for the second question: how to reach the goal? The answer is contained in the Pavlov's nervism theory and in the automatic control theory. If the unit of the stressogenic factor is conditioned reflex (M.P.Pavlov), then the law of the reflexogenesis is followed by the healing law - the "key" for the self healing according to Pavlov: permanent inhibiting of the reflex-symptom till zero – the disappearing.

The automatic system control theory formulates the same essence as the adaptation law—(the impatient client theory)—the request for the adaptation is satisfied immediately with the help of the feedback (look the step 1).(8)

Let us make a conclusion which is very important for the nervism theory and the practice of healing. The adaptation algorithm in the cibernetics and the health restoring law in the case of disadaptation diseases according to Pavlov coincide, as well as the laws of activities of the human and the technical automats according to N.Vinner.

Let us make the new ascent of the mystery.

MENTAL STEP 7

The system is considered to be completely informed if it is capable to answer all 5 questions. Let us try to answer the fourth question.

When should be BOA switched on? The answer to this question will help to realize why the whole ocean of modern healing the stressogenic diseases methods fail, especially the pharmacological ones.

The key of this mystery one can find in the researches of Pavlov, Bekhterev, Asratyan: in the reflexogenesis and the reflexosynthesis law. The stimuli and reaction combination principle dictates the algorithm of the BOA activity. It should be switched on immediately at the moment of the stress-signal activity and the disease reaction. Just in this case the disease reaction is inhibited and health reaction is brought up according to Bekhterev.(2)

The fifth question "what for" is clear: the contradiction of behavior, biological micro accident, which was born because of distress should be immediately removed according to the smart automates principle.

Let us make the ascent of the new mystery.

MENTAL STEP 8

The stress overcoming the level of personal durability destroys the organ safety system and the whirlwind mechanism occupies the pain receptors. The solitone—the nerve trace—disease conditioned reflex is formed by the swallowing up the pain psychoparticles as if it would be the dust particles. The disease is growing with the increasing of the solitones number.

The stress (emotional conditioned reflex) is very specific at the first stages of the disease. Superstress is switching on the pain, and further one watch the activity of the mechanism that could be called generalization of the conditioned reflex, following Pavlov's model. Now any stress and anxiety is flowing into the funnel-pain magnet. We have here the phenomenon of the emotional behavior dividing from the logical point of view. The stress becomes mad, in other words it looses the emotional way – the goal (for example while the querral or hard attempts to solve the problem) and falls down to the pain, giving its energy. That is why one could call the stress diseases the emotional schizophrenia.

How could one recover the emotional "breach"?

The answer is to put the psychozond, pushing the stress out of the pain place in the body. It will be conditioned reflex-psychoautomate, contrwhirlwind in the anti-disease system. It will put the stress away from the path of disease to the one of health. This "repairing" is made by the Mind with the help of the BOA according to the healing algorithm, which will be given in the methodical part.

Without the answer for the fundamental clinical question – what are the stress diseases symptoms we wouldn't be able to find the way to the rational healing. The antidote could be discovered only after the poison would be found by somebody.

Nervist's ideas have made great impact to the treasures of the collective thought - behavior science. The mighty dive of the Mind for the new level of the struggle with the stress diseases is possible only on the basis of the nervist's schools ideas, cibernatics, synergetics and development dialectics synthesis.

Nervist's research ideas have stopped near the doors to the clinic. There is still the gulf between clinical psychology and physiology of the higher nervous activity. Let us try to build a bridge from phenomena-symptom (for example, the pain) to its reflective essence.

Let us continue the ascent.

MENTAL STEP 9 (THE KEY FOR THE STRESS DISEASE FROM PAVLOV)

The image of the reflective brain activity having a lot in common with mechanism is coming from Decart to Sechenov-Pavlov. Considering that the acts of behavior are reflective and the conditioned reflex is the unit of the behavior and symptom is the unirt of the disease behavior, in other words the disease conditioned reflex, how can one overcome the disease symptom-reflex according to Pavlov? The answer is evident. Using the law of the inhibiting the disease till zero we eliminate the basic cause of the disease and prove the conditioned reflex essence of the disease symptom-phenomenon.

How is the pathological conditioned symptom-reflex formed finally? This mystery is guessed in the research of Asratyan. "Conditioned reflex is the synthesis of two unconditioned reflexes."(1) Where are they in the symptoms (for example, pain)?

This fact is basically important for the forming of the new healing paradigm.

Let us declare our own understanding of the problem. There is the reflexosyntesis of the first unconditioned emotionally-stress reflex and the second unconditioned one – pain reflex taking part in the symptoms of stress diseases. Anxiety-stress because of any reason – it is switching on the first half of the reflex, which radiated the pain reaction with the energy-information-substance.(1,14)

The synthesis of the stress-emotional reflex being combined with the astenic one gives clinical astenic symptom-complex, being combined with the mental one – symptom-complex of fixed ideas. We have here the system of disadaptation diseases. Having been inhibited systematically according to Pavlov with the help of the BOA they disappear. The declared idea was clinically proved.

How could one heal stressogenic diseases theoretically? It's necessary to make the ascent of the super knowledge. The "keys" for this problem belong to different nervist's schools. The synthesis of nervist's thoughts gives the great key for the mystery-rational flexible self-control.

Let us continue the ascent.

MENTAL STEP 10 (THE KEY FROM PAVLOV)

The might of Pavlov's ideas is the theoretical weapon of the new healing paradigm. One of them is Pavlov's perception of the inner and outer inhibition processes unity: "the inner inhibition is formed on the rest of the outer one".(7,8)

This declaration will help to go deeper in the understanding of the desadaption diseases sanogenesis etalon model. Pavlov has seen the reason of the conditioned reflex inhibition without strengthening in the inner inhibition mechanism, which he has called "the devil problem".

What is the deepest essence of the inner inhibition? This scientific problem is still the mystery of the reflexology. We will try to answer this question further.

The key from Bekhterev.

The genius of Bekhterev gives us the general peceipt equal to the law by the rank - how to cure the disorders with the psychogenic roots. "Reflexology is getting the way to the field of the nervous diseases therapy. The psycho reflective influence should have two goals: from the one hand to inhibit pathological combinatory reflex as possible, and from the other hand to bring up a new healthy one".(2) Our experience proves the universality of Bekhterev 's curing law towards any disadaptation disease.

The key from the Asratyan's school.

The powerful Asratyan's thought concerning the reflexosynthesis allows nervism to make a great theoretical dive towards clinic. According to Asratyan's and Rudenko's thought formation of the new reflexes is happening because of their uniting in such a way, that the first of them loosing its own reaction transforms into the complex signal, becoming the reason for switching on the second reflex.(2,9)

Let us make the ascent of the superknowledge.

MENTAL STEP 11 (THE LAW OF MAINTAINING THE BEHAVIOR IN CONDITIONED REFLEX)

Let us make an attempt to synthesize the ideas of the leading nervist's schools for the further ascent of the goal. The basic question of the behavior science: what is the determinant of the inner inhibition without strengthening of the conditioned reflex. The author has the following understanding.

The answer in the intuitive form is done by Bekhterev and Asratyan in the psycho-reflexsive therapy and reflexosynthesis (see above).

The unity of two contrary processes is taking part in the case of the conditioned reflex inhibition: the inhibition of the first one and the building in the same signal the second one as the result of their mutual penetration and mutual transformation one into the other on the basis of maintaining behavior law: the same quantity of the signal power and reaction is lost by the disease reflex and is added to its antisystem – healthy reflex.

The discovery of this law as the maintaining conditioned reflex behavior and flowing—transforming them one into another is equal to the possessing the inner eye, psychic view, widening of the consciousness. We are getting the goal of healing-adaptive behavior etalon model. The rational analitico-synthetical part of the BOA is formed. The Mind knows what to do at the moment of the disease appearing: to inhibit the disease symptoms till their disappearing with the help of the more strong BOA reflexes—in other words to transform "the disease reacts into the healthy reactions of non-suffering". The stress signals will switch on not the disease reactions, but contrareactions of non-suffering.

Here is our goal-possessing the self-adaptation to the distress. The anxieties would not switch on the disease. Stress breach of the body will be eliminated.

We have made the ascent up to the top of the hypothetical "mountain" from the side of its theory. The "head" of our "person" is formed, discovered.

Now we should make the ascent of the second half of the "mountain", where the methodic keys are situated, those that the system goal of the self-adaptation to the distress. We should discover the rest of the BOA parts – its "arms", "legs" and "body". We are talking now about the perepheric part of the antistress analyzer and neuro-psychic channel of the communication with the rational part of the brain.

Let us begin the ascent of the methodical top of the mental superknowledge mountain.

MENTAL STEP 12

Well, the controlling part of the BOA is formed. The Mind has enough information about the goal of the health self-restoring.

Let us repeat once again. The stress should be tough every time in the moment of switching on the disease: to enter not the "home" of the disease reaction, but on the contrary - the "home" of health reaction The person should do it consciously and every disease case can be suppressed-inhibited with the help of the more strong behavior dominant "Anti". "The brought up non-suffering reflex (Bekhterev) or realized sano-reflexosynthesis (Asratyan) – it is just Pavlov's "zero". The antistress automate – the homeostasis – the nervous immunity is declared.

Let us use the image of the stress signal "head", that should be cut of the disease "body" and put on the healthy "anti" reaction "body". The questions are coming. Where is situated antistress non-suffering reaction? Where will the Mind and the body get the flexibly controlled antistress health reaction, which would be capable to win every meeting with disease dominant? How is it possible to conquer the stress signal belonging to the disease and give it to the non-suffering reaction, so that it will serve light forces instead of the dark ones, switching on anti-disease, but not the disease.

We are talking here about the peripheric adaptation contour formation, about the discovery of the BOA "body-arms-legs" or the functional part of the system, realizing the goal-etalon program, self-healing algorithm. We are to answer the concrete questions about the reaching the goal methodically.

The key from the Kryzchanovsky's school.

"Activation of the antisystem" is one of the most important mechanism of the "healing and curing. The population of neurons that act separately and produce excess excitement we call the pathologically increased excitement generator. The radical method of neuropathological syndromes curing consists of the pathology main chain (it's determinant) elimination".(7)

Let us make the ascent of the new curing model.

MENTAL STEP 13

The ideas of Kryzchanovsky let us make great theoretical and methodical dive of the thought towards the superknowledge. The methodical way – ascent of the top of our hypothetical mental mountain.

It becomes clear that the Mind should make the stress to invert its energy for the curing. This goal should be realized by the disease "antisystem". It should become superdominant for this purpose, in other words it should have antistress energetics or more strong biomagnetism. According to Kryzchanovsky we should form anti-disease neurone population in the disease antisystem-antipain generator, that would produce healthy excess excitement at the moment when the stress-pain reaction is switches on.

Let us make the ascent of the new heuristic knowledge.

MENTAL STEP 14

Let us admit the coincidence of the Kryzchanovsky's view at the desadaptation process essence and synergetic health and disease whirlwinds model. The generator of the pathologically increased excitement, acting automatically, according to Kryzchanovsky is just the same whirlwind-psychosolitone of the stress and pain particles - microparticle of the conditioned reflex, which is born due to their combination-strengthening.

According to Kryzchanovsky actination of antisystem is one of the most important curing mechanism, thus the generators of the sanologically increased excitement, producing surplus healthy excitement should be created right there. This is the same phenomenon of the contrwhirlwinds-health and non-suffering solitones.

According to Kryzchanovsky neurogenic determinant is the pathology central chain. It's elimination is the radical method of neuropathological syndromes curing. The determinant of the psychosolitones, neurotic units of disadaptation, is the superstress signal, that creates the emotional whirlwind around the pain receptors. To eliminate the whirlwind energetics means to eliminate the pain, because the pain "bulbs" would not light without the energy.

According to Asratyan the same determinant of the disadaptation diseases is conditioned reflex (superstress), that "determines the destiny of any disease, opening one neuropathes and closing the others".(1)

Let us make an ascent of the new superknowledge.

MENTAL STEP 15

Let us make interscientific synthesis of the micro and macro reflective approaches for the understanding of the pathoneurogenic behavior unit - this "block" of the disease. The neuro-trace is understood as the memory or the disease pathostimul and pathoreaction unit of the learning in the psychology, or as the unit of strengthening and the microreflex unit formation - disease symptom, in the physiology. The same essence has the pathological neurogenerator formation according to Kryzchanovsky, and lso the birth of the pathopsychosolitone - disease neuro-whirlwind, microdominant according to Ukhtomsky. From the point of view of the psychocibernetics the unit of pathological neuro-feedback or the unit of the disease psychoautomat is born.

In this case the unit of the disease reflex inhibition by the healthy one means sanopsychogenesis, the unit of the healthy evolution – adaptation, the unit of the mutual transforming the substance of the disease and the health, in other words, the unit of the new self-organization and growing to the health (according to Pavlov, Bekhterev Asratyan).

The unit of the inhibition from the psychological point of view means "erasing" of the disease trace memory, the act of forgetting and the same time – the birth of the disease memory antitrace, getting the experience of non-suffering. The formation of the neurogenerator unit - psychosolitone - microdominant - psychoautomate - sanogenic neuro feedback of "antidisease", non-suffering has the same meaning.

Let us make an ascent of the new superknowledge.

MENTAL STEP 16

Let us try to examine the disease level and it's nervous power using as the example the clinical case of the patient R (Fig.1).

We can see at the figure that 264 cases of inhibition were used to suppress the disease reflex. It means that the disease system-symptom-dominanta includes 264 units of the reflexes - neurogenerators - psychosolitones - psychoautomates. The whole disease system works in resonance mode, like a whole unit, like one whirlwind - psychocyclone of stress and pain particles.

Pavlov's inhibition of pain reflex down to zero means that mental space of painful point amounts to 264 steps, the same number of cases of mutual transition and mutual transformations into the system of antidisease.

And now we will try to represent the measure of nerve power (the power of attraction and linking of pathostimules and pathoreaction, stress and pain) in the specified case. If one support of stress and pain gives rise to the unit of reflex-symptom, their neurogen-

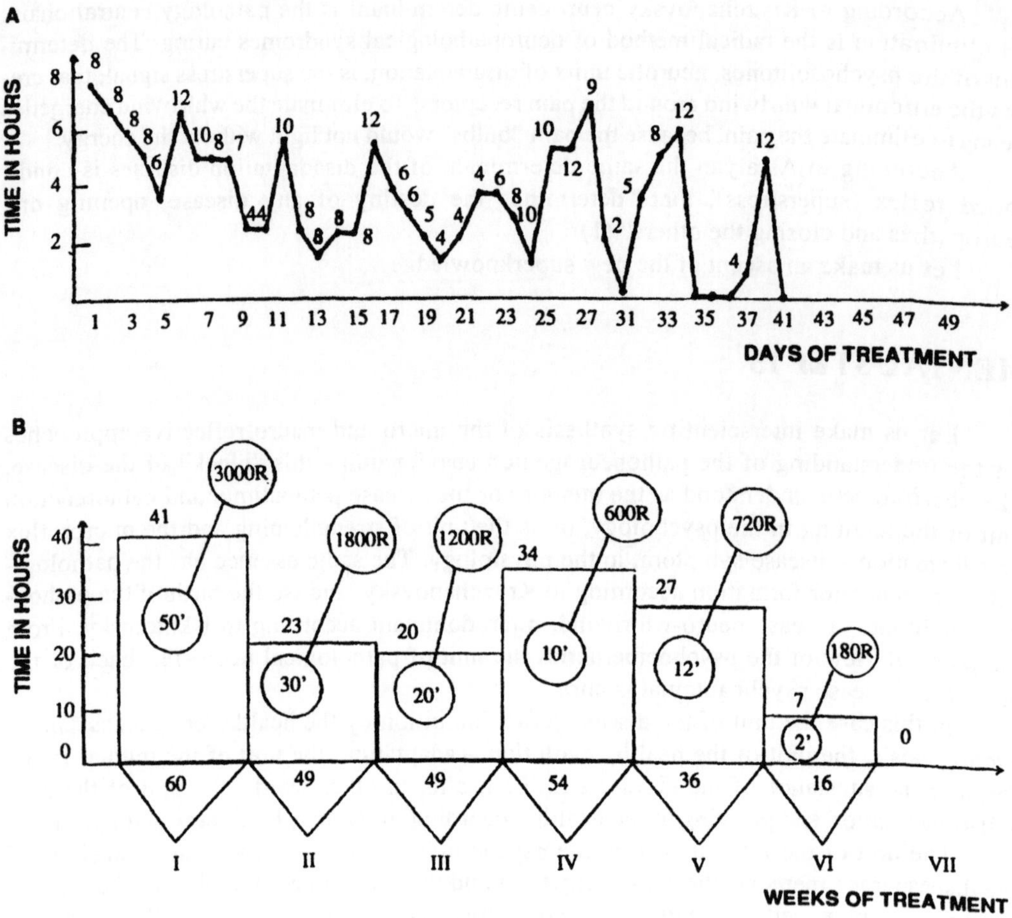

Figure 1.

erator causes pathological stimulation equaling the unit of the nerve flow, and psycosoleton–whirlwind causes the unit of biomagnetism. It means that bioelectromagnetical power of the whole system-dominant of the disease equals 264 bioelectromagnetical units.

In brief, the power of stress and pain attraction, their nerve connection equals 264 "biomagnets".

In the given figure we can see that on the 41st day the mutual transfer and mutual transformation of the system-disease reflex into its opposite was performed. The disappearance of disease pulsations means that stress was taught to activate an antidisease-antipsychocyclone of stress and pain particles instead of disease-psychocyclone of stress and pain particles. The power of the disease constitutes 264 biomagnetical units, and the power of antidisease outnumbers this quantity. We can make a very important for the human being conclusion: animal stress, this uncontrolled mental "tornado" is becoming human and harmless under the influence of thought. Thought directs it to the safe place which is called the system of non-suffering, the system of the disease self-liquidation. So human thought makes elements of nature work with the help of technology, and makes stress work with the help of psycho-methods.

Let us make an ascent of superknowledge.

MENTAL STEP 17

Let us ask the question. Where is the stress diseases "antisystem", for example, pain situated in the human body? We mean antistress homeokinetic system of non-suffering in the periferical systems (organs) of organism. This is a system of antipain receptors of bodily enjoyment, supercomfort.

In the projection of pain reflex where stress enters (painful point according to Pavlov) there are always receptors of antipain (healthy point). Antistress receptors giving rise to reaction of comfort and pleasure in the body represent mechanism created by nature to restrain stress and pain. This is its arresting device. It is here that nature laid mental "Archimedian" lever for the disease turning over and liquidation. Pleasure-system of antipain is a periferical part of BOA-system, antistress analyzer. This is "the body and the legs" of BOA.

Every time the stress initiates disease the Mind is to send a controlling command to the periphery to the antipain receptors for correction of the stress mistake and to teach it the culture of normal reaction according to the model of self-healing.

MENTAL STEP 18

To adapt the organism and the Brain to superstress it is necessary to find universal psycho-remedy in the organism itself. It is represented by antipain reflex through which (according to standard model) every case of activated pathologic stress-pain reflex-symptom is to be necessarily suppressed. We have already mentioned that standard model of self-healing is the principle of absolute suppression of pain reflex by normal one like an absolute liquidation of mistake in the automate's work, being a condition of the signal transformation and reaction of "anti" non-suffering.

There is no controlling communication channel between the Mind which is a commanding contour of adaptation and the system of observation which is a periferical contour of emotional homeokinesis. That is why brain can not arbitrarily switch pain off by purposeful influence on its "countra-lever"-system of antipain receptors. Here we face the problem of making periferical reaction of non-suffering more arbitrary by the Mind.

Through all human history collective thought has been looking for methods of combating the peripheral pain using switching-on antipain reactions. This aim was achieved through exerting local influence with medicines, physical methods. Another lever of influence is a general organizational, energetical one. If brain antistress center of pleasure is switched on, the center of negative emotions and energy supply of pain reflex will be switched off.

Experiments on autostimulation of pleasure center by Oldes and Millner (4), including the clinical experiment of electrode implantation and its chemical stimulation with drugs prove that human being possesses central pain control psycho-lever.

A crazy idea is proposed. Could not Intellect with the help of thought switch both central and peripheral "centers" of pleasure-antipain on?

We will ascend to superconsciousness.

MENTAL STEP 19

We propose general approach to the problem solution. We are using creative Pavlov's idea, namely, "repeatedly affecting the same excited cells (like falling drops) causes inhibition within them". And what about the inversed way? Let us affect repeatedly inhibited cells of antipain reaction of bodily pleasure at the moment when stress switches pain on. If we remove inhibition from them, the reaction of pain will be naturally switched off by the counterdominants mutual elimination mechanism. Besides, the second problem if solved as well—switching the brain center of pleasure on. The autostimulation of periphery from which impulses come to the cells of this center like falling drops gives such an opportunity.

Let us ascend to superknowledge and superskills.

MENTAL STEP 20 (THE HUMAN BEING MENTAL "NLC" DISCOVERY)

Presented project of making stress purposeful by means of conscious switching on the periferical system of "antipain-pleasure" and autostimulation of brain pleasure center can be solved with the discovery of universal antistress BOA. If, basing on mutual antistress energetics of all 6 antistress reactions we combine them into unified system, which can be launched independently by the Mind with the help of speech command, we will obtain the control channel of normal feedback communication contour between consciousness and all the antipain systems of bodily enjoyment in the organism.

The novelty of the approach is embodied in the fact that all 6 antistress reactions discharging simultaneously, synchronously in a single phase work rhythmically with pulse rate (1 impulse per second). The system of 6 reactions rhythmically and repeatedly affect inhibited cells of "antipain" reaction, i.e. the 7th reaction – "the body" of BOA until they remove inhibition from it. As soon as the reaction "antipain"—bodily pleasure gets rid of inhibition—the pain is eliminated, suppressed. The aim is achieved - the claim for adaptation is fulfilled.

But an unsolved methodical problem still exists at present. How can the energy of 6 antistress reactions (suborgans) be accumulated into one psychic ray, into one band of nerve energy and be transmitted from brain to periphery into receptors and antipain reaction? A special psychic, spiritual communication channel is required. This function is carried out by psychic NLC, this abbreviation means: "NC" - neurolinguistic, "C" - circle. It's a spiral-like rotating whirlwind—circle going from Brain—consciousness and bringing urgent spiritual help to the affected organ.

"C" – circle carrying information to the point of pain and energy – to the point of "antipain". This signal of consciousness "Pain is disappearing" is the language of communication with the cells of pain and stress. The Mind teaches them to communicate in this way making the cells of disease "human" of a kind. Pavlov's "zero" means that the language has been understood and the cells of disease have become cultural. So, Intellect is integrated with pain.

We are going to make ascent of superknowledge.

MENTAL STEP 21

We will make the idea as simple as common sense. In his time I.P.Pavlov with the help of saliva gland model had discovered the conditional reflex and its laws. He stimulated the saliva gland by means of sound and light and managed to control its functioning. Why couldn't we stimulate the periferical reactions by means of psychic image such as pulsating colorful circle of 6-reflex system in order to control the superstress and pain reactions?

The method mentioned above is not only curing but it also enables to lead the research of the human being conditional reflex. Pavlov was measuring the psychodynamics of conditional reflex in the drops of saliva. In the same way we can measure the power of disease dominant in terms of psychomechanical "quantum" depending on transfer of pain reflex into non-suffering reflex. We can see this at Fig.1 A and B.

Resonating work of the "band" of antistress rotating BOA-reactions is not linear as any single of them but it is ruled by the laws of whirlwind psychocentrifuge. All of them represent a powerful antistress neurogenerator, stimulating psychosolitone of antipain - the flow of psychoparticles of antisytress around the receptors of bodily enjoyment. The synchronous work of BOA is a kind of mental "magnifying glass" which multiplies the emotion of comfort to making it capable to "light a fire" of antipain receptors.

Let us make an ascent of a new superknowledge.

MENTAL STEP 22

There is new physiological interpretation of BOA activity. In order to make the stress purposeful let us make a synthesis of classical Russian physiological school of M.E.Vvedensky and A.A.Uchtomsky with the nervists' ideas. We mean the histeiosis - dominant phenomenon.(2)

According to M.E. Vvedensky's idea the histeriosis means rhythmical stimulation of afferenting nerve by incoming waves. The proectional nerve centers' irritability increases greatly. A.A.Uchtomsky's dominant is a major center of irritation suppressing all the other ones. Unlike the histeriosis phenomenon it is capable to accumulate irritation.

The rhythmic self-stimulation of antipain receptors with the help of psychorotating image BOA will increase their irritation to the level of antistress superdominant by A.A.Uchtomsky. Such histeriosis procedure can be called a "psychopumping" of antistress-antipain dominant. The synthesis of Vvedensky-Uchtomsky's theory with Pavlov's ideas creates a new approach to behavior control. A person gets a new property-volutional self-control over the superstress.

Let us make a new step to superknowledge.

MENTAL STEP 23

We have already discussed the constructive similarity of the technical auto-regulating automate and the antistress psychohomeostat - BOA. Taking into account that the latter is functioning as amplifying-transforming device and also provides the antistress systems' self-regulation we can make a conclusion about their functional similarity. Let us try to combine the "mental blocks" of BOA - the antistress analyzer performing a specialized function of antistress psychostat according to H.Delgado (6).

"The Head" – an informed consciousness: What to do - How-When-Why-What for.

The moment when the signal of biological accident comes, the Mind knows the aim - the standard model - and how to achieve it.

"The Body" – an antipain reaction which is caused by rotating resonance reflex by means of the histeriosis - dominant mechanism.

"The Arms" – a rotating system of 6 reflexes leading to cancel of stress-generated reactions in any part of the body. The thought becomes artificial with their help.

"The Legs" – is the reaction of extinguishing micro-transfer, compelling signal of stress "go away" from disease. So in the described clinical case 64 "steps" were needed to cover psychic space to health.

Psychical NLC – this rotating reflex functioning as a carrier of system signal-psychoreflexive quantum of information – energy-matter to the affected organ is psychotronic weapon of health invented by the Mind magnifying the power of thought millions times similarly just as electronic microscope, vehicle, rocket strengthening the power of eyesight, arms, legs.

Systematic antistress property of 6 reflexes of BOA provides man with new mental properties:

1. extramentality-ability to transfer disease to the side of non-suffering with the help of thought;
2. extrasensority-ability to switch on the "antipain" reactions by the thought;
3. extramotoric ability-ability to control superstress with the help of arms-eyes-breath.

Every person has antistress organ of adaptation. The task of medicine is to teach everybody how to use it.

Declaratory thesis—to treat a person not a disease—is filled with a new practical meaning. We'll use a metaphor. Why should you constantly fill a leaking barrel with water? It is necessary to eliminate the leak. So is the person with stress diseases. Why should we treat him during years and decades? It is necessary to show him the organ of biological protection, to teach him to use it and he will liquidate the"leak"by his own means. Stress will become harmless. That's why diseases of stress are first of all the diseases of ignorance, and only after this they can be called the civilization diseases.

Let us continue the ascent of superknowledge.

MENTAL STEP 24 (PREVENTIVE FUTURE STRESS CONTROL WITH THE HELP OF THE BOA)

The permanent suppression of pain reflex by normal one, i.e. BOA achieves the aim of self-reflexosynthesis performance according to Pavlov-Behterev -Asratyan - Vvedensky - Uhtomsky. The healing law of secondary (restoration) prophylaxis works here. Once again we ask question: what is Pavlov's "zero"? According to new understanding of the process of internal inhibition as a mechanism realizing law of behaviour conservation in functional system. The amount of decreased pain reflex equals the amount of normal one increased.

In our case, presented in figure "O" means that pain reflex has 264 biomagnetical units, and a normal one a few more. That is why it attracts stress-signal to its side. Conditioned-reflectoral biomagnetism and psychogravitation of stress-signal and antipain reactions have won. This is what we call mechanism of afferentation synthesis on unconscious level.

As we can see the balance of health is very unstable. Any superstress can break through "psycho-clamb" of health.

Let us control the problem. Can one prophilactically control future stress? We mean tertiary mental prophilaxis. The given problem can be solved positively. Let us present the theory. The signal of stress in the situation of performed sano-reflexsynthesis is "magnetized" by "antipain" and nerve passage to pain reaction is impathable under the principle of negative induction. Antipain system represents stereotype of conditioned reflexes - psychosolitones numbering supposedly 265. Under the law of conditioned stereotype (1,11) switching the first link on switches the whole reaction on. Hence, if with the help of BOA 2–3 rotations are performed with controlling command of intellect: "Healthy-No pains!", all system of "antipain" – 265 mental whirlwinds is activated. Within "antipain" system we performed stagnant self-stimulation, which still deepens the process of inhibition in the system of disease. That is why the coming stress can not activate pain.

We can make a conclusion. Using principles of conditional stereotype and conditioned back-feed communication according to Pavlov one can consciously and purposefully control the process of "pain-absence" and immunity of painful point to stress. In the given clinical case female patient during 2 years controlled the process of the disease spending 2–3 seconds for prophylaxis in the morning and before going to bed.

MENTAL STEP 25 (DISCOVERY OF THE THIRD BRAIN SIGNAL SYSTEM)

The third signal system represents the synthesis and further development of the first and the second ones. It includes not only informational, but also energo-evolutional component. "NLC" is the parapsychological psy-language of Mind communication and control for the stress and distress cells–signal of disease cancel.

One the see the humanizing of the neurotic cells at the Fig.1. "Zero" means that since the 41st day the word-thought are controlling the disease directly.

SELF-ADAPTATION TECHNIQUE TO STRESS-INITIATED DISEASES WITH THE HELP OF BIOLOGICAL SELF-REGULATION ORGAN

At the moment of the stress-initiated disease reaction (for example, pain) patient's personality consciously and deliberately activates 6 suborgans – the zntistress reflexes synchronously. We place the painful point in "psychocentrifuge"–psychowhirlwind of rotating reflex and speculatively while breathing out in accordance with pulse rate we perform 3–5–10 rotating comforting self-stimulations. Rotating circles are of rose-red shades.

There are two conditions of aim achievement. The first is: rotating selfirritations must be pleasant.

The second condition: in each case of symptoms manifestations should be made painless with the help of the above-mentioned method.

Actualization of BOA in the form of rhythmic rotating selfirritations of painful area carries out function of antistress energy generator. This energy being accumulated switches on disease "antisystem". Disease stops and naturally is eliminated in accordance with well-known physiological laws.

CLINICAL-EXPERIMENTAL CASE OF ADAPTATION TO STRESS DISEASE WITH THE HELP OF BOA

Patient R. - 50 years old. During 10 years she suffered from frequent headaches and heartaches being provoked as a rule by agitation on any reason. She was frequently examined and treated with traditional methods:pharmacology, physiotherapy, psychotherapy. The temporary effect after treatment was short-termed.

After we had formed BOA and taught her health self-restoration techniques, the above-mentioned pains disappeared and did not appear for 2 years. Absence of complaints was registered in clinical report during prophylactic medical examination. Technique of intertransition and healthy transformation of neurotic reactions (behavior) quantitative research by the method of wavelike conditioned reflex (quantum neurism).

The patient was taught to use method of painful reactions elimination with the help of BOA. During 24 hours within treatment course patient registered the time of pain elimination and their total amount during 24 hours. Then obtained information was presented graphically as 24 hours- and weekly psychodynamics. We obtained first time in the medicine's history visible picture not of a disease but of a recovery with the help of specialized organ of stress elimination as a result of stress adaptation.

PSYCHOPROGNOSIS OF RECOVERY OUTWARD APPEARANCE

At the figure 1A and 1B we can see outward appearance of biological rehabilitation process. Let us examine Fig. 1A. If pain symptom physiologically is a conditioned reflex (synthesis of stress and pain), then under continuos suppressing of it with the help of BOA it should disappear. It corresponds to Pavlov's law of suppressing reflex down to "zero". We observe the same tendency in Fig. 1B, where information is presented in more compact form. The pain life-term of the symptom is observed to be reduced with every week in accordance with a law of extinguishing. To eliminate the painful reaction 264 occasions of its inhibition with the help of BOA were necessary.

INWARD RECOVERY APPEARANCE

We will try to see behind the general picture of inhibition of neurotic reflex-pain symptoms the invisible process of natural transfer and mutual change of painful reaction into "Ántisystem"of non-suffering (normal reflex) basing on the law of behaviour preservation. The rate of painful reflex reduction corresponds to the rate of normal one increasing.

If every extinguishing of reflex produces psycho-whirlwind-solitone (biomagnet) of health which intercepts stress flying to the solitons of disease, we can observe a regular picture of disease reduction. This prognosis is proved. The life-term of pain reflex amounts to 41 hours during the first week. In two weeks after 109 extinguishing transformations the weekly amount of pains was only 7 hours. The pains disappeared with the 264-th extinguishing because now the stress is absorbed and intercepted every time by stronger biofield of 264 psychosolitons (the unit of pain lack). Now, let us consider the Fig. 1B again. Unshaded circles show average time of pain suppressing every week with the help of BOA.

Basing on the law of mutual transformations of pain reaction into reaction of pain-absence (antipain) we can authentically predict the development of this process. In circles behind squares the number of necessary rotating efforts necessary to be applied to suppress pains with the help of BOA is shown. If a reaction of antipain is formed with every pain reaction extinguishing, then it is antipain reaction that is "fired up" by rotation. The more reactions of non-suffering will appear, the less time it will take to suppress the pains with BOA. This tendency is totally proved if we study Fig. 1B. During the 1st week disease has 264 psychosolitons – biomagnet units. They demonstrate their power. Weekly time of pain suppression is 50 minutes. It took 3000 rotating movements to neutralize them. The strength of pain during the 1st week equaled 3000 biomagnet units.(one rotation produces one quantum of nerve flow "Ántipain" of which neutralizes one quantum of pain). During the 3rd week after 158 suppressions stress signal initiates 158 unites of pain absence and simultaneously 264 units of pain. The strength of disease is to reduce almost by half. Actually, to neutralize pain it took us not 3000 but only 1200 rotations.

During the 5th week after 248 suppressions the strength of disease was reduced even more because of transfer of stress-signal to psychosolitons of antipain. It is to be reflected in a faster disappearance of pains under their treatment with the help of BOA.

In fact, average period of pain extinguishing amounted to 720 rotations, and during the 6th week it comprised 180, i.e. only 3 minutes.

During two years complaints of headaches and heartpain were not registered and it was no accident. We observed here the results of prophylactic law action – the law of tertiary psychoprophylaxis. Every day accordingly to treatment instructions (the condition of recovery) patient consciously and purposefully suppressed the center of pain with the help of BOA, "demagnetizing" nerve flow of stress to pain and "magnetizing" the flow to antipain.

The given method was tested 25 years on 20 thousand patients with neurosis and psychosomatic diseases. It provides a high rehabilitation effect. In contrast to known methods of neuroses psychotherapy we called it a method of reflex psychosurgery.

RESUME

During thousands and thousands of years consciousness was involved in uncompromising war with subconsciousness – the kingdom of stress. Animal stress did not want to obey the intellect, to become obedient, and intellect did not possess enough power and knowledge for this. Antistress strategy of Nature-adaptation through illness and suffering appeared to be unnecessary. This is not the best way of survival. Peaceful strategy of intellect is supposed to exclude disease-war of stress by switching it from pain reaction to neutral one, its antisystem. This switch of pathologic stress into peaceful "trace" may be carried out by 7 antistress reactions, specialized biological organ of adaptation. Forming of BOA is intensive, consciously emotional improvement of a person allowing him or her to have intensive superstress way of life without suffering from different stress-imposed diseases.

The law of stress-imposed diseases prophylactics permit to carry out rational control and to recover emotional breaches of organism.

WHO slogan "Psychological protection from distress to everyone in 2000" may become a reality, if mass media inform everyone of his or her BOA and teach to be protected with its help from superstress affecting spirit, body and life of a person.

Making animal nature of stress more human and the development of emotional consciousness of person is the goal of evolution and culture.

REFERENCES

1. Asratyan E.A. Conditioned Reflexes Physiology Moscow, "Nauka", 1970, p.90,162, 175,184,191.

2. Bekhterev V.M. Selected Moscow, "Medgiz", 1952 p.3,383,433
3. Bykov K.M.,Vladimirov G.E., Konrady G.P., Slonim A.D. Textbook on Physiology Mosow, "Medgiz", 1954, p.624,626.
4. Valdman A.V. Emotional Stress Phamacological Regulation, Moscow, "Meditcina", 1979 p.78,147,148.
5. Gekht K. Psychohygiene, Moscow, "Progress", 1979 p.66.
6. Delgado H. Brain and Conciousness, Moscow, "Mir", 1971 p.240.
7. Kryzhanovsky G.N. Determinant Structures of Nervous System Pathology, Moscow, "Meditcina", 1980 p.315, 335, 353.
8. Moskatov G.K. Adaptive Systems Reliability, Moscow,"Sovetskoe radio", 1973 p.13.
9. Pavlov I.P. Unpublished Letters, Leningrad, "Nauka", 1975 p.29, 300.
10. Pavlov I.P. Twenty Years Experience of Animals' Higher Nervous Activity (behavior) Objective Research Moscow, "Nauka", 1973 p.151.
11. Rudenko L.P. Functional Organization of Elemantary and Complex Form of Conditioned-reflex Activity Moscow, "Nauka", 1974 p.10,13,24,85,200.
12. Selier G. Stress without Distress Moscow, "Progress", 1982 p.5.
13. Faidysh E.K. Changed States of Mind Moscow, "DEOS",1993 p.53.
14. Khantceverov F. Eniology Taganrog, 1995 p.129.

12

SEQUENTIAL HYDROTHERAPY IMPROVES THE IMMUNE RESPONSE OF CANCER PATIENTS

Gunhild Kuehn

Clinic for Natural Medicine of the Free University Berlin
Krankenhaus Moabit, Turmstr. 21
10559 Berlin, Germany
Telephone: 0049–30–3976–3410 / secr.:-3401; Fax: -3409

INTRODUCTION

Hydrotherapy is rather established in Germany concerning rehabilitation stays at spa villages. Good experience since at least 150 years taught us that the overall health status of patients with diverse chronical illnesses, including susceptibility to infection and degenerative diseases, will be strengthened.(1) Scientific evaluation, however, has not yet been able to differenciate the effectiveness of particular therapies, because many factors like diet, climate, rest and relaxation during a spa are working simultaneously and influence one another. Therefore in that context it was impossible to throw light on the principle mechanisms that would be able to explain the specific effectiveness of hydrotherapy, that was in general observed and described as global well-being and prophylaxis against infections.(2–4)

What we know is that cold water applied in hydrotherapy has to be looked at as an irritation of the autonomous nervous system mediated by the blood vessels of the skin. First, the cold stimulus to the body is followed by a sympathetic response of vasoconstriction in order to protect the body from losing inner warmth. Then, as a counterreaction, the vagotonus gains the upper hand and leads to vasodilatation to maintain circulation and nutrition of the peripheral tissue. This process to switch over the tone quickly and adequately can be trained by serial application. The balanced vegetative tone has consequences for all the organs like the heart, the bowel, the muscles etc.(5)

Pure medical research has identified structural connections between the nervous system and also the secondary organs of the immune system, where there exist receptors for autonomous transmitter substances. These explain the reagibility of the immune set-up. The power of which is depending on the degree of adaptation, reached by the vessels' reactivity to the cold. That also accounts for the fact, that untrained people will catch infections more easily when getting exposed to a high dose of cold irritation.(6,7)

Potentiating Health and the Crisis of the Immune System
edited by Mizrahi *et al.* Plenum Press, New York, 1997

Starting with these considerations, it seemed very interesting to find out whether these regulative processes are also working with cancer patients. This group makes up a big part of our in- and out-patients. On the one hand, resistance to infection is a very important aim for complementary therapy with those patients, who are under conventional cytoreductive therapies. On the other hand, deriving from many in vitro experiments, animal studies and clinical observations, one has to assume that the impact of the immune response, directed against tumour cells, can no longer be neglected for the prognosis of overall survival.(8–11) Until nowadays, there exist not a single scientific paper about immunologic improvement with hydrotherapy concerning cancer patients. This topic is solely investigated for sports.(12–18)

At the same time, we have to consider that quality of life is a dimension, which can be assessed exclusively by subjective jugdement.(19) The subjective outcome becomes by itself an important criterion for palliative therapy of cancer patients.(20) On an operational basis, quality of life can be seen in the context of the definition of comprehensive health, worked out by the WHO. That includes the cumulation of physical, mental and social well-being to bring people into the condition to realize themselves in an active way as members of the human society, suited to their wishes and abilities. But quality of life does not only mean a sum total of a bunch of circumstances in life.(21) Rather, we have to understand it as the product of a network of mutual interdependencies of these many factors. Quality of life cannot be calculated by machines, because the meaning of certain things varies a lot in consequence of people's different living and health conditions and their constitution, their degree of reflection and education, that determines their attitudes and habits, their past history and experiences, influencing their emotional and social preferences.(22) Even the numerous effords to assess the desirable quality of life for a self-contained group like the cancer patients revealed a very large diversity of statements of the affected people, that reflected not only the different kinds and stages of the disease but also their subjective going through, refering to individual personalities. The final agreement of the sociological research groups said that global estimation of well-being will bring the same amount of information as all other sophisticated instruments for its detailed assessment.(23) To be well or not to be well, that is the question.

For the purpose not to worsen quality of life by, e.g., chemotherapy without an acceptable benefit in a palliative situation for the patient remains a sense in the use of standardized questionnnaires.(24) The purpose to improve quality of life, which is deteriorated by the progressive disease, by some specific intervention is something else, and until now rather ignored and very little dealt in scientific community. In practice, most doctors give the patients the advice to keep on living just in the same way as before this life event of getting cancer.

Social research has proved that a high level of quality of life has been corresponding with longevity.(25) So we may put forward the hypothesis that measures to improve quality of life are able to prolong life expectancy. Former research of our medical department found out that certain forms of hydrotherapy were able to boost the patients' morale.(26) With cancer patients the psychological status is a main dimension of quality of life. The life threatening diagnosis brings fear of death to everybody affected, hopelessness to many, told not to have a curable stage of the disease, depression to those who do not know how to cope with that extraordinary situation. Chronically bad moods may aggravate the course of illness, as we know from psychooncologic science.(27) Most methods to help cancer patients to overcome their deep depression and comprehensible self-uncertainty after the diagnosis are based on intellectual digestion and bound to medical/ psychological institutional facilities.(28, 29) For many this cannot be the first step, because they are overwhelmed by their emotions and need to distance themselves from endless thinking.(30,31)

If, in this context, we complete the neuro- immunologic chain of physical reactions, provoked by hydrotherapy as mentioned above, with its supposed psychological, antidepressive effects, we are able to close the whole circle of psycho- neuro- immunology, taking this as the fundamental principles to explain a parallel outcome of an improvement concerning both the emotional and the immunological state.(32–36)

Until now, these interrelations are investigated first and foremost into the effects of life events and sports working on the body.(37–40) So it has been a challenge to demonstrate similar results with hydrotherapy. The focus of natural medicine in distinction to physical medicine is the dimension of the so-called ordering therapy, what means the additional aspects of emotional and social impact of a physical therapy. Not only the directly treated part of the body is questioned but also indirectly the whole person, seen as a complex information system of biological and emotional, intellectual and spiritual associations. In our western christian culture body and soul are separated and we find it difficult to abstract from this analytic thinking. To bring the unity of the material and immaterial body together, belongs necessarily to a holistic and integrative approach of human medicine.(41)

The group dynamics in educational lectures about hydrotherapy include thereby social life´s experience. The consistent communication, which is necessary between the patient and the doctor for explanation and motivation, feed back and good advice is part of the healing process as well.(42)

METHODS

Concepted as a pilotstudy we did an uncontrolled, prospective clinical observation with longitudinal intrapersonal comparisons. Starting from the assumation, that all cancer patients have at least temporarily an immunologic defect, even if not detectable with the test techniques available, we decided as a first step to work with patients without visible tumour activity, i.e. operated with a curative intention. We wanted to find out, whether in these less advanced cancer stages the immunological responsiveness, shown for healthy adults to hydrotherapeutical stimulus, could be repeated yet to the same extent. That implied a good physical state of the patients, so that the quality of life would above all be affected by mental well-being and functional fitness. In that cases of good risk, hydrotherapy got a sense of secondary prevention.

The doing itself

The feeling of refreshing

Topping up energy by overcoming present troubles

Boosting the whole faculty of sensation and sensitivity

Figure 1. Psychological effects and autosuggestion caused by hydrotherapy.

Implementation of Sequential Hydrotherapy at Home

Grading of the stimulus′ strength, raised every two weeks

1st step:
cold ablution of the upper or lower part of the body in the morning
poikilothermic /short cold immersion bath of the arms at noon
poikilothermic immersion bath of the lower legs in the evening

2nd step:
cold ablution of the whole body in the morning
cold immersion bath /cold affusion of the arms or the face at noon
cold affusion of the lower legs/whole legs in the evening

3rd step:
cold ablution of the whole body/cold affusion of the legs in the morning
cold affusion of the arms /the face/ the lower legs at noon
water treading in the evening

possibly:
variations using an even higher degree of stimulus, e.g.,
- whole body affusion
- cold immersion bath of the trunk
- cold immersion of the whole body

individual adaptation of the intensity of the therapeutic applications
- by the patient′s subjective experience of well-being
- by their evaluation in interchange with the physician

Temperature of the cold water about 15° Celsius

Figure 2. Hydrotherapy program for cancer patients.

For our study we won twenty women mostly with breast cancer after primary surgery and radiation and/ or chemotherapy, which had to be over at least three months before. They should be 30 to 65 years old. Immunologic affections or diseases or medications were excluded. They should not have had any former experience with hydrotherapy. Healthy lifestyle and additional substitution of mistletoe extract was not stopped but should not be changed in any way during the interventional period.

Among lots of different feasible applications of hydrotherapy we chose a not very sophisticated program of cold body ablutions, partial poikilothermic and cold immersion baths, cold affusions and water treading. As women, especially suffering from cancer disease, tend to state not to stand the cold well, we started very cautiously, attaching importance to the essential of feeling agreeably warm just after the refreshing cold, if not to warm themselves up before starting the procedures, in order not to endanger the patients to catch a flu.(43)

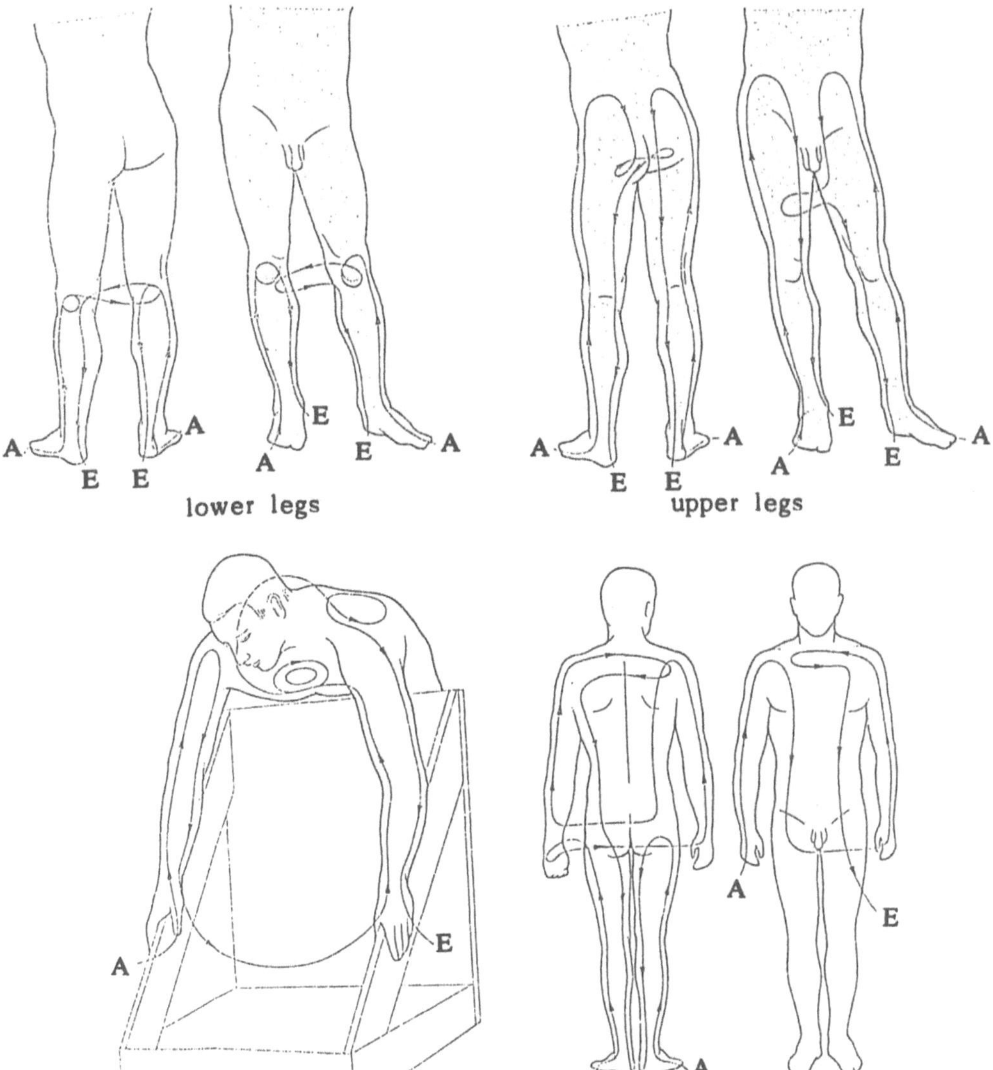

A: starting point - E: ending point
ending always by cold affusion of both the soles of the feet

Figure 3. Body models of kinds of affusions.

The ablutions of the body were done with a linen, folded several times. It was dipped in cold water and wrung out not to drip. It had to be turned over whenever it become warm by the contact with the body. Every other single application needed to be repeated three times, always ending with the cold. The right temperature was 15 rsp. 35°C for cold and warm water, the difference had to be at least 20°C. The immersion baths lasted 3–5 minutes warm rsp. 10–15 seconds cold. The cold affusions (in some cases preceded by warm ones) were done as quickly as possible in a fixed jet line (s.b.).(44)

The observation time was all in all six weeks. During this period we were confident to realise the expected effect estimated from following our experiences at the spas.(45) The patients got instructions always by the same physician and physiotherapist in small groups of ten patients, hold every two weeks for about two hours. After the demonstration, they were encouraged to do practical exercise and exchange of experience in order to find their individually tolerable dose of the cold. After every lecture they were demanded to repeat the taught applications for themselves at home, slowly increasing the intensity, corresponding to the gradual physiological adaptation. They were told to do it three times every day, in the morning, at noon or when coming home from work, and in the evening or just before going to sleep. They were warned not to exaggerate the intensity and time during the succession. The last meeting served only as an opportunity to give a final feed back to the therapists and to overthink it all perspectively.

Before and after the six week intervention period we took blood samples under standardized conditions (at 8–9 o'clock in the morning, with an empty stomach, after 20 minutes rest on a couch), to ascertain the white blood cell account and their differentiation. The neutrophils as well as the lymphocytic subclasses were studied by immunocytometry (Becton,Dickinson FACScan) using the marker molecules CD3, CD4, CD8, CD8-HLA-DR, CD19, CD16/ 56. We also determined the serum levels of antibodies against endemic virus (CMVand HSV by KBR and ELISA IgM, EBV by IFT). Besides, we asked

Statements about your sensibility to cold and warm

Please answer with YES or NO:

- I prefer to wash myself cold in the morning
- I tend to transpire in a warm climate or being strained
- Going to bed, I rather need one more blanket
- I easily get fever
- If the weather is cold, I always take care to get dressed in warm clothes
- I often suffer from cold feet and/ or hands
- Being together with other people in closed rooms, I am often getting hot
- I catch a flu very quickly
- I like to drink my coffee or tea quite hot
- In regard to other people, I think myself to be rather chilly

(most chilly = 20 points, least chilly = 10 points)

Figure 4. Chill questionnaire.

The Nottingham Health Profile

Listed below are some problems that people may have in their daily life
Please answer every question with YES or NO

I´m tired all the time
I have pain at night
things are getting me down

I have unbearable pain
I take pills to help me sleep
I´ve forgotten what it´s like to enjoy myself

I´m feeling on edge
I´m in pain when I walk
I lose my temper easily these days
I feel there is nobody I am close to

I lie awake for most of the night
I feel as if I´m losing control
I´m in pain when I´m standing

I find it hard to get dressed by myself
I soon run out of energy
I find it hard to stand for long
(e.g., in the kitchen sink,waiting in a line)

I´m in constant pain
It takes me a long time to get to sleep
I feel I´m a burden to people

I find it painful to change position
I feel lonely

I can walk about only indoors
I find it hard to bend
Everything is an effort

I´m waking up in the early hours of the morning
I´m unable to walk at all
I´m finding it hard to make contact with people

The days seem to drag
I have trouble getting up and down stairs
I find it hard to reach for things

Worry is keeping me awake at night
I feel that life is not worth living
I sleep badly at night

I´m finding it hard to get along with people
I need help to walk about outside
(e.g., a walking aid or someone to support me)
I´m in pain when going up and down stairs

I wake up feeling depressed
I´m in pain when I´m sitting

Now think about areas of Your life that may be affected by health problems.
IS YOUR PRESENT STATE OF HEALTH CAUSING PROBLEMS WITH:
- YOURWORK?
  (that is, payed employment)
- LOOKING AFTER THE HOME?
  (cleaning& cooking, repairs, odd jobs around the home, etc)
- SOCIAL LIFE?
  (going out, seeing friends, going to the movies, etc)
- HOME LIFE?
  (That is, relationships with other people in Your home)
- SEX LIFE?
- INTERESTS AND HOBBIES?
  (sports, arts and crafts, do- it yourself, etc)
- VACATIONS?
  (summer or winter vacations, weekends away, etc)

Figure 5. Quality of life questionnaire.

Table 1. White blood cells and their subclasses of cancer patients before and after hydrotherapy

Cell classes/ul	Before Mean	SD	SEM	After Mean	SD	SEM	2-tailed t-test significance p<	
WBC	4900	1200	275	5500	1600	365	0,022	*
neutrophils	3102	799	183	3572	1301	296	0,042	*
monocytes	343	95	22	377	94	22	0,061	
lymphocytes	1401	440	101	1507	409	94	0,210	
CD 3+	935	344	79	1040	334	77	0,015	*
CD3+/CD4+	643	233	53	724	236	54	0,003	**
CD3+/CD8+	480	491	49	541	186	43	0,208	
CD3+/HLA-DR+	155	123	28	171	118	27	0,065	
CD3+/CD16+/CD56+	84	54	13	112	93	21	0,234	
CD19+	157	83	19	194	93	21	0,001	***
CD3-/CD16+/CD56+	309	135	31	347	124	28	0,279	

about the patients' sensibility to the chill, using a self-made questionnaire about everyday behaviour.

We inquired their quality of life, using the Nottingham Health Profile questionnaire, in Germany widely evaluated with cancer patients under conventional therapies.(46–49)

After the intervention we interviewed on general well-being and functional complaints.

Statistical computing was done by microsoft windows SPSS, using the two-tailed t-test for bound pairs.

RESULTS

The mean age of the patients was 51 years, the last cytoreductive therapy was accomplished at least four months to six years ago. All of them were members of self-help groups. Only one dropped out just before the beginning because of a severe feverish infection. Two of them had signs of tumour activity at the time of intervention. Except one woman with a colon cancer and one with a multiple myeloma, they all had breast cancer, one with pulmonary metastases at the time. A higher sensibility to cold was confirmed with 15 of 19 women (answer score > 15). After the therapy only 12 of them remained in this chilly condition. Yet in the group´s mean, the thermal protective behaviour did not change.

The mean values of all blood cell counts of the unspecific and specific immune response showed an augmentation.

Significant improvements were found with the amount of the overall white blood cells, the neutrophils, the overall T-lymphocytes on a 95% level. The T-helper- lymphocytes (p<0,003) and the B-lymphocytes (p<0,001) increased significantly even on the 99% level. The other fractions of lymphocytes,including the activated T killers and the natural killer cells increased only in direction, the same with the monocytes, the precursor cells of the macrophages. The viral antibody levels were lowest normal at the beginning of our study, so that they could be further lowered by therapy.

The quality of life measured by the NHP questionnaire could not sensefully be evaluated by mean scores. Only one patient had stated a rather bad quality of life (score= 24 out of 38 possible negative points). She reached an improvement over 40% in the end.

Seven more patients showed a change in direction. Single items noted individual effects like, e.g., less pain when strained, less despondency, less touchiness, feeling oneself no longer a burden for other people, and having less difficulty to get along with others.

The statements in the open interviews at the end were in line with the physiological adaptation. For the majority of the participants, the hydrotherapeutic applications became more and more a pleasant procedure. They felt no longer as chilly as before. The rather sensitive constitution had made it necessary to intensify the cold stimulus very slowly as expected. Our patient group did not come to large affusions of the trunk or whole body within the sceduled time of the six weeks. In the course of the conditioning process the resulting adaptation aimed in a higher degree of well-being. This subjective feeling must vice versa be recognized as the clinical guideline for the intensity of any natural irritation therapy.(50, 51) Sticking to this elementary rule, facilitated to dose in an appropriate way in order to avoid hyperstimulation with the countereffect of immune suppression. Nobody of the patients was falling sick.

Noticeable was a decrease of functional complaints like sleeping disorders, constipation, recurrent headache, and menopausal syndrome. Indeed, everybody was happy about the raising kick up, they got at a more positive attitude towards future and the joie de vivre.

DISCUSSION

The presented results from this pilot study, comprising an intraindividual comparison of immunological data and dimensions of quality of life with cancer patients before and after sequential hydrotherapy at home, give evidence of a simply feasible contribution to palliation, if not actually to secondary prevention. The main characteristic features of this kind of natural medicine are the independent activity instead of reactivity and the chance to strengthen the ego, i.e. empowerment, necessary for patients to succeed in coping with a severe and threatening illness. Therefore it seems correct to name this therapy,-working through body-experience for getting awareness of oneself-: *body-related psychotherapy.*(52, 53)

There is no sense in calling this placebo, because it is not replaceable with any other measure or inert pill at all. It is not the often quoted human care and bodily touch, that is part of either psychologic dialogue or body massage. On the contrary, in our case people become more independent from help of professionals.(54–56) They treat themselves and assess the adequate dose, only counselled by the doctor. So, they feel self-competent and self-determined.

What we did not take into account, were the immunologic changes which were thinkable to take place in the marching season by the increasing light exposure in march leaving wintertime on the northern hemisphere and west wind zone.(57) These effects would have been able to be excluded by a control group, which we did not record in this pilot study. Nevertheless, to interpret the significant rise of the blood cell capacity to be nothing but spontaneous, seemed extremely unlikely, too. Thus, we feel allowed to attribute the improvement of the immunologic response to the hydrotherapy. Though all absolute figures lay in the normal range before starting, some rose even significantly after the therapy.

As we know from oncological literature, the activity of the unspecific neutrophils as well as of the specific T lymphocytes is of important value in the anti-tumour fight of the body itself. In these cell classes we found a significant augmentation, especially concern-

Advantages of Natural Therapies:

Coincidental influence on physical and psychological and social aspects of quality of life

- **Physical education**
- **Body training**
- **physiological adaptation**
- **Immunological strengthening**

- **Improvement of the psychological status**
- **Activation and self-assurance**
- **Body sensitivity and self-assessment**

- **Self-competence concerning illness and therapy**
- **Emancipation in medical institutions**
- **Self-determination**
- **Socialisation by continuous communication**

Conditions for life expectancy in chronic disease?

Figure 6. Palliation of cancer disease - advantages of natural medicine.

ing the T helpers, which play a crucial role in the immunologic network. The normal level of the natural killers did not move up significantly, maybe because all patients had continued to substitute mistletoe extract. So thereby the NK cells might have been pushed already in a matchless way before the intervention, while the neutrophils have a very high turn over rate, and so their proliferation, stimulated by, e.g., mistletoe will not last in the long run, and needs new impulses continously.(58, 59)

The numerical increase of B lymphocytes as the cellular representatives of the specific humoral pathway of the immune system, standing as its reserve in the background, may be seen as a sign of better resistance to viral or bacterial infection. In analogy the antibody titres remained low, as well as there was no remarkable increase of T killers in action, expressing HLA-DR receptor molecules. One might imagine that these specific tools against infectious enemies needed not to be ordered, because the unspecific first-line force of the phagocytes worked sufficiently against any microbic invasion. And of course this relaxed situation at the microbial front, created by the 'hardening' with hydrotherapy, may be regarded to be best for keeping a keen eye on latent cancer cells' formation. The very highly significant rise of the B lymphocytes stands for a high protection against microbian overrun.

This deduction may be true, but another hypothesis could also be, that part of those people, who do not suffer from occasional feverish inflammations, as told from many oncologic patients in their medical history, may have some weakness in controlling cancer, for whatever reason, perhaps associated with this extraordinary fitness against infection. The research field about details of the immune mechanisms against cancer is still underdeveloped. So these questions remain unresolved.

The standardized questionnaire, we used to ask about quality of life, the Nottingham Health Profile, turned out to be rather insensitive for our special patient group. Certainly, most of the women were physically not in a bad shape, because neither the stage of the disease was rather late, nor were they undergoing harming therapies. So, complaints about pain, fatigue and other handicaps, mentioned in the NHP, did not hit their reality. Single items, they answered to be a problem, were drowned in the mean score. That one, who had lung metastases causing a reduced physical state, could raise it clearly.

Besides, all of them belonged since several times to self-help groups, where they had already trained to overcome reactive depression and to manage crises, learned to love themselves, pay attention to and act after their own desires. Over and above that, they created a very friendly and facing atmosphere during the lectures, an example for their social competence. But nevertheless, one can imagine that these factors altogether led to the content mood found already at the start of the interventional trial.(60)

The statements of relief from diverse functional sufferings (s.a.), must in this context also be submerged under the topic of quality of life. Again we found positive correlations between physical and emotional symptoms, meaning that, e.g., headache or sleeplessness disappear while feeling good.

We found it interesting, that the review of the thermic sensitivity brought some evidence for a autonomous habituation to cold. Yet, the behaviour did not change in a corresponding manner. A reason for that might be the cold weather of springtime in our temperate climate zone and the ingrained habitudes to get dressed.

A further feature of quality of life is the joie de vivre, in this study expressed by enthusiasm with the applied procedures themselves. A postal reconsultation one year after the study had closed, informed us, that 16 of the 19 patients were continuing to do hydrotherapy regularly at least twice every day. They told us doing so not only for the reason of a high health motivation, but also of the good feeling itself, they got while and after doing so.

Whether a strong immune system is responsable for the well-being or whether the emotional well-being reinforces a fit immune system, remains a matter of academic dispute.What we do know is that quality of life and self-determined activity raise the people's vitality and overall survival: by having a critical look at one's being sick as well as by directing one's attentiveness to different subjects and distractions.(61–63) Hydrotherapy being a typical example of ordering therapy means regularity both physiologically in a successional counterbalance of the vegetative tonus and mentally in spending time for one's own disposal and getting distance to the everyday wrangling.

The model of the psychosomatic salutogenesis has been developed to a theory of psychooncological research in the last ten years.(64, 65) We believe to supply some scientific material, that brings forward the idea of the salutogenetic sense of hydrotherapy as an

Psycho -	Neuro -	Immunology
psychological well-being	autonomous regulations	defence against infection
quality of life	functional troubles/ diseases	influence on the course of cancer disease (survival)
vitalization	less sensibility to cold	
improvement of the mood:	disappearance of headache	sign. numeral increase of the WBC:
antidepressive effect	improvement of sleep disturbances	neutrophils (unspecific)
social emanzipation	regulation of menopausal-syndrome	T lymphocytes (cellular) T helper cells (crucial role)
	lessening of constipation	B lymphocytes (humoral)

Figure 7. Psycho-neuro-immunology of hydrotherapy.

alternative or complementory essential for anti-cancer therapy and prevention: what's good, is good for you!

The psycho-neuro-immunology seems to explain some interrelations of specific effects of hydrotherapy, we found in our interventional trial, though the number of sounded patients was too small to derive final knowledge out of it.

SUMMARY

In an uncontrolled trial of 19 female cancer patients (17 suffering from breast cancer), an improvement of the immune response could be achieved by a sequential daily hydrotherapy over six weeks at home. The neutrophils and the overall T lymphocytes increased significantly ($p<0,05$), especially did the group of T helpers ($p<0,005$) and the B lymphocytes (0,001). All values were calculated in the two-tailed t-test for bound pairs. At the same time several functional complaints like sleeplessness and constipation decreased. The quality of life was on a very high level already in the beginning, so that the mean score of the NHP, we had chosen from literature as the instrument of assessment, brought no relevant information for the whole group. The theory of psycho-neuro-immunology is taken for a possible explanation of physiological and mental changes, that go parallelly in the course of the intervention. The placebo effect is excluded critically. This study may be seen as a first contribution to evaluate specific natural treatments, known from the rehabilitation praxis, but not applied separately. The meaning of ordering therapy concerning hydrotherapy is worked out by the special situation of cancer patients.

REFERENCES

1. Reinhold D.Kurorttherapie als ganzheitliche Präventions- und Rehabilitationsmassnahme in:Rehabilitation im Kurort (Heipertz W. Editor) Braun Karlsruhe 1996.
2. Kuehn G. (1993) Kurbehandlung aus medizinischer Sicht. Heilbad und Kurort **45**, 78–79.
3. Gaertner U.et al (1996) Beschwerden, Belastungen sowie Lebenszufriedenheit onkologischer Patienten. Medizinische Klinik **91**, 501–508.
4. Friedrich K.-W. Buehring M. Behandlung mit Kalt- und Warmreizen in: Naturheilverfahren (Buehring M. Kemper F. Editors) Springer Berlin 1996.
5. Brueggemann W. Kneipptherapie Springer Berlin 1980.
6. Felten D.L. et al (1995) Noradrenergic and Peptidergic Innervation of Lymphoid Tissue. JourImm **135** (2), 755–765.
7. Buehring M. Abhaertung durch kaltes Wasser in: Naturheilverfahren und unkonventionelle medizinische Richtungen (Buehring M. Kemper F. Editors) Springer Berlin 1996.
8. Beverly P. Tumour Immunology in Immunology (Roitt I. et al Editor) Mosby London 1996.
9. Rosenberg S.A. et al Principles and Appliktions of Biologic Therapy in:Cancer (DeVita V.T. Hellman Jr.S. Rosenberg S.A. Editors) Lippincott Philadelphia 1989.
10. Chang A.E. Shu S. (1992) Immunotherapy with Sensitized Lymphocytes. Cancer Investigation **10**(5), 357–369.
11. Dénes C. Szende B. Hajós G. et al (1990) Selective restoration of immunosuppressive effect of cytotoxic agents by thymopoietin fragments. CancerimmunolImmunther **32**, 51–54.
12. Findeisen D.G. Sport, Psyche und Immunsystem. Frieling Berlin 1994.
13. Shephard R.J. (1990) Physical activity and Cancer.Int.J.SportsMed.**11**, 413–420.
14. Crist D.M. Mackinnon L.T. Thompson R.F. et al (1989) Physical Exercise Increases Natural Cellular-Mediated Tumor Cytotoxicity in Elderly Women. Gerontology **35**, 66–71.
15. Keast D. Cameron K. Morton A.R. (1988) Exercise and the Immune Response. Sports Medicine **5**, 248–267.
16. Mackinnon L.T. Tomasi T.B. (1986) Immunology of exercise. Annals of Sports Medicine **3**(1), 1–3.

17. Kohl H. LaPorte R.E. Blair S.N. (1988) Physical Activity and Cancer. Sports Medicine **6**, 222–237.
18. Peters C. Loetzerich H. Schuele K. et al (1995) Auswirkungen eines Ausdauertrainings auf die Phagocytoseaktivitaet von Monocyten bei Brustkrebspatientinnen. Phys Rehab Kur Med **5**, 42–45.
19. Gill T.M. Feinstein A.M. (1994) A Critical Appraisal of the Quality-of-Life Measurements. JAMA **272**, 619–625.
20. Bowling A. Measuring Health. Milton Keynes Philadelphia 1991.
21. Schwarz R. Bernhard J. Flechtner H. et al Lebensqualität in der Onkologie in:Aktuelle Onkologie **63**, 1–145.
22. Tuechler H. Lutz D. Lebensqualität und Krankheit. Deutsche Aerzte Koeln 1991.
23. Singer E. Garfinkel R. Cohen S.M. et al (1976) Mortality and Mental Health: Evidence.
24. Tchekmedyian N.S. Cella D.F. Aaronson N.K. et al (1990) Quality of Life in Current Oncology Ptactice and Research.Oncology **4**, 22–208.
25. Kuehn H. Krankheit und soziale Ungleichheit in: Healthismus. Sigma Berlin 1993, from the Midtown Manhattan Restudy. Soc. Sci. & Med. **10**, 517–525.
26. Kuehn G. (1995) Physical therapy and quality of life: design and results of a study on hydrotherapy. Complementary Therapies in Medicine **3**, 138–141.
27. Helmkamp M. Paul H. Psychosomatische Krebsforschung. H.Huber Bern 1984.
28. FawzyF.I. Kemeny M.E. Fawzy N.W. et al (1990) A Structured Psychiatric Intervention for Cancer Patients. Arch Gen Psychiatry **47**, 729–735.
29. Jany L. Csef H. Rueckle H. (1992) Active Living with Cancer- ASupport Program for Cancer Patients and Their Families. Onkologie **15**, 405–408.
30. Koch U. Potreck-Rose F. Krebsrehabilitation und Psychoonkologie. Springer Berlin 1990.
31. Greer S. (1991) Psychological response to cancer and survival. Psychological Medicine **21**, 43–49.
32. Solomon G.F. (1987) Psychoneuroimmunology: Interactions Between Central Nervous system and Immune System. Journal of Neuroscience Research **18**, 1–9.
33. Jamner L.D. Schwartz G. E. Leigh H. (1988) The Relationship Between Repressive and Defensive Coping Styles and Monocyte, Eosinophile, and Glucose Levels: Support for the Opiod Peptide Hypothesis of Repression. Psychosomatic Medicine **50**, 567–575.
34. Brenner G.J. Felten S.Y. Felten D.L. et al (1992) Sympathetic nervous system modulation of tumor metastases and host defense mechanisms. Journal of Neuroimmunology **37**, 191–202.
35. Kiecolt-Glaser J.K. Glaser R. (1986) Psychological influence on immunity. Psychosomatics **27**, 621–624.
36. Kropiunigg U. Psyche und Immunsystem. Springer Wien 1990.
37. Evans D.L. Folds J.D. Pettito J.D. et al (1992) Circulating natural killer cell phenotypes in men and women with major depression. Arch gen Psychiatry **49**, 388–395.
38. Kukull W.A. McCorkle R. Driever M. (1986) Symptom distress, psychosocial variables, and survival from lung cancer. Journal of Psychosocial Oncology **4**(1/2), 91–105.
39. Levenson J.L. Bemis C. (1991) The role of physiological factors in cancer onset and progression. Psychosomatics **32**(2), 124–132.
40. Redd W.H. Silberfarb P.M. Andersen B.L. et al (1991) Physiologic and psychobehavioral research in oncology. Cancer **67**, 813–822.
41. Haberzettl A. Wenn das kalte Wasser kommt... Lang Frankfurt 1996.
42. Buehring M. Kuehn G. (1993) Das physiotherapeutische Prinzip. Heilberufe **45**, 84–87.
43. Brock F.E. Hydro-/ Thermotherapie, Allgemeine Grundlagen in:Naturheilverfahren und unkonventionelle medizinische Richtungen (Buehring M. Kemper F. Editors) Springer Berlin 1993.
44. Guenther R. Jantsch H. Physikalische Medizin. Springer Berlin 1986.
45. Hildebrandt G. Gutenbrunner C. Wissenschaftliche Grundlagen der Kurorttherapie in: Rehabilitation im Kurort (Heipertz W. Editor) Braun Karlsruhe 1996.
46. Raspe H.H. (1990) Measures to Assess Quality of Life. Medizinische Forschung **2**, 297–307.
47. Westhoff G. Handbuch psychosozialer Messinstrumente.Hogrefe Goettingen 1993.
48. Hunt S.M. McKenna S.P. McEwen J. The Nottingham Health Profile. User´s manual. Galen Research and Consultancy Manchester 1989.
49. Kohlmann Th. Bullinger M. Hunt S.M. et al (1992) Zur Messung von Dimensionen der subjektiven Gesundheit - Die deutsche Version des "Nottingham Health Profile" (NHP) Gesundh.-Wes **54**, 56 (abstract des Arbeitsberichtes Luebeck).
50. Gutenbrunner C. Hildebrandt G. Der Prozess der adaptiven Normalisierung in: Rehabilitation im Kurort (Heipertz W. Editor) Braun Karlsruhe 1996.
51. Kuehn G. Wirkungen der Hydrotherapie in: idem 1996.
52. Weber H. Laux L. Bewältigung und Wohlbefinden in: Wohlbefinden (Abele A. Becker P. Editors) Juventa Muenchen 1994.

53. Braehler E. Koerpererleben Psychosocial Giessen 1995.
54. Kuehn G.Psychische Wirkungen von Naturheilweisen - Wissenschaft und Medizin in der klinischen Forschung - alternativ und ehrlich? in: Forschungsmethoden in der Komplementaermedizin (Hornung J. Editor) Schattauer Stuttgart 1996.
55. Stange R. Die gefaellige Natur - Das Placebo-Problem und die Naturheilverfahren. in: idem.
56. Craen deA.J.M. Kleijnen J. The Role of Non-Specifc Factors in Randomized Clinical Trials - Proposals for the Future. in: idem.
57. Buehring M. Kuehn G. Wirkungen der Heliotherapie. in:Rehabilitation im Kurort (Heipertz W. Editor) Braun Karlsruhe 1996.
58. Scheer R. Becker H. Berg P.A. Grundlagen der Misteltherapie. Aktueller Stand der Forschung und klinische Anwendung. Hippokrates Stuttgart 1996.
59. Pike M.C. (1990) Reducing Cancer Risk in Women Through Lifestyle-Mediated Changes in Hormone Levels.Cancer Detection and Prevention **14**, 595–607.
60. Petzold H. Schobert R. Selbsthilfe und Psychosomatik. Jnfermann Paderborn 1991.
61. Coates A. Gebski V. Signorini D. et al (1992) Prognostic Value of Quality-of-Life Scores During Chemotherapy for Advanced Breast Cancer. J Clin Oncol **10**, 1833–1837.
62. Fallowfield L.J. (1995) Assessment of Qualitzy of life in Breast Cancer. Acta Oncologica **34**, 689–696.
63. Weeks J. (1992)Quality of Life Assessment: Performance Status upstaged? Journal of Clinical Oncology **10**, 1827–1838.
64. Antonovsky A. Unravelling the Mystery of Health. Jossey-Bass San Francisco 1987.
65. Heim E. (1995) Coping-based intervention strategies. Patient Education and Counseling **26**, 145–151.

13

ADAPTOGENS – NATURAL PROTECTORS OF THE IMMUNE SYSTEM

Ben Tabachnik

PrimeQuest International
2354 Garden Road
Monterey, California 93940

I. ABSTRACT

Modern medicine tends to place an increasing emphasis on disease prevention. This involves a search for biological active substances with preventive properties. Furthermore, there is a need for specific agents to correct the disturbances of the organism's homeostasis associated with environmental impacts and increasingly stressful lifestyles.

The field of preventive pharmacology in Russia was pioneered by the prominent Russian scientist, N. Lazarev, in the 1950–60's. Lazarev and his colleagues were the first to study a group of plants named "Adaptogens." Adaptogens were shown to have the ability to correct a wide range of disturbances of homeostasis and protect our immune system.

This presentation will provide an over view of studies that show the supportive effects of Adaptogens on the immune system.

II. INTRODUCTION

Modern civilization makes high demands of the physical and mental abilities of man.

During this century, unlike anytime in history, man's activity has resulted in a wide spread change to our living and working environments. Although there is no contesting the benefits of our modern technological age, there is no denying the stresses it has put on mankind andthe planet. Some of the negative results are pollution from industrial wastes, foreign chemicals compounds (drugs, medicine, food additives, etc.). Additionally, noise, vibration, radio waves, artificial illumination, and radiation are constantly increasing. In addition, to physical and chemical stressors, modern man experienced effected significantly greater mental and emotional social stress.

Evolution has not prepared man to cope with modern conditions of life. And, technological development is hampered by man's biological characteristics. The gap between the rapid development of civilization and man's natural abilities grows continually. It is believed that 70–80 percent of disease is caused by physical, chemical or emotional stress.

Potentiating Health and the Crisis of the Immune System
edited by Mizrahi *et al.* Plenum Press, New York, 1997

Stress-related health problems include but are not limited to: headaches, ulcers, heart disease, sleep disorders, depression etc. The National Coalition on Immune System Disorders has reported that about 65 million Americas are suffering from inadequate immune system function. this is due in part to environmental and life style factors such as chemical pollution, drug therapy, malnutrition, smoking, radiation and stress. Unfortunately, in modern, industrialized, fast paced society many of these factors are not avoidable. Scientists are now proving that the best way to strengthen ones immune system is through proper lifestyle habits including diet, exercise, relaxation. Therefore it is critically important that proper nutritional intake is maintained on a day to day basis.

III. TEXT

The ability of certain nutritional plants to produce general holistic effects has been known to herbalists for thousands of years. Their philosophy is based on the belief that lasting vitality is achieved through keeping the body healthy, rather than treating it for disease. The Orientals were so proud of their special natural remedies that they named them "kingly" and "elite" herbs. For thousands of years these plants were utilization by people in China, Russian, Japan, Korea and finally in Europe. Nonetheless, the effectiveness of these plants from a scientific standpoint was not confirmed until 45 years ago. It was then that the Russian physician and pharmacologist, Dr. Israel and his mentor, Prof. Lazarev named them "Adaptogens. This name was chosen due to the scientific proof of their effectiveness in helping the human body to "adapt" to changes in the environment. Scientific studies have shown that humans and other organisms are able to adapt better ad survive longer when using these adaptogenic herbs. In fact, these special plants have managed to survive in harsh environments for centuries due to their unique composition of biologically active substances.[3]

The group of plants, which were classified by Soviet scientist as "Adaptogens" includes: Ginseng, Eleutherococcus, Aralia mandshurica, Schizandra chinensis, Rhodiola Rosea, Rhaponticum carthamoides, etc. A large number of research studies on adaptogens were made by Dr. Brekhman and his scientific school. They demonstrated that adaptogens put the organism into a state of non-specific resistance in order to better cope with a variety of unfavorable factors and adapt to extraordinary challenges. Protective properties of adaptogens stimulate some body processes which prepare the organism to endure damaging influences, consequently ensuring a more rapid restoration of homeostasis.

A few of the most important actions of adaptogens are: 1) Anti-Stressors: increases tolerance and endurance to stress, both physical and mental, as well as decreasing the incidence of the harmful side-effects of stress on the body. 2) Normalizing: has regulating effect on body functions which tend to fluctuate. It has been shown that adaptogens normalize abnormalities of function: adrenal, thyroid, blood pressure, cholesterol, blood glucose, etc. Scientific date indicates that changes produced by adaptogens in the organism are due to stimulation of several systems: nervous system, endocrine system, tissue metabolism and immune system. 3) Prophylactic: adaptogens have demonstrated a wide range of therapeutic actions: influenza, acute respiratory disease, atherosclerosis, hypertension, diabetes, skin disorders, sexual disorders and cancer. 4) Anti-toxic: increases the body's tolerance of many chemical, biological and radioactive toxins.

An important aspect of the actions of adaptogens is their influence on the immunological responses of the body. It appears that adaptogens stimulate and regulate the immune system directly and indirectly.[1, 2, 3, 8, 12]

It is well known that stress decreases the activity of the immune system. Stress hormones made by adrenal glands actually interfere with the immune system and can make us vulnerable to disease.

It has been proven that adaptogenes increase the body's ability to resist stress which automatically protects the immune system. A close link was demonstrated between the development of stress reaction and the functional condition and biochemical composition of natural. In Germany, a placebo-controlled study of the effect of Eleutherococcus extract on the immune system was performed with 36 healthy volunteers utilizing quantitative multi-parameter flow cytometry and monoclonal antibodies directed against specific surface markers of human lymphocyte subsets. The most salient feature in the verum group was a drastic increase in the absolute number of immunoconpetent cells, predominately of the helper/inducer type, but also cytotoxic and natural killer cells. In addition, a general enhancement of the activation state of T-lymphocytes was observed. No side effects were observed during the trial or afterwards (observation period 6 months). The authors concluded from the obtained data that Eleutherococcus senticosus exerts a strong immunomodulatory influence in healthy normal subjects and can be considered a nonspecific immunostimulant and an immunoregulator (immunomodulator), or better yet, a biological response modifier.(2,4)

It was shown that Eleutherococcus and other adaptogens stimulate the immune system to produce defensive proteins. A large number of studies, mostly in the former USSR but also in Germany, China and the US, have shown how important this effect is when the immune system is damaged.

Clinical studies(5,6,11,15) indicate that Eleutherococcus acts as an anti-influenza agent when used prophylactically. This activity may be explained both by the adaptogenic properties of the extract, which indirectly enhances the body's resistance to viral infection, and by its ability directly to retard the growth of viruses (demonstrated in mammalian cell culture by Wacker and Eilmes, 1978). Several experiments on the prolonged use of Eleutherococcus root extract for prophylaxis of illness revealed a significant decrease in both the incidence of illness and work loss due to illness. The extract was administered over a period of ten years among workers at an automobile factory. Between 13,000 - 15,000 workers at the factory received a liquid form of the extract, resulting in a 30–50% decrease in cases of flu and a 20–30% reduction in lost work time due to absence or disability. In the same study, 3,000–4,000 factory drivers received the extract over a period of six years, revealing a significant and steadily decreasing incidence of flu, hypertension and ischemia. The net decrease in illness as a result of the extract's use at the factory provided for improved health and working conditions, constituting an important economic effect (Table 1).

Another example was a study conducted in the winter of 1972–73, where approximately 1,000 workers of a North Siberia mining and smelting plant received daily adaptogen supplements for two months. The influenza and HBP incidence reduced almost 2.4 times versus the same number of workers engaged at a shop with the same working conditions (Figure 1).

There is sufficient evidence(7,8,9,12,13,14) that adaptogens in oncology is especially effective for cancer prevention and in those cases when tumor mass is small, i.e. at early stages of cancer development, and also after a primary tumor has been removed — to prevent recurrence an development of metastases. Adaptogens are also highly effective in stimulating recovery, tissue reparation and regeneration after radiation therapy, chemotherapy or surgical removal of the tumor.

The extract of Eleutherococcus senticosus (Siberian ginseng), Schizandra chinesis, Rodeola rosea have been used as a part of oncological regimens. The combination of the

Table 1.

Years of consuming adaptogenic supplement	Sickness on 100 drivers					
	Flu		High blood pressure		Ischemia	
	I	II	I	II	I	II
1st Year*	42	286	6.6	171	6.7	282
2nd Year	30	171	3.2	46	5.3	210
3rd Year	33	188	2.6	23	1.6	51
4th Year	24	144	2.7	24	1.5	35
5th Year	13	74	1.2	19	1.3	21
6th Year	6	32	2.1	21	1.5	12
7th Year	3	11	0.5	5	0.2	3

Description: I — Number of Cases; II — Number of Days of Sickness.
*Adaptogens were not given during the first year of the experiment.

Eleutherococcus extract and chemotherapy was reported to be more effective than chemotherapy alone, both in the short and long term, in breast cancer patients while the combination of Eleutherococcus and radiotherapy was superior to radiotherapy alone in lower lip cancer patients. [8] The addition of the Eleutherococcus extract to the regimen for gastric cancer patients improved the short term results of the treatment Tzeitlin (1981) reported the improved results of treating Hodgkin's disease in children when using the Eleutherococcus extract in preoperative period and during recovery from surgery. There is also evidence of Eleutherococcus modifying the immune response in cancer patients.[9,10]

It appears that clinical application of adaptogens is highly promising. At present there is sufficient evidence to conclude that adaptogens can be effectively used in preventing development and recurrence of cancer, suppressing metastases, and decreasing the adverse effects of conventional methods of cytostatic therapy. On the basis of the existing data one can design further clinical trials of adaptogens within various treatment regimens and different phases of cancer treatment, and ultimately develop comprehensive recommendations for the clinical use of adaptogens in oncological practice.

One can roughly divide the treatment of a cancer patient into four phases: 1) preparation for treatment; 2) specific treatment; 3) treatment to prevent recurrence and development of metastases; 4) further prevention (14). Adaptogens can be used in each of these phases for the following purposes: In the first phase - to induce SEGR (the State of En-

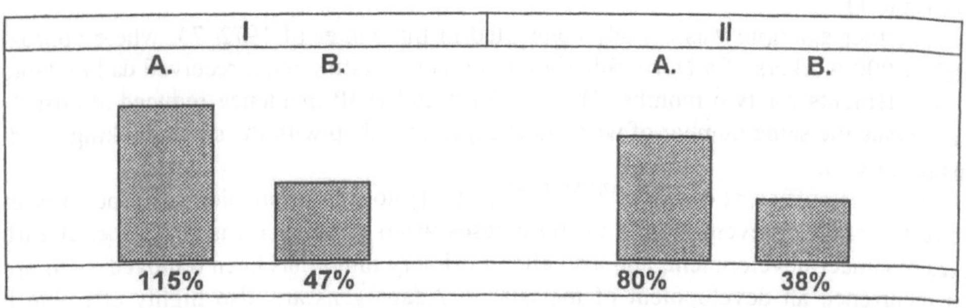

Figure 1. Effect of adaptogens on the disease incidence and manpower losses in the shops of the Norilsk Mining and Smelting Plant. **I**: Variation in influenza incidence. **II**: Variation in the total manpower losses due to influenza incidence. %: The data obtained on year before experiments are taken for 100%. **A**: Control (Shop 1); **B**: Experimental (Shop 2. Workers received adaptogens).

hanced General Resistance) in the organism, thereby making a patient better prepared for specific treatment; In the second phase - to optimize the development of stress reaction during specific anti-cancer therapy, and prevent the complications of sytostatic treatments; In the third phase - to prevent recurrence and development of metastases; In the fourth phase - to optimize the homeostasis and maintaining SEGR in the patients' organism.

IV. SUMMARY

Years of studies have demonstrated that protective properties of adaptogen herbs stimulate some processes which prepare the organism to endure stressful or damaging influences and subsequently ensuring a more rapid restoration of homeostasis.

The discovery of adaptogens holds great promise for preventative medicine.

V. REFERENCES

1. Barenboim, G. M., Sterlina, A. G., Fisenko, A. P., et al. 1984. Studies of Natural Killer Activators with Eleutherococcus Extract. 2nd International Symposium on Eleutherococcus. All-Union Cancer Research Center, Moscow.
2. Bohn, B., Nebe, C.T., and Birr, C., "Flow-Cytometric Studies with Eleutherococcus Senticosus Extract as an Immunomodulatory Agent." Arzneimittel-Forschung 35 (10), 1987, 1193–1196.
3. Brekhman II Eleutherococcus, Nauka Publishers, Leningrad, 1968, p. 85.
4. Dardymov, IV "Some Aspects of the Mechanism of Action of Eleutherococcus Extract," Eleutherococcus: Strategy of the Use and New Fundamental Data, Medexport, Moscow, 1984
5. Gagarin, I.A., "Eleutherococcus Prophylaxis of the Disease Incidence in the Arctic." Adaptation and the Adaptogens. Far Eastern Scientific Center, USSR Academy of Sciences, Vladivostok.
6. Galanova, L. K., "Eleutherococcus in the Prophylaxis of Influenza and Relapses of Essential Hypertension." Adaptation and the Adaptogens. Far Eastern Scientific Centre, USSR Academy of Sciences, Vladivostok.
7. Gvamichava, A.R., Khatiashvili, T.M., Khudzhadze, R. G., Lezhava. G. G., Tsulaya, G. E. "The first results of the use of eleutherococcus in the combined treatment of breast cancer In: Materials for Studying Ginseng and Other Medicinal Agents of the Far East." VII. Eleutherococcus and other Adaptogens from Far Eastern Plants. Far Eastern Book Publishers. Vladivostok, 1966, p. 231–235.
8. Khatiashvili, T. M., "Experience of Liquid Eleutherococcus Extract Application in the Complex Treatment of Patients with Labial and Oral Cavity Cancer. In: Materials for Studying Ginseng and Other Therapeutic Medicines of the Far East." 7: Eleutherococcus and Other Adaptogens Derived from Far Eastern Plants. Far Eastern Book Publishers, Vladivostok. 1966, p. 237–241.
9. Kupin, V.I., "Biologically Active Modifiers of Immunity in the Complex Treatment of Malignant New Growths." Author's Synopsis of the Dissertation for a Scientific Degree as a Doctor of Medical Sciences, Moscow, 1985, p. 46.
10. Kupin, V.J., Blanko, F. F., Sorokin, A. M., Comparative Evaluation of Some Biologically Active Adaptogens and the Possibility of their Application in Experimental Oncology. Paper presented at the 2nd International Symposium on Eleutherococcus, Moscow, 1984.
11. Shchezin, A. K., Zincovich, V. I., Golanova, L. K., Eleutherococcus in the Prophylaxis of Influenza, Essential Hypertension and Ischaemic Heart Disease in the Drivers of the VAP. In: New Data on Eleutherococcus and Other Adaptogens. Far Eastern Research Centre. USSR Academy of Sciences.
12. Wacker, A., Eilmes, H. G., Virushemmung mit Eleutherokokk Fluid-Extrakt. Erfahrungsheilkunde, 27(6), p. 346–51.
13. Yaremenko, K.V., Adaptogens as a Means of Prophylactic Medicine, Tomsk, 1990, p. 94.
14. Yaremenko, K.V., Pashinski, V.G., "Preparations of Natural Origin as Remedies for Prophylactic Oncology." New Medicinal Preparations from Plants of Siberia and the Far East. Tomsk, 1986, p.171–172.
15. Zykov, M.P., Protasova, S.F., Prospects of Immunomodulated Vaccine against Influenza Using Eleutherococcus and Other Drugs of Vegetable Origin. Paper presented at the 2nd International Symposium on Eleutherococcus. Moscow, 1984. All-Union Cancer Research Centre of the USSR Academy of Medical Sciences.

14

POTENTIATING THE IMMUNE SYSTEM ACCORDING TO THE TRADITIONAL CHINESE MEDICINE

Y. Morgenstern

International Holistic Center
13 Harav Zirlson Street 62302
Tel Aviv, Israel
Tel: 972 3 604533293

1. INTRODUCTION

This article will deal with the traditional Chines view on the strengthening of the immune system. This tradition is based on the notions of YIN and YANG, and JING, which is the vitality. We will view the concept of the immune system as being a part of the Oriental system of the 'defense function' and the ways to strengthen it. This is a meeting point between East and West, between ancient healing traditions and new notions of bodily malfunctions. I hope that this article will open and clarify both the new and the ancient paths to better living in our time.

2. THE CHINESE TRADITION

The Chinese tradition is very ancient. It started a few thousand years ago, based on experience, and on the direct observation of phenomena and conscious listening to nature.

Later on, the observations were supported by wisdom and philosophy which offered various models through which all the information could be better used.

The Chinese tradition is effected by the specific qualities of the Chinese language. This language is visual, and it contains images, symbols, and signs which allow the reader to experience directly the described phenomena. The Chinese culture allows a combination, of curiosity, the observer and his observation of phenomena, comprehension and utilization of nature through various models and the collection of data in books and manuscripts. The Chinese point of view is holistic and points towards Man's task to be a link between Heaven and Earth. It concerns Man's different aspects (body, mind, soul, spirit) and his relationship with the environment, through the couple, the family, the generations, the society, and nature.

Potentiating Health and the Crisis of the Immune System
edited by Mizrahi *et al.* Plenum Press, New York, 1997

As a holistic point of view, this method has two aspects: 1. The whole is the sum of different parts. 2. Each part is connected in space and in time to the whole and it contains the information relevant to it. Just as the does the genetic material in the body's cells. In human life this link can be materialized in two ways:

- The link with space by consciousness in one's relationship with the Other, with Nature and with the whole Cosmos. A Chinese proverb says: "When a leaf falls on the earth a star in the sky is changing it's orbit."
- The link with time is explained by consciousness of the preceding and following generations, the history, and the tradition.

These two attitudes makes the difference between two states of consciousness. That of human beings who are linked, belonging, in connection and in a relationship with themselves and with others. They live in their environment as open systems. And secondly that of human beings, who are separated, isolated, alone, who live in the world as a closed systems. What kinds of experience allows Man to make his conscious grow from a closed to an open state?

What makes the difference between the two? Between the connected, generous eternal Man and the separated, temporary selective Man? What is the different between the body of one and the body of the other? Between a merely living body and a vital one? Does this difference have an influence on the human immune system? What is the role of this vitality which is the basis of our health and our capacity to adapt and to protect ourselves?

We will try to understand these questions and what is behind them.

3. A FEW CONCEPTS FROM THE CHINESE TRADITION

In order to try and understand the Chinese response to these life questions, I will introduce two essential notions that assist in understanding Chinese thought. One notion is universally present throughout Chinese history and serves as link between the different aspects of this culture. This notion become very popular recently in the West. It is the YIN and the YANG or rather the YIN\YANG. The YIN\YANG is a simple model that involves the observation of cosmic phoneme and the attention on the functioning of life in a penetrating manner. In order to understand the nature of this model I will describe how a traditional Chinese painter creates the following emblem.

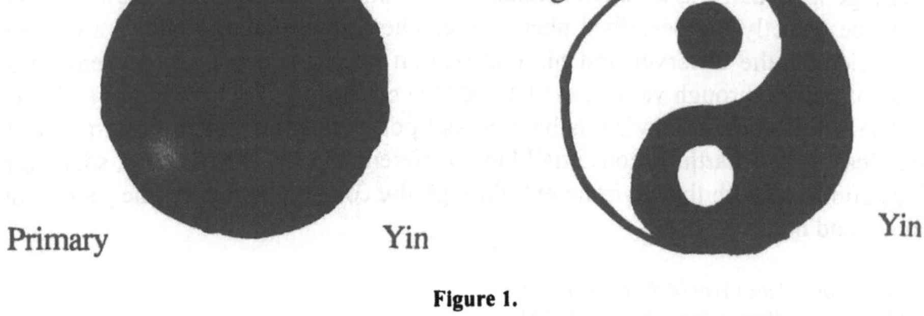

Figure 1.

He uses a piece of wood that is circular which symbolizes unity, totality, the perfect. The wood symbolizes life, the living. He paints the whole surface black in order to create the mysterious space that is without light and without order, yet contains everything. It is called the primary YING. On the black surface he actively draws the red fish with the black eye, this is called the YANG. Thus the black fish is drawn passively, and with a red eye added it becomes, and is called the secondary YIN.

What is the resemblance between the two fish?
What distinguishes between them?
What are the different natures?

There is a large oriental teaching, and many books, that have discussed these questions. One aspect may help our understanding. Under the (YANG) red fish is found in the black color, which symbolizes the first YIN. Under the black (YIN) fish there is the original YIN. The YANG is supported by the primary YIN. The secondary YIN receives its form from the YANG that left its mark in the primary YIN. The primary YIN holds up everything. It is the foundation for all, the beginning, the chaos, all matter lies within it, it is the source of vitality.

The second concept is Vitality – JING in Chinese. In observing the Oriental written character for this concept one finds:

1. The tree is above right and signifies movement, life, the transformation between Heaven Earth. It is a fruit bearing tree with all the relation between the fruit and the tree. Its color is green.
2. The field of Cinnabar is in the lower right and it is bright red. Its role is to allow the subtle transformation of matter and to make it come alive.

The two symbols indicate the link in space between earth and heaven, and its relationships through transformation and change with living matter. The bright colors emphasize and are the outer expression of the profound nature of being.

This JING vitality has an innate aspect and also an acquired one. Its contained in the primary YIN, in the inside, in obscurity, in chaos where life is very intense but without light. It emanates from the mystery of life, all the beginning of the universe. At the same time it exists in the present. It is a great and powerful source of energy which cannot be controlled and dominated. This is the reason why we are scared of it and we compress it and imprison it inside us. It is an inextinguishable energy that we can create through our life by subtle transforming the external energies within us. We can create it through the food we eat, physical exercise, our emotions, our love, through our conscious relations with the world. This vitality is like a river that is in flood, like an army of chariots that charge across the country, covered in a cloud of dust. It is strong, wild and uncontrollable, the basis of our life, basis of our capacity to defend ourselves.

Jing

Figure 2.

4. THE FUNCTION OF DEFENSE IN THE BODY ACCORDING TO CHINESE VISION

The Chinese make many analogies in order to understand different systems. For example they have often compared the human body with a country. We will attempt to understand the immune system of the body by comparing it with the defense of a country. What is contained in the notions of defense of a country?

There is the army, soldiers, officers, weapons, property (buildings, vehicles etc.), knowledge, information. In addition it needs controlling by a:

- Defense Ministry that fabricates and purchases weapons, and is responsible for providing the army with all it might need.
- Health Ministry that is responsible for the well being of the population and for providing medicine.
- Education Ministry which is responsible for educating and training soldiers.
- Transport Ministry.

And what else? Morals, communication, culture, faith, to make the people responsible and humane citizens, and other concepts that are difficult to describe. The human body is the same. The defense systems contains all the physical and non physical aspects of a human being. It includes an immune system working within the body and all other psycho spiritual levels.

According to the Chinese, the immune system is contained within:

- Eleven internal organs. The internal organ that is responsible for external protection is the liver, the spleen is responsible for internal protection.
- Internal reservoirs of energy.
- Orifices that serve as antennae for external information. The senses pass through the orifices: seven in the head and two in the lower body.
- Paths of energy named Meridians. The external Meridians follow through the body on the same pattern as the defense memory of the body.

Among the energies of the body are including the specific defense energy which called in Chinese WEI QI. This in a wild, rapid, powerful but unstable energy. It is created from the nutrition that enters the body. The origin is in the pelvic area, and from there it spreads to all the body like soldiers on the front line, and like guards it is based in the meridians but circulates from there through the fascia.

From the psycho–spiritual point of view, the strength of resistance comes from the preservation of unity which includes the two aspects: The YIN and the YANG. The symbol of this is the word WO, which in Chinese means I. It is described by two swords in peace which are connected together. Inside us there are two aspects that are living together in gentle balance. An inner separation weakens the body which have less energy for his own protection. In a country, if there is an inner rebellion against the central government the external defense becomes weak. In the same way, children and animals stop to eat when they are sick to preserve their energy to be able to attack the external enemy.

From the moment that our inner world is divided into two, there is a struggle between the two parts in which one part (usually the mental) wants to dominate the other one (body and emotions). Then the unity stops and is replaced with the duality which is the kingdom of the EGO. The oneness is governed by the spirit of Heaven -SHEN- as it called by the Chinese. SHEN is the conductor of an orchestra, the force of organization of the being. The emotions should be in harmony and be expressed. There are specific emotions connected to each inner organ, and they participate in building the organs themselves. If they are too intense, they can destroy those organs. For example, sadness is related to Lungs, and fear is related to the Kidneys.

When the emotions are under pressure or their inner paths are blocked, the vital forces can't circulate in the body and the process of destruction begins. To stop this process we have to make a big effort towards the gathering together of our scattered selves and acceptance of reality exactly as it is. We have to use our spirit, soul and body to make a good base for our protection system. This is the base of our energy, our health (physical, emotional and spiritual), and defensive and immune systems.

5. STRENGTHENING OUR IMMUNE SYSTEM

Traditional Chinese medicine considers it of great importance to reinforce the immune system by preventive measures on a day to day basis. They advise an activity of awareness which is connected with the seasons, the constitution, the age and personality of the individual. The diet and use of herbs must be adapted to the seasons, the activities of life, the typologies and pathologies of the person. For example: cinnamon, which is heating in its nature should be taken during winter. Mint which is cold in nature should be taken during summer.

The habits of life should be related to the person. For example: sleep early in the winter, go for walks in the spring. Physical exercise is intended to recruit the energies of nature and to transform them by inner process, to store what is necessary and to diffuse what is left over. The basic body work in Chinese medicine is called QI KONG. QI means energy QI KONG means working with energy. This physical activity brings many benefits to the body, toning, coordinating and permitting access to subtle energy, which is available to defend the body. By those exercise, the QI energy is created and transformed to JING- vitality which is stored in the Kidneys, for further utilization or for creating the essences of the body (including sperm and ovule).

The Kidneys are in the lower part of the body where WEI QI - the defense energy is born. There are specific points related to the immune system, that can be notified by exercises or by acupuncture. For example, the point DA - TUEI on the seventh vertebra in the place where the YANG is able to penetrate the body, and is responsible on the immune system of the body.

In every Meridian there are few points that tonify the immune system and direct this energy towards the injured place. There are also specific points which can help to produce, spread and use WE, QI - the vitalization energy.

In order to strengthen the immune system, we need to insure free flow of the emotions, which allows to the energy and vitality go every where. From spiritual point of view, the Chinese doctor attempts to create a spiritual contact with the SHEN-spirit of the patient and to guide him towards living in unity which will help him to find his way to a real health.

Traditional Chinese medicine gives the tools to integrate all different levels of the human being by methods of prevention or healing when sickness comes. This method of working with the whole body was demonstrated at this conference by Shin Terayama, from Japan. By his stories he shows how to be cured by the healing power of integrating love of life.

"I was an engineer with cancer for 11 years. I have been free of metastasis for the last 7 years. I transformed myself and recovered from cancer with cello playing. I confess how I loved my cancer, instead of fighting it. The most important thing which I learned was that the cancer was part of my body, it is not an enemy. It is still my body.... I used to say so to my body, I made a mistake, I'm very sorry, you are here and I love you. I'm grateful to my cancer because it made me to become a different person."

Only the relationship with the unknown, with the uncontrollable, with the mysterious in life, with the vitality of life, can give deep confidence in the world. This confidence can be transformed to hope, hope based on real experience and not the unreal.

6. CONCLUSION

The immune system is active from day one of the human being, from birth. Breast feeding is vital for reactivating the immune system. Children's sicknesses are involved in building this system which exists in the body until old age and death.

The immunity is a defense system which is in a state of constant awareness and depends on excellent functioning of the body. Serious illness can reach the body when the immune system is weak. A strong immune system is able to return the body very quickly to a balanced condition. The correct action of the immune system depends on balancing the emotional and spiritual energies, and the ability of man to choose to be responsible and to love. Only by gathering all our physical and non physical energies can we protect ourselves and become really healthy!

15

NON-CONVENTIONAL CANCER THERAPIES

Criteria for Evaluation and Rationales for Implementation—Helping Patients Decide

Susan Silberstein

Center for Advancement in Cancer Education
Suite 100, 300 East Lancaster Avenue
Wynnewood, Pennsylvania 19096

1. ABSTRACT

As the failure of radiation and chemotherapy to cure most cancers has become more apparent in recent years, dozens of alternative cancer therapies have commanded increasing attention. However, the complex world of unorthodox modalities can be confusing to patients and practitioners alike. Overwhelmed with information from the broadcast and print media, and torn by conflicting recommendations from conventional and unconventional sources, patients left to their own devices often make bad choices or no choices.

Yet there is a place in oncology for unorthodox approaches as either adjuncts or alternatives to traditional treatments. Three different therapeutic avenues in oncology with three distinct goals illustrate the shortcomings of conventional therapies targeting tumor reduction, the ways in which selected adjuvant treatments support host resistance, and the potential for biological repair with use of appropriate alternative modalities.

Review of the major unconventional treatment classifications reveals an intriguing parallelism between experimental mainstream therapies -hyperthermias, hormone therapies and immunotherapies — and the best of the "unproven" therapies — clinical nutrition, botanical medicine and psychoneuroimmunology.

Criteria for evaluating unorthodox therapies need to be stringent, with the ideal program focusing on the essential distinction between a host-oriented and a tumor-oriented approach. When tumor reduction remains the primary goal, results with alternative therapies have been inconsistent and sometimes not significantly better than with orthodox methods. The programs which have produced the most consistent results are those which recognize the body as a self-healing mechanism by giving it the essential tools for immunocompetence and self-repair.

By respecting the common ground upon which conventional and nonconventional approaches can meet and by seeking an appropriate marriage between the body's own natural healing potential and the best that medical science has to offer, health profession-

Potentiating Health and the Crisis of the Immune System
edited by Mizrahi *et al.* Plenum Press, New York, 1997

als can play an important role in helping their patients to design individualized intervention plans and to make the most effective therapeutic choices.

2. INTRODUCTION

Today's cancer patients are suffering from much more than cancer—they are suffering from a syndrome called "information overload." Bombarded with data on treatment options courtesy of the broadcast and print media, and offered conflicting recommendations from doctors, well-meaning friends, neighbors and relatives, modern patients are simply overwhelmed. As a result, the passive ones are only too happy to abdicate their participatory role to the first oncologic opinion that comes along; and those aspiring to make informed decisions as sophisticated medical consumers often end up researching themselves into totally indecisive confusion.

The heads of patients and their advocates spin with information about Naessens' 714X, Clark's herbal parasiticides, Contreras' Laetrile injections, Burzynski's anti-neoplastins, Burton's IAT, Dannopoulus' urea therapy, Lane's shark cartilage, Sopcak's Cancell, Pauling's vitamin C, Simonton's guided imagery, Mexican clinics, macrobiotic diets, coffee enemas, ozone treatments, and the list goes on and on.... Asking one's doctor about these programs is of little help. Most physicians are either ignorant or scornful of anything not mainstream, and they generally tell patients they are wasting their time.

To a certain extent, that is true. The average patient, when faced with a cancer diagnosis, has neither the time, energy, money, nor knowledge to sift through such a multiplicity of resources. How is a patient to make sense of that muck of information which fills the void between the shortcomings of the conventional therapies and those of the non-conventional world? On the one hand, some unorthodox centers are so taken with being anti-establishment and holistic that they ignore the importance of a competent medical work-up, proper diagnostic tools, and careful medical supervision; the cancer patient becomes easy prey to quacks and opportunists. On the other hand, conventional cancer centers are so overly concerned with tumor reduction that host resistance and quality of life are often sacrificed in the process.

Yet there is a place in oncology for alternative and complementary medicine, and that is what we are here to explore. The questions most cancer patients are asking us at the Center for Advancement in Cancer Education are, in fact, the very inquiries of the present Conference: Is it possible to bridge the gap? Is it possible to get the best of both worlds? Is marriage possible between the body's own natural healing potential and the best that medical science has to offer? And if so, how can we help them decide on a program that is relevant and purposeful for them, while at the same time protective of their precious physical, financial and emotional resources?

The answer to these questions lies in completely shifting the perspective from which we view cancer treatment. The resolution of the conflict between conventional and non-conventional cancer therapies lies in one essential distinction: the difference between a tumor-oriented and a hostoriented approach.

3. TEXT

3.1. Negative Aspects of Orthodox Therapy

To understand the difference between a tumor orientation and a host orientation to cancer, one must start with a brief examination of three therapeutic avenues and the yardsticks for measuring the success of each:

Therapeutic Approach	Goal, Primary Measure of Success
Traditional/Orthodox/Conventional Therapy	Tumor Reduction
Adjunctive/Adjuvant/Complementary Therapy	Host Resistance
Alternative Therapy	Biological Repair

Now, before we explain these further, I'd like to focus for a moment on the negative impact of the three primary orthodox therapies upon biological function. This is neither a value judgment nor an indictment of these therapies. Tumor reduction is an appropriate goal, certainly, in lifethreatening situations and when significant functionability or quality of life depends on timely intervention. Furthermore, when the tumor is huge and/or rapidly growing, debulking is often highly desirable. However, surgery, chemotherapy and radiation are not without their severe shortcomings.

Let us look at chemotherapy first. A patient may rightfully be uncomfortable about hair loss and nausea, but these are clearly the least of his or her problems. The immunosuppressive activity of chemotherapeutic agents has long been recognized. Bone marrow depression, decrease in white blood cell counts, and general inability to fight off infection are common occurrences among patients treated with long-term chemotherapy. Further, the cardiotoxicity, renal toxicity and hepatotoxicity of certain drugs are unfortunate iatrogenic complications which compromise the body's overall biological integrity. Dr. Ulrich Abel, a German biostatistician who performed a meta-analysis reviewing thousands of research studies on chemotherapy, concluded that its long term use actually promotes increasingly rapid proliferation of cancer cells.[1]

Surgery, as we know, can be curative. But for many patients in whose cases it has been labelled "curative," disease recurs—much to the surprise not only of the patient but actually of the surgeon as well. And so we hear comments like, "I don't understand; (s)he said it was encapsulated, (s)he got it all out, (s)he cut way beyond the borders of the tumor—I guess a microscopic cell must have somehow escaped." This common scenario is all too familiar to many of us. It is perhaps not so well known that procedures such as surgery, general anesthesia, and blood transfusions lower immune function. *The British Medical Journal* has reported that postoperative depression of tumor-directed immunity is quite common.[2] In recipients of blood transfusions, furthermore, a diminished helper/suppressor cell ratio and decreased natural killer cell activity—both key paramaters of immune function—have been noted.[3]

Finally, let us look at radiation therapy. It is well known that radiation can cause local mutations and malignancies in cells it does not destroy.

But radiation therapy also has its systemic effects. The journal *Radiology* reported on research examining the T-cell subsets of cancer patients before and after radiation treatment. Researchers noted decrease in the total number of T cells, reduced proportions of T helper cells, and increased ratios of T suppressor cells, with some of the disturbances in immune status and lymph flow lasting long after treatment was discontinued.[4]

The implications of such findings are two-fold: First, regardless of which system you use to destroy tumor, tumor reduction alone does not generally represent a cure. If you do not go back to the cause of the cancer—to the environmental exposures, to the toxic lifestyles, to the imbalance in body chemistry, to the weakness in immune defenses, to whatever allowed the body to produce abnormal cells in the first place—and somehow make the correction at that point, then you have an enormous likelihood that the disease will recur down the road. And second, if, in the process of reducing or destroying tumor, you use a system that actually promotes cancer or suppresses immune function, then the likelihood of a recurrence is hastened.

3.2. Value of a Host-Oriented Approach

For these reasons, then, we are interested in looking at an approach that focuses more on the body than on the tumor. Most cancer patients do not die directly of their tumors, unless the malignancy is in some way suffocating a vital life function. Rather, patients die of malnutrition, toxemia, and/or opportunistic infections. The adjunctive therapies can supplement the traditional yardstick of tumor reduction through any or all of the following:

1. supporting host resistance
2. enhancing immune response
3. potentiating the positive effects of chemo- or radio-therapy
4. helping to minimize the negative effects of these treatments
5. picking up where surgery leaves off (or after any treatment protocol is discontinued)
6. actualizing patient empowerment
7. enhancing quality of life, even among the terminal

For patients who are not candidates for, have failed on, or simply refuse conventional treatment modalities, we must look towards alternative therapies. The word "alternative" is used, for want of a better term, to represent a treatment approach employed instead of, rather than along with, conventional treatments. (It should be clearly noted, however, that this is an umbrella term covering hundreds of different treatments, many of which we do not support. Criteria for evaluating alternative cancer therapies will be elaborated upon later.) The important issue is that with the alternative therapies the traditional yardstick of tumor reduction must be replaced by the goal of biological repair. This means repairing the body to every extent possible so that we achieve optimal function of every organ, gland and system.

3.3. Experimental Mainstream Therapies

Now, what are the adjunctive and alternative therapies? Because of time constraints, I can only offer an overview of an extremely diversified list of treatments. For our convenience, I have grouped the major nonconventional treatment classifications into two subsets: Experimental mainstream therapies and the officially "unproven" therapies.

Among the experimental mainstream therapies are the hyperthermias, hormonal treatments, and the immunotherapies.

3.3.1. Hyperthermia. The hyperthermias—also called thermotherapy, heat treatments or fever therapy—have been known from the time of the Romans to reduce tumor and are mostly used today in conjunction with other therapies for their synergistic effects. The heat may be applied locally, regionally or systemically, and the heat source may be thermal blanket, wet suit, microwaves, high frequency radiowaves, heated probe or extracorporeal heat exchanger. For the patient choosing a biological system, or in the case of widely disseminated disease, we prefer carefully administered whole body heat therapy because it most completely resembles a natural fever, now known to be part of the innate human healing response. Problems with hyperthermia include the fact that treatment is not readily accessible, especially for deeper tumors, it is costly, and is not generally covered by insurance.

3.3.2. Hormone Therapy. Hormonal treatments involve administration of substances which inhibit production or uptake of hormones which feed certain hormone-dependent tumors like breast or prostate disease. They can be very effective and are less toxic than chemotherapy, but they can have uncomfortable side effects, tend to cease working, and, in the case of tamoxifen, for example, can cause ocular and hepatic toxicity and even uterine cancer.

3.3.3. Immunotherapy. The third category of experimental mainstream therapies is the immunotherapies, aimed at enhancing the body's ability to break down tumor from within. Many of these therapies have originated with the Biological Response Modifiers Program of the National Cancer Institute, whose mission it is to develop and test biological agents to boost immune defenses. I have divided the immunotherapies into two main types: those that I call "immuno-augmentive" and those that I label "immuno-stimulatory."

The latter type involves administration of a small amount of toxic material in an attempt to wake up the immune system. It includes the BCG tuberculin bacillus, Coley's toxins (a bacterial cocktail concocted by a 19th-century New York surgeon, Dr. William Coley), Rhus Tox (poison ivy) first applied by an oncologic surgeon named Lenaghan, and others. These therapies tend to be unreliable, sometimes reducing tumor and sometimes producing no effect.

The "immuno-augmentive" type of immunotherapy involves administration of substances innate to a properly functioning immune system. Lymphokines, interferons, interleukins, tumor necrosis factor, and thymosin, for example, are representatives of this group. They may be directly toxic to tumor cells, partially suppress neoplastic cell transformation, or induce protective T-cell or natural killer cell activity. However, clinical trials are still very limited, and toxic side effects such as spleen congestion, lung inflammations, high fevers and other complications have not been managed with total success.

3.4. The "Unproven" Therapies

The "unproven" therapies—and I hasten to add that "unproven" does not necessarily mean "disproven" or inefficacious—fall into many categories. The three which tend to produce the most consistent results -clinical nutrition, psychoneuroimmunology, and botanical medicine—all aim at enabling the body to enhance its own immunocompetence. Each of these fields is so rich with information that each could alone constitute a presentation of several hours. More important, each of these fields, properly applied, has the potential to accomplish endogenously what the three experimental mainstream approaches try to accomplish exogenously -that is, generating fevers, triggering hormonal changes, and enhancing immune response.

3.4.1. Clinical Nutrition. Let us look at nutrition first. You are probably familiar with macrobiotic miracles and megavitamin mania. While often helpful, neither of these do I consider to offer necessarily the best that clinical nutrition can contribute to cancer therapy. And contrary to prevailing opinion, literally tens of thousands of articles have been published in hundreds of biomedical journals on the scientific relationships between diet and cancer survival -not just cancer prevention. An enormous body of research has documented the immuno-stimulatory effects, the hormonal effects, the impact on tumor growth factors, and the influence on tumor angiogenesis (the ability of the tumor to develop its own blood supply), all mediated by certain nutritional factors. Further, nutrition is the best tool we have for producing new healthy cell growth.

I will give you just a few examples of this fascinating field. For example, lowering fat intake to 15% or 20% of total calories, increasing the omega 3 to omega 6 fatty acid ratio, and reducing oxidized fats can dramatically alter the tumor cell proliferative process. Consumption of large amounts of concentrated mixed carotenoids increases T cell, macrophage and NK cell function. Cruciferous vegetables contain a chemical called indole-3-carbinol which stimulates important liver enzyme systems helpful in reducing levels of circulating hormones and immuno-suppressive compounds.

Dozens of other foods or phytonutrients, including sea vegetables, garlic, shitake mushrooms, raw green tea, and flax seeds have been correlated with specific immune-enhancing effects and tumor inhibition.[5] Although our Center has helped thousands of patients by distributing information on a generic diet, it is especially desirable for the patient to work with a practitioner who can design a specific nutritional program of vitamins, minerals, enzymes, glandulars and other supplements, tailored to the patient's biochemical individuality.

3.4.2. Psychoneuroimmunology. For about two decades, the fascinating field of psychoneuroimmunology (PNI) has been studying the powerful relationship between psychological and emotional factors, cancer and the immune system. Scientists in the emerging field of behavioral immunity have actually been able to document in the laboratory the influence of certain mental states on various aspects of immune function, and have been able to delineate the specific biochemical pathways by which the nervous, endocrine and immune systems communicate.

Findings in this field have been applied to cancer treatment by researchers and clinicians in three basic areas: 1) relaxation and guided imagery, 2) support groups and 3) counseling and psychotherapy. For example, Gruber and Hall published a pilot study of changes in immune parameters in metastatic cancer patients using relaxation and guided imagery in 1988. Le Shan, an experimental psychologist doing research with cancer patients for over four decades, has concluded that psychotherapy, besides improving copeability, can also help stimulate a patient's compromised immune system and enhance his self-healing abilities.[8]

3.4.3. Botanical Medicine. Botanical therapy, an integral part of oriental medicine for centuries, has recently begun coming into its own among practitioners of western medicine as well. The use of herbs or plants in cancer treatment is generally not directed specifically at reducing the tumor, but rather at biological repair of the host through a variety of functions. They include stimulating circulation, flushing the kidneys, detoxifying and regenerating the liver, relieving lymphatic congestion, stimulating grandular function, activating eliminatory organs, and potentiating the immune response.

Let me give you a few illustrations. Astragalus root, for example, can enhance productions of interferons, significantly improve macrophage activity, and strengthen the body's resistance to illness, especially in wasting diseases like cancer. Echinacea also has been demonstrated to activate macrophages, stimulate T-lymphocytes, increase phagocytosis and stimulate interferon production. Eleutherococcus, or Siberian ginseng, can significantly enhance phagocytic activity and is a helpful immunostimulatory adjunct during radiation therapy.[9] Viscum album has been used in Europe since 1921 in the treatment of human cancers, especially for its ability to cause proliferation of thymus cells. It has also been demonstrated to accelerate the growth of healthy bone marrow and spleen tissue in irradiated animals. The list of excellent botanicals for cancer includes not only Iscador from Switzerland, but also Essiac from Canada, the Hoxsey herbal tonic from Mexico, pau d'arco from Brazil, carbo betula polaris from Finland, and the Jason Winters formula, among others.

3.4.4. Bioenergetic Medicine. I would be remiss if I did not at least mention one additional class of unconventional therapies, the bioenergetic approaches. This group includes acupuncture, acupressure, massage, lymph drainage, yoga, tai chi, chi gong, therapeutic touch, reiki, and reflexology. All of these techniques are very helpful post surgery and during chemotherapy or radiation for restoring vitality and energy, promoting a state of relaxation, reducing pain and anxiety, improving immune function, enhancing circulation and elimination, and stimulating glandular efficiency. This class also includes a number of other treatments which show promise and which we are still watching, principally the bioelectromagnetic therapies and ozone or other hyperoxygenating therapies.

3.5. Criteria for Evaluating Alternative Therapies

In operating a referral agency for adjunctive and alternative cancer treatments, my colleagues and I at the Center for Advancement in Cancer Education have had occasion to examine dozens of therapies, communicate with hundreds of practitioners and receive feedback from over ten thousand patients over the last twenty years. That feedback, in particular, has placed us in a unique position to develop criteria for evaluating nontraditional cancer treatments and for determining which types of programs produce the best results for the most patients over the longest period of time. Any time we learn of a patient who has "been successful" using an "alternative therapy" we use the following evaluation criteria:

1. CONFIRMED DIAGNOSIS (We understand that some patients may not even want a biopsy — that is their prerogative, but we cannot evaluate a therapy based on a probability that there was a malignancy.)
2. NO ANTI-PHYSICIAN PLATFORM (We do encourage patient participation but not do-it yourself programs, and we are interested in marriage with, not divorce from, organized medicine.)
3. USE OF SAFE, NON-TOXIC MATERIALS (Primum non nocere, said Hippocrates. We are comfortable with substances with which the body is comfortable.)
4. NO MAGIC BULLETS (We do not rely on one single key substance to repair what we consider to be a multiplicity of body malfunctions)
5. BIOCHEMICAL INDIVIDUALITY (The program should be adjustable and applicable to individual body chemistry, not "one size fits all." This includes sensitivity to individual pace and especially to toxicity patterns that tend to wax and wane.)
6. RESULTS FOR MAJORITY (The vast majority of patients treated, not a few isolated anecdotes, should show significant improvement.)
7. RESULTS LASTING MORE THAN FIVE YEARS (We do not believe that five year survival represents cure, especially in view of the fact that our increasingly excellent diagnostic tools can set back the clock a couple of years.)
8. QUALITY LONGEVITY (Mere survival is inadequate anyway. Our biological yardstick measures not only life's length, but also its width and depth. In some

cases, excellent responses are achieved even though complete cure may not be possible.)

9. NO QUICK FIX (The time frame for recovery should last one to two years. Anyone who claims he can cure cancer in three weeks or three months with an unconventional cancer therapy is no more impressive than an oncologist who can get rid of a tumor which then comes back in a few weeks or months. As the body chemistry breaks down gradually over years, we expect complete repair and recovery to take a couple of years.
10. MIND-BODY CONNECTION (Attention to destructive emotional or behavioral patterns is essential for wellness; without it, no other therapy is likely to be effective for any length of time.)
11. ATTENTION TO IMMUNE PARAMETERS (Balanced function of the body's defense mechanisms must be regained and maintained.)
12. FOCUS ON BODY METABOLISM (Harmonious and efficient function of all organs, glands and systems should always be the goal.)
13. HOST ORIENTATION (Summing up all these factors, the most basic of all criteria is that the focus of the treatment be the host, not the tumor, and that the treatment emphasis not be on mere tumor destruction but rather on biological repair.

3.6. Individualizing the Program

Once we evaluate a therapy for these criteria, we also evaluate it for its appropriateness for a given patient. Our approach is highly individualized and patient-driven. It is much less important to us what kind of cancer the patient has than what kind of patient has the cancer. We look at patients' physical, financial and geographic limitations. We look at their belief systems, work patterns, family dynamics, lifestyles, and goals. We look at their mental or emotional readiness for a certain treatment approach. We never tell patients what they should or should not do. We ask them what their doctor has told them, what their doctor is offering them, how they feel about what the doctor is offering them, how they feel about the doctor, what they feel comfortable doing in the conventional or unconventional medical worlds.... Then we start designing a game plan and offering them resources.

3.6.1. Making a Needs Assessment. Here are some of the questions that are asked during a typical needs assessment:

- Is there a confirmed diagnosis of cancer? When was the diagnosis made? What is the primary site? Are there any metasteses?
- What treatment has already been done? Over how long a period? Has it been discontinued? How long ago? Why?
- Is any treatment currently being administered? What kind? How do you feel about it?
- Is any treatment being proposed? What kind? What has your doctor told you about the treatment, its purpose, and his or her expectations of it? (Have you dared ask?) How do you feel about the proposed treatment? Why?
- Are you looking for adjunctive support or a strictly alternative approach? Do you know what type? Do you wish to learn more about any particular approach? Would you be receptive to taped or printed educational materials?

- Is a referral requested? What type of professional? Do you need help deciding on the type of professional?
- What financial limitations exist? What type of insurance do you have? What resources can you draw upon for treatments which insurance will not cover?
- What geographic limitations exist? How is your mobility?
- What is your quality of life? Your functionability? In what particular way would you most like to see the quality of your life improve?
- What is your work schedule or other responsibilities? How flexible are they or could you make them?
- What is your family situation? Who is in your support system? Would you like more support? From what type of source?
- What stress patterns were present prior to diagnosis? Which of these are not resolved?
- What are your other current stressors? Would you like help managing them?
- What were your eating habits prior to diagnosis? To what extent have you changed them? To what extent are you willing to change them? What is your weight/height ratio? Who can shop for/prepare your food?
- How efficient is your digestion and elimination? Would you like to learn more about improving them?
- Finally—and most interesting—Do you have any ideas as to why you got this cancer? What might have been your risk factors? Could any lifestyle factors like smoking, alcohol, drugs, diet, work or stress have played a role? Which of these factors are still operative?

3.6.2. Qualities of Successful Patients. Those patients who are most open and willing to dialogue on all of these issues place us, as support personnel, in the best position to be of real help. And those patients who manage to achieve a balance between cognition, emotion and behavior are the most likely to outlive their prognosis with quality longevity. From that group of successful patients emerges a clear pattern of psycho-social-behavioral qualities:

1. ACCEPTANCE OF THE DIAGNOSIS/REJECTION OF THE PROGNOSIS
2. PARTICIPATION/INITIATIVE/COMMITMENT (Active participation means choosing a treatment plan according to one's own wishes, understandings and beliefs. This may involve disagreeing with one's doctors or loved ones in favor of one's intuitive sense of what is right for oneself — and then persevering.)
3. INTROSPECTION (Use of illness for personal learning and growth, resolving previous losses, completing griefwork, and self-actualizing)
4. TRANSFORMATION IN INTERPERSONAL RELATIONSHIPS (This especially includes learning to receive, making oneself a priority, reconciling conflicts (even with those deceased!), and purging toxic relationships.)
5. LIFESTYLE CHANGES (Developing new patterns of diet, exercise, work and play, particularly changing jobs or living arrangements)
6. EXPRESSION OF EMOTIONS (This includes both positive and negative emotions that may have been repressed: externalizing anger and resentments, and digging deep for the joie de vivre of early childhood.)
7. LIFE PURPOSE (Finding life purpose means clarifying one's goals, creating meaning in life, and developing a sense of self-worth.)

3.6.3. Helping Patients Decide. I would like to focus briefly on the role of the health-care practitioner or counselor in helping cancer patients decide on appropriate alternative and complementary therapies. The most basic advice is LISTEN, LISTEN, LISTEN.

1. Help the patient DETERMINE REALISTIC GOALS. (That does not necessarily mean full recovery from cancer. It may be more energy, less pain, more mobility, weight gain, living until one's son graduates or one's daughter marries, better family communication, or greater spirituality.)
2. Find the patient's particular way to DEFINE QUALITY OF LIFE. (For some patients, quality of life is meat and potatoes (so a macrobiotic diet would not work); for others, it is going on a cruise (forget diet here too). Others will take enemas and drink awful-tasting teas around the clock as long as their energy improves.)
3. DIRECT ATTENTION TO THE HOST, NOT THE TUMOR. (Patients can live many productive years with a tumor in homeostasis with the immune system, not disappearing but not going anywhere — so it is better to focus on repairing the body and the spirit.)
4. START WITH THE BASICS OF HUMAN NATURE AND NURTURE. (Look at patients in terms of what they're eating and what's eating them. Appropriate nutritional and emotional repair is often all that is needed.)
5. GIVE OPTIONS NOT OPINIONS. (Put out a smorgasbord of ideas, accept that you do not need to be an authority. Don't fall into the "what should I do?" trap. Respond by asking, "What are you comfortable doing?"
6. BE SUPPORTIVE NOT JUDGMENTAL. (A patient may not choose what you know to be best for him or her. The fact that (s)he is making choices at all is more to the point.)
7. WORK WITHIN THE CLIENT'S BELIEF SYSTEM. (Help design a game plan that makes sense to him. Don't try to talk a patient out of chemotherapy if she believes her doctor knows best, and vice-versa.)
8. NEGOTIATE AND COMPROMISE. ("If you won't give up ice cream, could you at least cut down? If you won't change your diet, will you take some herbal supplements? If you can't travel one hour, would you consider 30 minutes? If you can't afford a $300 juicer, would you buy an $80 one?"
9. HELP PATIENTS ACHIEVE BALANCE IN THEIR LIVES. (Help them avoid becoming professional cancer patients 24 hours a day. Help them choose a program that leaves them room for play and fun, family and friends, and any other choices they may make. Make certain their life is worth living.)
10. GEAR CHOICES TOWARDS COMFORT AND CONTROL. (Pay attention to what fits with the logistics of their lives. The "Big C" does not necessarily stand for "cancer"; comfort and control are really what cancer patients lack.)
11. HELP PATIENTS DETERMINE WHEN TO BECOME IM-PATIENTS. (Over-eager patients can push the body too far too fast; biological repair takes time with many ups and downs, and patience while the body heals is crucial. On the other hand, waiting until a recurrence before the doctor has a treatment to offer is much too passive; patients want to take a pro-active, pre-emptive role, and they should.)
12. SHOW THE DIFFERENCE BETWEEN CURING AND HEALING. (The terms are not synonymous. Stress the difference between removal of symptoms and repair of the host, between superficial and often transient external responses, and deeper, more lasting internal responses. We have seem many examples of treat-

ments that make tumors disappear at the expense of the patient. Even in death, much healing can take place. Often emotional and spiritual transformations end up producing physical transformations and remarkable recoveries.

13. TEACH THE KEY QUALITIES OF SUCCESSFUL PATIENTS.

3.6.4. Ultimate Individualization. Over the last 20 years, thousands of patients have contacted our Center for alternative or complementary treatment resources. Many have left with referrals to medical nutritionists, second opinion surgeons, third opinion oncologists, clinical immunologists, acupuncturists, herbalists, osteopaths, naturopaths, homeopaths, yoga teachers, support groups, hypnotherapists, colon therapists, psychotherapists, or family therapists. Many more have walked out with a recommendation to paint, sculpt, sing, act, write, play an instrument, move out of town, change a job, adopt a baby, or get a divorce. So there you have it—the ultimate host-oriented approach!

4. SUMMARY

The reason conventional treatments fail is because they concentrate on killing cancer cells but neglect or abuse the host in the process; that is, they address the symptom rather than the root cause. But we must not substitute alternative treatments in the same allopathic equation. Many socalled "alternative cancer therapies" simply mimic orthodox medicine by focusing on eradicating the symptom rather than repairing the host. It is well known in the medical community that tumor reduction does not represent a cancer cure; the controlling factor is really how the body functions.

All therapies, conventional or non-conventional, which make tumor reduction their goal, ultimately fail—even if they do buy survival time, improve quality of life, or temporaryily relieve symptoms. When tumor reduction is the primary goal, results with alternative treatments have been inconsistent, unpredictable and often not much better than with conventional treatments. The programs which have produced the most consistent results are those that recognize the body as a self-healing mechanism and give it the essential tools for immunocompetence and self-repair.

It is our responsibility as oncologic caregivers and support personnel to help patients wade through the proliferation of cancer treatment options by fostering a better understanding of appropriate criteria for their evaluation and rationales for their implementation. I would like to close with quotations from professionals with far more impressive credentials than my own, professionals who deeply understood the host-oriented approach.

Hippocrates, the Father of Medicine, said: "Natural forces within us are the true healers. Let they food be thy medicine...."

Famed medieval physician Moses Maimonides wrote: "The physician should not treat the disease but the patient who is suffering from it."

Florence Nightingale stated: "Nature alone cures...and what nursing has to do...is put the patient in the best condition for nature to act upon him."

And finally, the great Albert Schweitzer philosophized: "Each patient carries his own doctor inside him. We are at our best when we give the doctor who resides within...a chance to go to work."

5. REFERENCES

1. Abel U: Cytostatic Therapy of Advanced Epithelial Tumors: A Critique. Hippocrates Verlag. Stuttgart.

2. Cochran et al. (1972) Postoperative Depression of Tumor-Directed Immunity in Patients With Malignant Disease. British Medical Journal *4*, 67.
3. Kaplan et al. (1984) Diminished Helper/Suppressor Lymphocyte Ratios and Natural Killer Cell Activity in Recipients of Repeated Blood Transfusions. Blood *64*, 308.
4. Yang et al. (1988) Changes in T-Cell Subsets After Radiation Therapy. Radiology *168*, 537.
5. Sparandeo J: Scientific Relationships Between Diet and Cancer Survival. Comprehensive Nutritional News. Easton, PA 1991.
6. Gruber B, Hall N, Hersh S and Dubois P. (1988) Immune System and Psychological Changes in Metastatic Cancer Patients Using Relaxation and Guided Imagery. Scandinavian Journal of Behavior Therapy *17*, 25.
7. Spiegel D, Bloom J, Kraemer H and Gottheil E. (1989) Effect of Psychosocial Treatment on Survival of Patients With Metastatic Breast Cancer. The Lancet *October 14*, 888.
8. LeShan L: Cancer as a Turning Point. Dutton NY 1989.
9. Duke J: Ginseng: A Concise Handbook. Reference Pub. Algonac, MI 1989.
10. Bloksma N, Schmiermann P, De Reuver M, et al. (1982) Stimulation of Humoral and Cellular Immunity by Viscum Preparations. Planta Medica *46*, 221–227.

16

COPING WITH CANCER AND THE SEARCH FOR MEANING

Ton Staps[1] and William Yang[2]

[1]Department of Medical Psychology
University Hospital Nijmegen
P.O. Box 1901
6500 HB Nijmegen
The Netherlands
[2]Taborhuis
Postbus 9001
6560 GB Groesbeek
The Netherlands

ABSTRACT

Psycho-Energetic Therapy (PET) has been developed to aid patients mobilize their physical, psychological, social and spiritual resources in order to cope with the many and often drastic changes in their lives. It includes regular counseling/psychotherapy and a variety of mind/body interventions. The latter have been developed to assist in the release of physical and emotional tensions and in the buildup of a reservoir of psychic energy. The mind/body interventions can also play an important role in the search for meaning.

PET is based on a five dimensional model. The five dimensions are: attention, communication, emotions, body, and meaning. Each of these five dimensions has its own characteristics and can be taken as the starting point for counseling. The dimensions are closely interrelated so that change in one dimension can also elicit changes in the other dimensions. PET seeks to create a dynamic balance among the five dimensions considered vitally important for any patient coping with cancer.

Although the five dimensions are generally of equal importance, meaning takes the lead role in coping with cancer. The basis of the coping process is the search for meaning, which takes place in this dimension. Corresponding to the partly different tasks of the cerebral hemispheres, the distinction is made between the logical mind and the metaphorical mind. In the activities of the logical mind the self is placed opposite the non-self. In the activities of the metaphorical mind in contrast, the self is integrated with the non-self. The search for meaning also respects the distinction between an active, logical process of giving meaning to the illness, on the one hand, and the experiential process of recognizing

Potentiating Health and the Crisis of the Immune System
edited by Mizrahi *et al.* Plenum Press, New York, 1997

that one is a part of a meaningful whole that includes the illness, on the other hand. PET helps the patient to find the dynamic balance between the two and thereby cope effectively (i.e., accept the irreversible and nevertheless behave autonomously within this restriction). In the course of effectively coping with an illness, the patient uses two mechanisms in his search for meaning. The first mechanism is changing the time perspective in which the patient tries to direct his attention to the here and now rather than the future. In such a way, the seriousness of his situation can be relativized. The second mechanism is changing the reference point. In doing this, the patient accepts the irreversible changes and starts organizing his life accordingly. Each change of reference point is made possible by a mourning process and by placing his own situation in a larger meaning context. Different changes of reference point, may take place with the progression of the illness, and also bring the patient to the acceptance of death.

1. INTRODUCTION

Cancer creates a sudden and drastic change in the lives of patients. Cancer is not just a confrontation with another life situation. It is a confrontation with the fragility of life and reality of death. While it is difficult enough to relate to life and cope with it in a healthy and effective manner, relating to death is the ultimate challenge for every human being.

Psycho-energetic therapy (PET) has been developed to help patients mobilize their physical, psychological, social and spiritual resources in order to cope with the many and often drastic changes in their lives. It includes regular counseling/psychotherapy and a variety of mind/body interventions. The latter have been developed as relaxation techniques and to assist in the release of physical and emotional tensions and the buildup of a reservoir of psychic energy.

In this paper, the practical and theoretical background to PET will be presented along with the basic principles. Special attention will also then be given to the role of meaning in coping with cancer.

2. BACKGROUND

The authors and developers of PET have worked almost exclusively with cancer patients. Our initial need, was to bring our experiences together in a model that could perhaps explain the working of the different aspects of coping and also be used as a lead for interventions with patients who are having problems in coping with their cancer. The general system theory was a fruitful theoretical framework for the development of our ideas (1). According to this theory, living nature is composed of systems. These systems have a hierarchical structure. Each system is composed of several subsystems, and also itself a subsystem of a system one level higher in the hierarchy. Each subsystem has its own functions, but all of these functions may not be used. Which function is used at what moment and for how long is determined by the regulating center of the system, which is in turn influenced by a higher system. A cell, for instance, has the capacity to divide but it does not do so until it gets a signal from the surrounding tissue. Similarly, the cell will only stop when it gets a signal to do so from the tissue. The signal to divide may come when I cut myself, and the signal to stop may come when the wound is closed. In this light, it should be noted that once a cancer cell starts to divide, it does so in an autono-

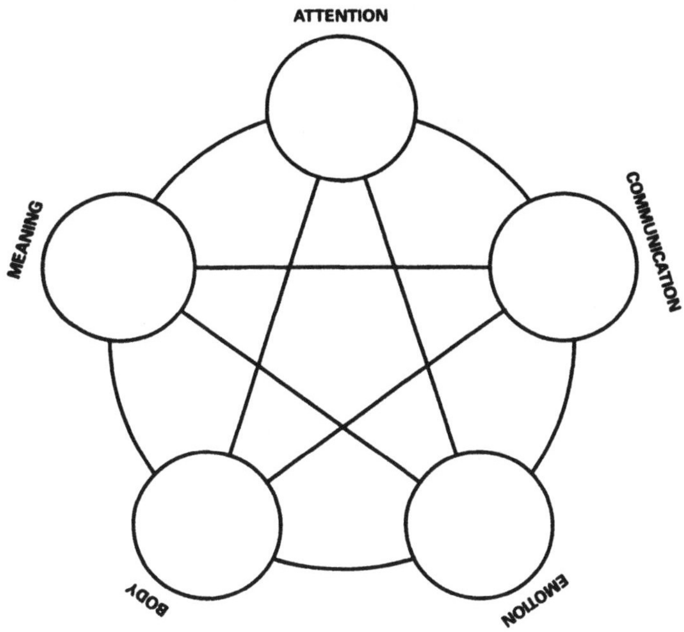

Figure 1. Psycho-energetic therapy: basic model.

mous manner without listening to the signals from its surroundings. This is what makes the cancer cell so devastating.

According to general system theory, a system is subject to a everlasting stream of changes: internal as well as external. The system tries to survive in this stream of changes, and struggles to do this by trying to make the changes undone, changing its reaction to the changes, leaving the changed situation, or by utilizing a combination of these reactions.

3. THE MODEL

In the model underlying PET, we distinguish five dimensions considered vitally important for any patient coping with cancer (figure 1).

The dimensions—which should be seen as physical-emotional-mental "spheres of influence" — are:

- attention,
- communication,
- emotions,
- body, and
- meaning.

These five dimensions are closely interrelated as visualised by the various lines connecting them in the accompanying figures. A blockage or release in one dimension can either directly or indirectly affect the other dimensions. The model reflects the relatively equal importance of every dimension. Psycho-energetic therapy seeks to establish a dynamic balance between the different dimensions. Therapeutic intervention can start from any specific dimension.

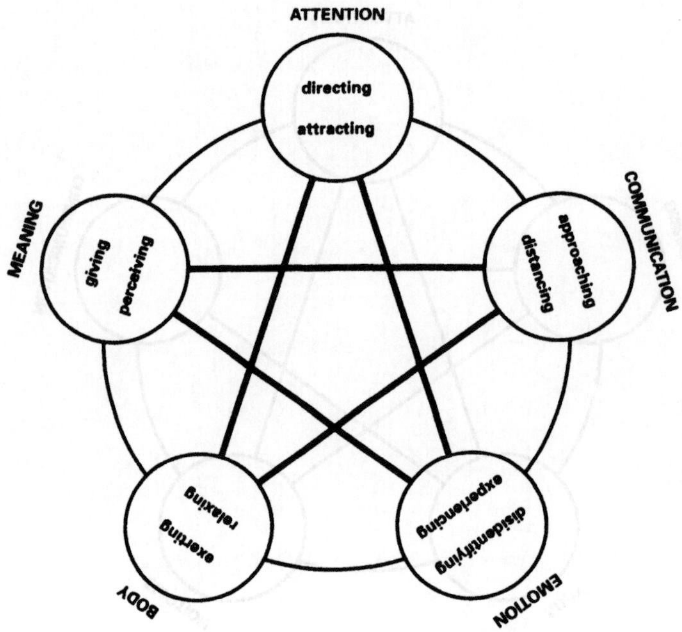

Figure 2. Psycho-energetic therapy: five interdependent dimensions and their polarities.

Healthy coping is viewed as a dynamic balance between these five dimensions within the energy system of the particular individual. The different dimensions work together harmoniously (i.e., support and control each other). There also needs to be a balance within each dimension. Both centrifugal and centripetal forces are at work within each dimension. Active, outward flowing forces are called centrifugal and passive, inward flowing forces are called centripetal. The polarities characteristic of each dimension are depicted in figure 2. A balance must always be found between the different polarities.

Attention can be attracted by a particular external object or internal experience. Attention can also be actively directed towards or away from an external object or internal experience. The effective balance between these two polarities is called mastery of attention.

Communication fluctuates between approaching and distancing. The balance is supportive exchange. The dimension emotions ranges from experiencing the emotions and to disidentifying one's self from them. The balance is the cognized feeling. The body can fluctuate between activity (exertion) and rest (relaxation), with eutonia as the balance. The dimension meaning can range from giving meaning in which the self is seen as separate from the nonself, and perceiving meaning in which the self is perceived as integrated into a larger system of life and ultimately the whole of life and death. The balance here is called realistic appraisal.

When a patient says that he cannot cope with his illness, we presume the balance both between and within the various dimensions to be disturbed. This results in psychological and spiritual distress with one of the polarities in a dimension taking a dominant role for an extended period of time. The threat of an illness (progression of the cancer, suffering, death), can attract attention with such power that one is totally fixated on or obsessed by the illness and one's ability to direct one's attention to some other external object or internal experience is reduced. The opposite imbalance develops when the thread of illness is so much feared that the person uses all his energy to avoid attention to the thread and direct his attention to everything but the illness. With regard to communication approaching can turn into a claim on others for company, listening, and support with the threat of illness as the only theme for communication. The opposite imbalance develops when distancing predominates and the patient withdraws from contact and doesn't communicate what he is going through. The diversity of emotion can also be reduced to anxiety, rage and/or sorrow. When the experiencing of emotions predominates, the patient may be overwhelmed by the negative emotions in particular. When disidentification with the emotions predominates, we may see heroic stoicism, apathy or depression. With regard to the body healthy exertion can turn into hyperactivity or relaxation can turn into its extreme: inertia. With regard to meaning, the threat to the autonomous self may predominate and the patient become the victim of intrusive thoughts about the evil things that can happen to him or extremely preoccupied with thoughts of how to stay in control. In such a case there is no place for the perception of meaning. When the integration of the self and the non-self or surrender fully predominate, self-denial can be the end result.

4. COPING WITH CANCER

Cancer is an active process within the body and brings major changes in the existence of the patient. Among these changes are:

- the consequences of the illness and its treatment including the symptoms, side effects, regular visits to the hospital, etc.
- a disruption of the normal course of life, either temporarily or permanently. Tasks and plans cannot be fulfilled.
- a possible threat to one's future (i.e., the progress of the illness with possible deterioration, pain and death).

The patient has to deal/cope with these new challenges, and this coping process is largely governed by the search for meaning. The normal sequence is initial bodily change, followed by coping. The bodily change isn't asked for; it is simply there. The patient has to cope with it whether he wants to or not, and it may therefore be experienced as forced upon him.

For effective coping, the patient must:

- come to accept the irreversible changes,
- get a grip on intrusive thoughts,
- find out which aspects of the situation he still has control over and use this control,
- re-evaluate his existence with the relativity of his existence as the point of reference, and
- keep an open mind for the positive aspects of his existence.

The coping process may not be effective when the individual:

- cannot accept the irreversible changes,
- feels helpless and hopeless,
- is overwhelmed by intrusive thoughts,
- clings to the premorbid situation, and
- only experiences the negative aspects of his existence.

The way in which a patient copes with his illness is strongly determined by his appraisal of the situation or the particular meaning that the situation has for him.

4.1. Meaning in General

The Jewish philosopher Martin Buber (2) states that man has a dual way of relating to himself and to the world. Buber uses pairs of words to refer to these ways of relating and speaks of the me-it relation and of the me-you relation. In the me-it relation, I experience myself as different from the perceived object or subject. I can give a description of them. In the me-you relation, I and the object or subject are together in the relation. As there is no distance, no description is possible.

In everyday life, we experience our self as an individual separate from the non-self. At the same time, we can experience ourselves as a fellowman or as a part of nature.

Based on the partly different functions of the cerebral hemi-spheres, Rossi (3) speaks of the logic mind and the metaphoric mind.

As psychologists, we prefer to speak of giving meaning as an act of the thinking or autonomous self and perceiving meaning as an act of the perceiving or integrated self. These selves let us know the world in different ways. Characteristic for giving meaning is the mental activity of thinking and thereby knowing the world in small logical steps. There is separateness. Thinking also offers us the possibility of travelling in time. We can think about the here and now, but we can also go back to the past or go forward to the future. The appraisal of the here and now is then colored by comparison with the past or the future. In the case of an illness where the past was better and the future is insecure, this can lead to the rejection of the here and now.

Characteristic for perceiving meaning is knowing the world directly. There is no separateness, only relatedness and integration. Objects and events are part of a pattern which itself is part of a larger pattern (4). The perceiving takes place in the here and now; it is unconditional (5). The here and now is perceived as it is, without judgement. What is, is and is neither good nor evil. This leads to a choiceless existence, with no clinging to just one aspect of reality.

For coping with cancer, both of these functions of the mind are necessary.

We can speak of giving meaning as an active mental process that strengthens autonomy and perceiving meaning as a passive mental process that strengthens integration.

In our Western world, there is a strong tendency towards secularisation and individualisation. This leads to an overemphasis on the autonomous self and a relative neglect of the integrated self.

4.2. Cancer and Meaning

Cancer implies a loss of control. For the autonomous self it is difficult to give a rational meaning to these irrational aspects of the cancer process. For the integrated self, however, the cancer can be perceived as a small aspect of a larger meaning context.

In an effective coping process, we see increasingly greater weight given to the integrated self and relatively less weight given to the autonomous self. By placing the personal illness in a larger context (6) the threat of the illness for the autonomous self can be relativized. This process of relativation can also result in spiritual development.

5. MIND, MEANING, AND MECHANISMS FOR COPING WITH CANCER

Coping with cancer means abandoning the old situation, accepting the new and often restricted situation, and deal with a future that is threatened. In most patients, we see two mechanisms used to do this. The first mechanism is changing one's perspective on time. The second mechanism is changing one's point of reference.

5.1. Changing One's Perspective on Time

The patient cannot control the future. If his attention is concentrated on the future for a longer period of time, the awareness of being out of control will be given a negative meaning. This, in turn, can elicit negative emotions that inhibit his functioning in the actual situation. This inhibition is why patients commonly tries to avoid thinking about the future and direct their attention to the here and now. This can be seen as a mild form of denial or dissociation, but in doing so, the threat of the future is reduced and more room is made for perceiving meaning. This mechanism can be used for coping with the illness in general and also for different aspects of treatment. With respect to the illness in general, the patient often helps himself, on the mental level, by thinking that the illness is just a temporary situation and that he will soon—after a possibly difficult period of treatment—be the old one again. This shift in time perspective also expresses itself in the form of fewer plans for the future and less long-range plans.

The treatment can be very difficult to endure. If the patient is simultaneously aware that the treatment may be without effect, he will simply not have the energy to deal with it and the illness. The patient coping effectively may therefore direct his attention to the treatment and postpone the confrontation with the possible consequences of such treatment until it is completed. On the mental level they may say: "let us first deal with the treatment and then we can see what else is coming".

This dispersion into time fragments can be seen in every therapy modality including surgery, radiotherapy and chemotherapy.

It is an effective way of coping, and keeps inhibiting emotions under control. A consequence, however, is that part or all of the emotional reaction may come later when the patient is mostly not prepared for it: the treatment is over, the health condition is good, friends say that he has reasons to be glad, and that is what the patient; nevertheless, he feels depressive and cries easily.

That the patient is able to block out the confrontation with threatening thoughts and the largely negative emotions corresponding to these thoughts does not mean that he is not aware of them. The actual thoughts and feelings are locked up somewhere but may nevertheless physically express themselves as tension, a constant state of arousal, or sleeping problems. When attention is directed towards the future, these thoughts and the corresponding emotions can be brought out. On special occasions, they may also flare up (for example, when the patient feels a bodily change, when the patient has to go to the hospital for a follow-up visit).

This shift in time perspective can be seen as an effective coping strategy. It is advisable therefore not to disrupt this strategy. Even if a patient is suffering as a result of such a strategy, it can be dangerous to break it. As an example, we mention the patient who cannot stand the confrontation with nearing death. Pain often signals such a situation. The pain is in the here and now. To avoid facing the thought of his impending death, a patient may completely fixate on the pain experience. A resistance to therapy may also occur when an attempt is made to manage the pain. For this patient then, pain relief may only bring greater suffering.

There is a group of patients that says they cannot live with the insecurity of their cancer. Their attention is continuously drawn to on the negative aspects of the future. They are convinced that something negative is going to happen to them, and they reject this possibility. Their emotional life is determined by negative feelings and the bodily expressions of these. Also in their communication, there is only space for the threatening.

Most patients can cope effectively with their cancer. In them you see flexibility. They can direct their attention to the threatening future, think about it, experience negative emotions with their bodily correlates, and talk about the threatening future. This situation doesn't last for these patients, however. They have the force to change the direction of their attention to the possibilities of the moment or something quite different. Their thinking and experiences also change accordingly. There can now be positive feelings in addition to negative feelings, and this variety also exists in their communication.

5.2. Changing One's Reference Point

As we already noted, many patients relativize the threat of illness by telling themselves that it is just temporary. Many of them however—about 50 percent—will discover that the illness will bring permanent changes and that they will never be the same old one again. When they start to realize this, the mechanism of changing their reference point and resigning themselves to the situation has started to operate. The patient leaves the premorbid phase behind as his point of reference to surrender himself to the new situation and starts to arrange his life with a new reference point. The patient starts to admit to his thinking self that certain changes are irreversible and that he may have to give up certain aspirations. This process may cause distress for the autonomous self, and the patient may, ask himself: "Why does this have to happen to me?" Letting go of the old situation elicits an emotional reaction and a process of mourning. The expression of these emotions is very important as they represent a further step in the direction of acceptance. Their expression also reduces their emotional loading (7). It will be gradually possible for the patient to think about the new situation with less emotion. Room for the perceiving self to view the situation in a neutral way and perceive its integratedness, will also be created. This new balance makes it possible for the thinking self to broaden its perspective. It can start with the realization that he is not alone, that other family members have had cancer, and that there are fellow patients in the hospital. For many patients, the surrender to the new situation is facilitated by the realization that the situation of these other patients may be worse. On a larger scale, the religious patient may broaden his perspective by viewing his illness as an inscrutable act of the deity; the humanist may perceive his illness as an act of nature. The acceptance is mentally completed when the patient no longer asks: "Why does this have to happen to me?" and can say: "Why should it happen to only others and not to me?" Once this acceptance has been done, the patient can arrange his life according to his new reality and no longer needs to fight it. This can provide a tremendous sense of relief. An example is the lung cancer patient who goes upstairs every evening to sleep in his bedroom. Every day it gets more and more difficult; by the time he has to stop on almost every step, it has become a torture. When this man accepts that he is better off sleeping downstairs, it will come as a relief. The little energy that he has left can now be used for communication and experienced as a positive side-effect of his decision.

At the beginning of an illness process, the idea of gradually having to give up everything, can be very threatening to a patient. At that point, the patient only sees what he has to give up. When actually in the illness/treatment situation, the patient may realize that it is not just giving up but also the relief of no longer having to fight, and the possible gains from cherishing his blessings more in the here and now. The new situation still has a meaning for the patient, thus. This implies that a patient can never tell what will be acceptable for him in a latter stage of illness and has consequences for the euthanasia wish. In our country, many patients, will—when they hear that they have a metastasis—ask their doctor if he is prepared to help them towards an end when they can no longer cope. The group that later actually asks for such help is very small. Similarly at the beginning of an illness it is not uncommon for patients to declare that life can have no meaning for them when they are bedridden or suffering severe pain. At that moment however, they have more options. When the options are reduced to being bedridden and having pain or being dead, however, most patients evaluate the first option to be preferable over the latter option.

With the progression of the illness, the patient goes through a series of reference shifts and mourning processes. Each shift is, in fact, a giving up (abandoning) of functions, possibilities, plans, etc. In this process, the patient grows towards an acceptance of death. As stated at the beginning of this paper, this growth is experienced by the patient as being forced upon him by the progression of the illness; nevertheless, he knows to give it a meaning and perceive meaning until the very end.

This is not possible for all patients. There is a small group of patients for whom the last phase of their illness is most devastating. They have been exhausted by the long course of their illness; their bodily complaints and discomforts demand all their attention. We refer to this as the endpoint of a centripetal process in which their is only the body. They have no more energy to open up to a broader perspective and the positive side-effects that such an opening can bring. There is only suffering. For these patients, neither their life nor there suffering has any meaning. They scream for death, which means an end to their suffering.

These observations suggest that man can endure a tremendous amount of suffering as long as he can give and perceive some meaning for it. To us, this means as long as he can place the suffering within a larger context and in some way still derive some positive experiences from his life.

I have been told that there are patients capable of undergoing such devastation in a very serene way, that is, they are still able to give and perceive meaning.

REFERENCES

1. Vries M.J.de Choosing Life. Swets and Zeitlinger Lisse 1993.
2. Buber M. Das dialogische Prinzip (The dialogical principle). Verlag Lambert Schneider Heidelberg 1965.
3. Rossi E.L. The psychobiology of mind-body healing. W.W.Norton & Company, Inc. New York 1986.
4. Leshan L. The medium, the mystic and the physicist. Ballantine Books New York 1976.
5. Levine S. Healing into life and death. Anchor Books New York 1987
6. Frankl V.E. Man's search for meaning. Pocket Books New York 1971.
7. Greenberg L.S., Rice L.N., and Elliot R. Facilitating emotional change. The Guilford Press New York 1993.

AN INVESTIGATION INTO THE EFFECTS OF A COUNSELLING AND GUIDED IMAGERY PROGRAMME ON THE OUTCOME OF PEOPLE WITH STABLE ANGINA AND FOLLOWING MYOCARDIAL INFARCTION

Carl Stonier

Pontefract General Infirmary
Friarwood Lane
Pontefract, West Yorkshire, WF8 1PL, United Kingdom

ABSTRACT

For valid, albeit pragmatic, reasons, orthodox medicine is founded on the erroneous Cartesian philosophy that the mind and the body are separate and therefore, that what happens with one of the two has no effect on the other. Whilst the self-evident fallacy of this standpoint is now being recognised in some areas, there has traditionally been much resistance to the notion that one can alter disease processes through psychological means, despite the work that has been done in areas of chronic and/or life threatening diseases such as cancer. Indeed, this work is frequently dismissed as anecdotal or methodologically unsound, rather than acknowledging that there is an effect, and seeking to develop and investigate a hypothesis to explain the effect.

This study is an attempt to redress this imbalance by investigating the hypothesis that there is an intimate relationship between psychological health and physical health, and that psychological factors are significant in the aetiology of (in this case) ischaemic heart disease (IHD) and that the disease process can be arrested and possibly reversed by addressing those psychological factors.

The method chosen to investigate this hypothesis is a randomised, controlled study of people with a diagnosis of IHD. Eligible people are randomised into the control or experimental sub-group appropriate to their diagnosis (angina or MI). The experimental group receive a series of counselling sessions, usually about 10 although the number is not critical, and a specifically written and recorded audio-cassette tape to guide them through a visualisation of the heart damage being repaired. The style of counselling is humanistic, essentially person centred, and the focus of the counselling is to identify the stressors that are relevant for that individual.

Potentiating Health and the Crisis of the Immune System
edited by Mizrahi *et al.* Plenum Press, New York, 1997

Evaluation of effect is through the scores on 2 questionnaires which are administered before randomisation, after 3 months (or 2 weeks after completion of counselling for the experimental group), after 6 months and after 12 months. Records are also kept of hospital admissions for cardiac related conditions, referral rates for coronary artery bypass graft surgery, medication usage and patient self-report.

The study is currently approximately half way through, so full results are not yet available, but early indications are that a definite improvement in clinical status occurs as a result of the experimental programme, with subjects experiencing a marked improvement in all parameters eg symptoms, wellbeing, medication useage. There also appears to be a trend emerging with regard to the broad life experiences of IHD subjects (as distinct from personality.

INTRODUCTION

This paper is essentially the story so far of my research into Ischaemic Heart Disease (IHD) and into whether the disease process may be reversed by addressing psychological factors. I will present briefly the background to the study, the formal hypothesis, the methodology and a sample of the evaluation. Whilst the study employs rigorous (hopefully) scientific methodology to investigate the hypothesis, the real message is about the value of taking a whole person approach to health care; of focussing less on the science and technology of symptoms, physical manifestations and chemical or surgical manipulations and more on the deeper, human aspects of what is happening for that person. This is the Art of Medicine as opposed to the Science of Medicine and both are needed.

As Voltaire said, "The Art of Medicine lies in keeping the patient amused whilst Nature effects a cure".

BACKGROUND

Orthodox medicine is founded on a philosophical error! The Medieval Church would not allow further explorations of how the body functioned to be made by the medical scientists of the day for fear of offending God who had, according to the Scriptures, created Man in His own image. The philosopher Rene DesCartes saved the day by producing proof, through philosophical argument in 'Meditations on First Philosophy, in which the Existence of God and the Real Distinction between Mind and Body are Demonstrated' (1641), that the God-image part of man is the mind, that the body is nothing more than a mechanical vehicle to contain and animate the God-image part, and that therefore God would not be offended if this mechanical part was dissected and explored. This argument was accepted by the Church.

Pragmatically, whilst this philosophy was necessary before further progress could be made, its influence was such that it has underpinned medical education since that time, despite the evident fallacy of the assertion that the mind and the body are separate and that neither has an impact on the other.

On the other hand, so called traditional medicine is based upon the philosophy that man is a whole being formed from the integration of the different parts of mind, body, soul and spirit. As such, all parts are intimately connected, and will thus have an effect on each other, rather like a hanging mobile decoration, all is in balance and in a constant state, but if one part is altered, the whole is affected and moves out of balance.

The winds of change are now blowing more strongly, and there is an increasing number of health care practitioners, of all disciplines, who acknowledge the self-evident strengths of orthodox medicine, but recognise the need for, and the validity of, the holistic approach.

One of the forefathers of this change was Hans Selye (1) who developed the General Adaptation Syndrome. The General Adaptation Syndrome shows the relationship between the mind and the body, and clearly illustrates the effect on the body of stress.

As will be seen from figure 1, stress has a direct effect on the immune system, the gastro-intestinal system, the endocrine system and the cardio-vascular system, and Selye's pioneering work has since been developed by many other workers who have further clari-

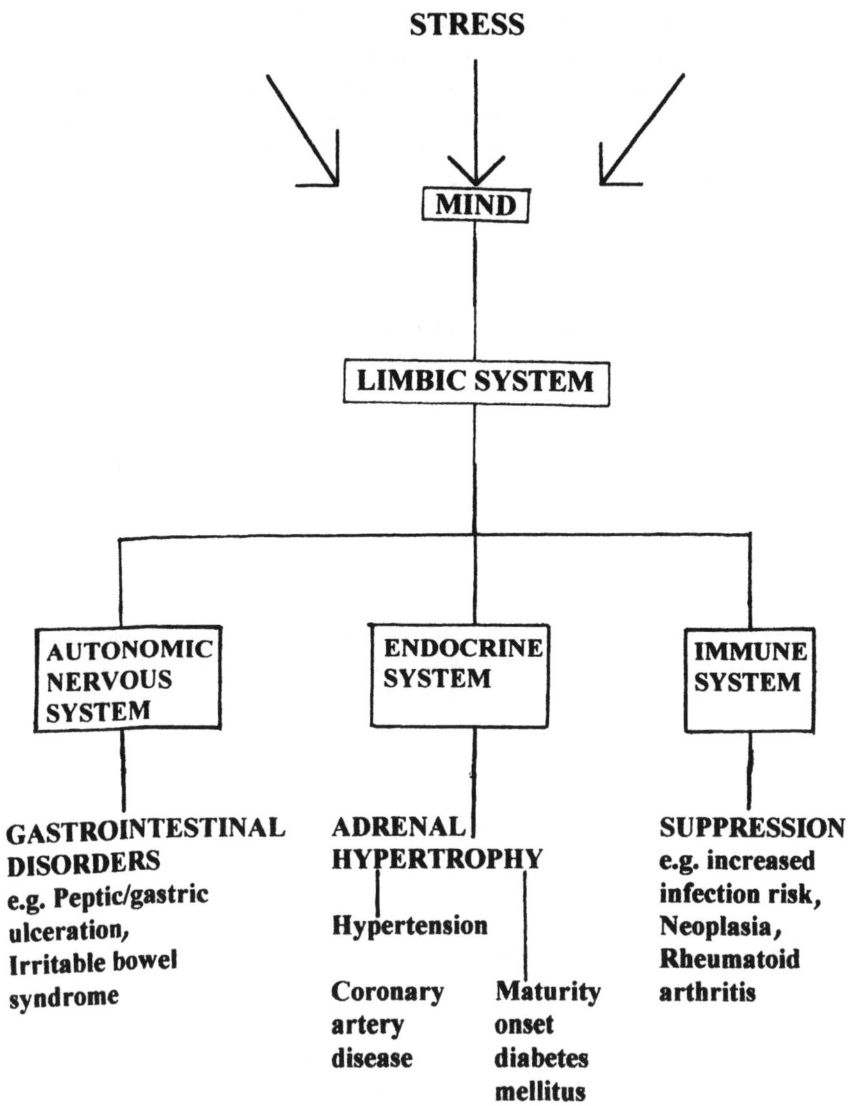

Figure 1. Simplified representation of Selye's General Adaptation Syndrome, showing the effects of 'stress' on different parts of the body.

fied the links between the mind and the body and who have further strengthened the evidence supporting these links.

As the paradigm changed some 350 years ago from the whole person view of traditional medicine to the reductionist view of the Cartesian duality, so is it now starting to change again. Indeed, whilst the heritage of the traditional wisdom is shown in the statement of Paracelsus "As Man imagines himself to be, so shall he be, and he is that which he imagines", the scientific pedigree is shown in the statement made by Einstein "Whatever we can visualise, we can actualise", showing that one of the finest scientific minds of all time recognised the profound effects that can be achieved by using visualisation and the power that can be harnessed by learning to exercise control over the mind.

Since Selye's work in the 1950's, many people have recognised the importance of attending to the psychological aspects of ill health, but few have actively used a psychological approach as a discrete therapeutic modality in it's own right, and those that have have either been dismissed as charlatans or have been condemned for producing methodologically unsound research. Yet if there is a mind-body link, and what happens with the mind affects the body and vice-versa, then correcting the psychological aspects that are causing distress should create the optimum internal milieu for the healing of the physical disease. This then is the hypothesis for this study.

HYPOTHESIS

That psychological factors are significant in the aetiology of Ischaemic Heart Disease and that the disease process can be arrested and possibly reversed by addressing those psychological factors.

THE STUDY

The method designed to investigate this hypothesis is a randomised, controlled study of people with diagnosed ischaemic heart disease.

The study design calls for a cohort of 200 eligible people, eligibility being determined by a diagnosis of a) stable angina, with or without previous myocardial infarction or b) recent myocardial infarction. People with IHD but who are over the age of 70, have concurrent life threatening disease or current psychiatric disorder are excluded.

The study size was determined following statistical advice based on the results of an uncontrolled pilot study that I made in 1992 with 10 subjects.

All recruitment to the study is from the Cardiology service of Pontefract General Infirmary, UK., with eligible people being invited to join the study following an explanation both verbally and in writing of the study demands.

Those agreeing to participate then complete two baseline questionnaires – the 60 item General Health Questionnaire (GHQ60), which is well documented for it's validity in the field of cardiology, and a 10 item Visual Analogue Scale (VAS) of my own design, the validity of which was tested during the pilot study.

Following completion of the questionnaires, participants are randomised into either the experimental group or the control group, with randomisation being effected by the drawing of a numbered ticket. Odd numbers are allocated to the control group and even numbers to the experimental group.

In addition to the standard medical care, as determined by the Consultant in Cardiology, people in the experimental group receive a series of counselling sessions, ideally

Please mark with a cross the position on the line that you feel most accurately represents you now for each of the following statements.

e.g. I feel very well.

Not at all ____ As well as I can imagine being.

1. I feel very well.

Not at all ____ As well as I can imagine being.

2. I feel short of breath.

All the time ____ Not at all.

3. I experience chest pains

Very frequently ____ Never.

4. I am as active as I have always been.

Very inactive ____ As much as ever.

5. I can tolerate as much effort as I ever could.

Not at all ____ As much as ever.

6. I feel overworked.

All the time ____ Not at all.

Figure 2. The visual analogue scale.

7. **I feel very fit.**

Not at all ______________________________ As fit as I can imagine being.

8. **I feel able to cope.**

Not at all ______________________________ All the time.

9. **I feel tense.**

All the time ______________________________ Not at all.

10. **I easily get tired/fatigued.**

Very easily ______________________________ Not at all.

Thank you for completing these questions.

Figure 2. *(Continued)*

once per week, though this may vary according to either the patient's or the counsellor's constraints. Each session lasts for a 'therapeutic' hour i.e. between 50 and 70 minutes. The number of sessions is nominally 10, but this may vary according to the needs/problems of the patient. Currently the range is 3 to 22 sessions. This variable is of no importance to the validity of the study, since if there is an effect attributable to attending to psycho-emotional issues, the length of time taken is irrelevant. Participants also receive an audiotape which contains a specifically written and recorded guided imagery meditation. This takes them firstly to an 'Ideal Place', a place where they can be just as they wish to be, then to an inner control room, from where they can control any of their bodily functions or processes then finally to a visualisation (mental picture) of the damaged heart muscle and blocked arteries responding to the measures taken in the control room and being healed and restored to optimal functioning. (See appendix for full script). The duration of the tape is 20 minutes and participants are asked to use it at least once per day and preferably three times per day, since previous studies indicate that this is the optimal frequency (personal communication from Stephanie Simonton).

The counselling is of an integrative Humanistic style which is heavily influenced by Carl Rogers' Person Centred Therapy. The participant is initially invited to tell his or her own story in as much detail as he or she can manage, with the counsellor being alert to any cues which may indicate areas of dissatisfaction or conflict in life. These are then explored at the time and in subsequent sessions.

Two weeks after completion of the counselling the GHQ60 and the VAS are re-administered and scored, at which time the initial questionnaires are also scored. Feedback on the results is then given to the participant and comments invited. This administration of the questionnaires is done in person.

The questionnaires are then repeated after a further three months and again six months after that, on these two occasions, being sent through the post with a prepaid return envelope. A completed set of questionnaires for each participant thus comprises four pairs - the initial ones, three months, six months and twelve months. Records are also kept of cardiac related hospital admissions, referrals for coronary artery bypass graft surgery, medication usage and patient self-report.

Control group participants receive the standard medical care as determined by the Consultant in Cardiology, and receive further GHQ60 and VAS questionnaires at three, six and twelve months from completion of the initial questionnaires. With the twelve month questionnaire is included an invitation to then join the experimental group. This was included for ethical reasons so that randomisation into the control group would not result in exclusion from what may be a potentially beneficial intervention, simply a delay.

The questionnaires measure a number of different aspects of illness compared to wellness, namely:

GHQ60	Somatisation of affect	Questions 1 - 10
	Anxiety	Questions 11- 20
	Social Dysfunction	Questions 21- 42
	Severe Depression	Questions 43- 60
VAS	Quality of Life	Questions 1,4,5,7,10
	Physical Symptoms	Questions 2,3
	Perception of being under Stress	Questions 6,8,9

The study is currently approximately half way through, so results are not yet available, and I have deliberately avoided looking at all the data so far collected to prevent possible contamination of my thinking or approach. Having said that, a quick glance at some of the questionnaires indicates that the results concur with the hypothesis, but this data is evidence more of a well-being effect rather than necessarily a definite clinical effect, which would be evidenced, for example, by a difference in mortality rates between the experimental and control groups. In other words, the questionnaire data is somewhat 'soft', and the 'harder' data has not yet been collected, and therefore for both this reason and to avoid the accusation of a lack of methodological robustness (such as I have levelled at some other studies) I have chosen not to publish interim results at this time, but to await the completion of the study and to then publish the full results and data.

However, it is relevant to publish the pilot study data, since this phase of the study is complete, and it was statistical advice based on the pilot study which determined the cohort size of the main study.

Analysis of the data collected during the pilot study indicated that a GHQ60 score differential of four points would be statistically significant using the cohort numbers of this study, (Table 1).

Finally, there are two broad themes emerging with regard to the broad life experiences of the experimental group subjects (I can comment only on the experimental group subjects, since these are the only ones to have undergone the in depth counselling). Everyone with whom I have thus far worked has experienced in some way either a) poor parenting, e.g. from the death of a parent, from physically and/or emotionally abusive parents

Table 1. GHQ60 averaged percentage scores for finishers and non finishers from the pilot study

Finishers	Somatisation Average % Pre	Somatisation Average % Post	Anxiety Average % Pre	Anxiety Average % Post	Social Dysfunction Average % Pre	Social Dysfunction Average % Post	Depression Average % Pre	Depression Average % Post
↑	50	7.5	60	15	42	4	31	3
↓	5	20	30	60	39	52	28	39
All	35	13	50	30	41	20	30	15
Non-finishers	37.5	–	50	–	45	–	49	–

and even from such things as being evacuated to a place of safety during wartime at a critical time in the development of the sense of self and esteem or b) a sense of being on the wrong life path i.e. a life's work in the wrong job, a life spent with the wrong partner, a life spent with the early dreams, aspirations and/or perceived potentials remaining unfulfilled.

All participants thus far interviewed can be categorised into one or the other of these categories, and some fall into both. Whilst this is too prevalent not to be significant, it does raise the question of whether there are people without ischaemic heart disease who also fall into one or other of these categories, and if so, what is it that protects them from IHD, or conversely, what is it that leaves the IHD group without protection. It is also possible that subjects will be encountered who do not fall into either of these categories, which will then beg the question of what factor(s) have compromised their hardiness or IHD 'immunity'.

SUMMARY

This presentation has briefly highlighted the philosophical background to the primacy of Mind–Body dualism in orthodox Western medical thinking. It has postulated a hypothesis that this Mind–Body duality is in fact non-existent, and that the Mind and the Body are in reality closely and intimately connected, such that each has an impact on the other, and that, therefore, addressing long term psychological conflicts will have a beneficial effect on the disease process. A study is described which has been designed to investigate this hypothesis, and some interim results are presented. These results appear to support the hypothesis.

ACKNOWLEDGMENTS

I would like to acknowledge the support and assistance of Dr RV Lewis, Consultant in Cardiology, Pontefract General Infirmary, and the support of the Pontefract Hospitals NHS Trust.

REFERENCES

1. Selye H (1936) A Syndrome Produced by Diverse Noxious Agents. Cited in Selye H., (1956) *The Stress of Life*. New York, McGraw Hill.

APPENDIX

The Script for the Guided Imagery Tape

I want you now to take a deep breath in and hold it for as long as is comfortable, and then gently breathe out. Continue to breathe in this manner, and as you breathe in you can feel the relaxation start to spread through your body, starting with the soles of your feet. As you breathe out you can feel all tension leaving your body, to be replaced by the next wave of relaxation as you breathe in.

The relaxation continues to spread up your legs; up your back to your shoulders; down your arms right to the tips of your fingers and finally up through your neck, over your scalp and down through your face. If your eyes are still open, you may let them gently close now, and if there is any tension left in your jaw, allow it to leave, so that your mouth is loose and slightly open.

You are now deeply relaxed, perhaps more relaxed than you have ever been before. Already, you are starting to feel better, and I want to enhance that feeling now by taking you on a journey—a journey to your Ideal Place—a place where you can be just as you want to be.

This place may be somewhere that you know—perhaps from a holiday, or it may be entirely from within your imagination. It may be on a mountainside, in a forest, on a beach, by a lake or in a garden—anywhere at all where you feel at peace with yourself and with the world.

Wherever this place is, it is your Ideal Place; you feel perfectly safe and secure in this place; you can be just as you want to be, and only you can go to this place, unless you choose to invite someone else.

Get a mental picture of this place now, as strong and as clear a picture as you can make it, and if you choose, allow yourself to go there. Experience this place in its totality. Kick your shoes off and feel the texture of the floor in this place. Smell the smells and hear the sounds. Feel the warmth of the sun's rays as they touch your body. Taste the purity of the air as you breathe in. Look around and see the sights and the landmarks of this place.

Recognise that you are feeling absolutely wonderful. Any aches and pains that you may have had have disappeared. Any ill health or disease is cured; any anxieties, fears or worries cease to exist. you are in a perfect state of physical, mental and spiritual health and I want you now to explore this place. Walk or run around. Get to know it intimately so that you will be able to recognise any part of this place anfd return to it at any time in the future. Continue now to explore your Ideal Place, whilst I remain quiet and the music continues to play for a few minutes.

As You explore your Ideal Place, you come upon an entrance. This can be any kind of entrance—a door, a cave, a hollow tree, a waterfall—how it appears is irrelevant. The important thing is that this is the doorway into your mind.

Choose now to go through that entrance and into the corridor beyond. There you will find many doors leading off the corridor, and I want you now to find the door marked 'Control Room'.

When you find this door, open it and go through into the room beyond. Inside this room, you will find row upon row of taps. Each of these taps controls a function of your body, and since you are in total control in this reality, you are able to control any of these functions that you wish. Each of these taps is labelled, so choose now to look around until you find a tap marked 'heart'. Know with every fibre of your being that you are in control, and that the tap marked 'heart' controls the blood supply to your heart. Choose now to turn on the tap marked 'heart'. Turn this tap on as tightly as you can. Look around and find a spanner or some other tool and use it to make certain that the tap remains open. As you turn the tap on, you are aware that your heart is receiving all the blood that it needs to work effectively. The arteries are becoming unblocked and able to supply all the blood that your heart needs. Know with every part of your being that this is so. Know that you have total control.

Now that you have turned on the tap marked 'heart', you can deal with any other problems you may have. If you have pain, you can turn off the tap marked 'pain'. Notice that as the tap is turned off, the pain lessens until it disappears altogether.

If you are taking any medication, you can turn on the tap marked 'medication', knowing that as you do so, the medication is directed to the place where it is needed, so that only the desired effects are achieved. Any side effects are eliminated, and your medication becomes wonderfully effective.

Any other health problems that you may have can be dealt with similarly. You have total control over your body and its workings and you know this with very cell of your body, every aspect of your mind, every fibre of your being.

Now that you have adjusted all the control taps that you wish to adjust, take a look up on the wall of the control room. There you will see a monitor screen that you can focus on any part of your body. Focus the monitor on your heart, and you will be able to view the effect that you have had by adjusting the control taps.

As you look at the monitor, you are aware that the screen is dissolving, so that you are now looking directly at your heart. You can clearly see the arteries becoming unblocked. You can see the heart working efficiently. Notice how this appears to you.

Simply allow the image to form, regardless of how outrageous it might be. Censorship is unnecessary in this place. Simply allow the image and its symbolism to be and to exist without hindrance.

You can see the nerves calming down as any pain subsides; you can see any medication that you are taking flowing in and helping to create the best internal environment possible so that your body can heal itself. As clearly as you can, experience all of this taking place. Experience the arteries in your heart becoming unblocked. See it happening, however that may be, and know that it is happening because of your directions, because you are in control. Experience any pain decreasing, then disappearing. Experience any other problems resolving, and know that your experience is valid. You are in control of all of these processes, and you are now exercising that control to enable your body to become totally healthy.

As you experience these events, you reinforce the knowledge that you have control, and with each repeat of the experience, this knowledge becomes more deeply entrenched into your being, so that every cell, every atom, every burst of energy, every fibre of your being knows that your reality is now one where ill health and disease have left. You know this to be so with a knowledge that transcends simple belief.

Continue to experience this now, for a few minutes, whilst the music plays in the background and I stop talking.

It is almost time to return your focus to the control room, but before you do so, give directions to ensure that the unblocking of the arteries in your heart continues, even whilst your attention is elsewhere. Gently come back through the monitor screen and return your awareness to the control room.

Check that all the taps are still in the position that you want them to be in, then leave the control room. Walk back along the corridor to the exit door, and back into your Ideal Place. Know that you can return to this place at any time, and then, when you are ready, gently come back into this room; becoming aware of the surface beneath you; becoming aware of any background noise; then, as you are ready, coming back fully into this room, eyes open and feeling awake, alert and bursting with health.

18

REALITY AND MYTHS ABOUT BREAST CANCER

A Personal Journey

Galia Sefchovich

CADIR
Creatividad y Desarrollo Integral
Socrates 128–5 Col.Polanco C.P. 11560, México, D.F.
Tel.: (525) 3958839; Fax: (525) 3825729

First of all I should like to thank Dr. Avshalom Mizrahi, and THE DEAD SEA CONFERENCES organizing committee for the opportunity of participating in this forum of specialists on health, medicine and alternative medicine. The paper I am going to present is not the result of theoretical knowledge in any of these fields; my need to participate stems from personal experience in the struggle against cancer.

In 1990, I was diagnosed with a cancerous breast tumor (I don't wish to use the "malignant" word as this has unhealthy symbolic connotations). This event changed my life at a profound level, altering the way I saw, felt, thought and related to myself and others. These changes may not be apparent at first sight, since I am still the same person I used to be, with the same tastes and needs, the same passion for my family, my friends and my work. However, the experience of living close to death for a long time enhanced my levels of sensitivity and perception.

I am not interested in discussing my case from a medical point of view (I have been treated by the best physicians in my country, Mexico, and also at the M.D. Anderssons Hospital in Houston, Texas, one of the finest hospitals in the world for cancer treatment.)

I have also resorted to various forms of alternative medicine and healing, and been looked after and guided by extremely ethical and professional experts. I should add that I am now in excellent health and that I am able to lead a normal life.

Why then, do I feel a deep need to share part of this experience at this particular forum? Why have I devoted my time, energy and my dear friend Luna's time to reviving this event from the past?

I think it is because during the battles that one has to fight in the war against illness, some things are hidden between the cold, clinical lines of the medical files and records of curers, healers and therapists, which are either too empirical or else non-existent, and I think that these findings may prove useful to other people.

Potentiating Health and the Crisis of the Immune System
edited by Mizrahi *et al.* Plenum Press, New York, 1997

Some of these people are here today and prepared to listen so that they may reflect once again and continue to advance, as a human group, in the search for new solutions to old problems.

After I discovered that I had a tumor, and went through the initial stages of anger against God, myself and the world, I began to focus my attention on the search for those who had managed to conquer illness and restore their health, often in spite of doctors' prognostications.

I kept asking myself why I was in that situation?, what it was I had to learn?, what others had to learn through me?, and how it had all begun?

These questions led me to make concrete decisions which were easy to turn into conscious, deliberate actions which made me feel I was in control of the situation and my body, instead of continuing in the initial state of anger and depression in which I perceived myself as a powerless victim.

I began to be fully aware that my body was mine and that everything that happened to it was profoundly linked to my conscious or unconscious decisions, so I decided to live and run the risks that this decision implied.

It may seem paradoxical that I posited the fact of living as a great risk, but the fact is that in the face of an illness that threatens one's very existence, there are only two choices available; to become paralyzed with the fear of the pain and struggle involved so in this case you surrender and can let your self die, or to become paralyzed with the fear of the unknown, i.e. death, and decide to fight and hang on to life. Both positions require a great deal of courage, both are equally valid, and in my view, should be fully respected. In most cases, these decisions are buried deep in the patient's unconscious, and may not be acknowledged as such without the help of appropriate therapy.

It is also obvious that if these decisions are made conscious, then the struggle for life or calm in the face of death becomes easier.

Indeed, it is possible to assume that the origin of illness is a complex combination of genetic, physical and biological factors, and also the unconscious somatization rooted in the origins of the life history of each patient or perhaps even earlier.

In my case, I decided to live and focus my energies on achieving this.

My first concrete actions were expressed in the way I related to the doctors. I decided not to ask any questions about life statistics or life prognoses, which may be necessary for science and research but only caused me to have horrific, useless fantasies that drained the little energy I had, because I am not a scientist.

I felt that most doctors acted extremely arrogantly. They felt they had the power to decide whether their patients would live or whether they should issue them a death sentence.

In our Western culture, we have given scientific knowledge too much power. It is not that the former does not warrant this, but we should separate scientific truth from our "deification" of the bearers of this truth.

In the situation of fear and vulnerability in which people who have recently been diagnosed with life-threatening diseases find themselves, doctors' words are not only heard, but become self-fulfilling. Consequently, I had to make a conscious effort not to accept medical opinions as absolute truths and learn to view doctors as highly competent, extremely professional persons who deserved my trust but were only human, fallible counselors.

During my stays at the hospital, I heard accounts of patients' relatives who responded like automata to the dictates of this occasionally overbearing knowledge, as they unconsciously programmed themselves to respond to doctors' expectations and bear out research figures instead of living like autonomous people in control of their bodies and capable of influencing their destiny.

I now know that feeling oneself and trusting in one's body facilitates a patient's recovery, because it prevents depression, but in today's medical system, the patient is virtually alone and it is difficult for him to feel like an integrated person. Much of his energy is dispersed and wasted in vain, whether in reading and signing documents related to the struggles of doctors fearful of lawsuits and lawyers who demand that all risks be "clarified" to all patient without exception before any form of treatment is begun, without taking into account the amount of stress this produces in the patient who begins to perceive and fantasize about dangers or risks that often do not warrant being explained in such detail.

And even in cases where illness for whatever reason proves incurable, the process of death is experienced with greater acceptance and dignity if the person does not become a number or, even worse, an object of experimentation, divided into bits of body which are handed over to each specialist and sub-specialist, causing her to enter a state of unbearable confusion and ambiguity, and lose her sense of unity and autonomy.

I believe that each person's life span is a sort of conscious or unconscious "secret" or "private agreement" between God and the person's free will.

I have found it difficult to meet people who have recovered and are prepared to discuss their experience. It is as though the magic or miracle of curing or healing might be destroyed if it is shared.

Most of the accounts with which I was familiar are recorded in the now-classical books and works of Dr. Berny Siegel, Dr. Deepak Chopra, Simonton, Louise Hay, and Norman Cousins, among others. However, I felt a great need to see, listen to and touch someone who could tell me about their experience with "spontaneous remission" or something similar. This curiosity led me to talk to many of the volunteers who offer their services at the Hospital. The flotilla of volunteers form a buffer between the doctors and patients. It is always possible to talk about many issues with them, in a discreet, unofficial way.

Talking to the volunteers is a sort of "open secret". They have all had cancer or accompanied a loved one in this situation. This is a condition for being able to work as a volunteer at the Hospital.

In my talks with them, I was able to prove my hypothesis that being diagnosed with cancer is not a death sentence. There are many types of solution and indeed it is worth making personal combinations regarding the use of scientific as well as alternative contributions. Delving into these two worlds may be a fascinating experience but one which also sows confusion. Where does one begin to look? Who should one pay attention to? What combines with what? My need to talk to people who had recovered also led me to find a Jungian therapist who had suffered cancer of the womb over 20 years earlier. She had received treatment at the same hospital where I was, and been "given" six months to live. I decided to be in individual therapy with her as long as necessary for two reasons. The first was that I had never felt comfortable in the support groups organized by the hospitals. I found it extremely difficult not to get involved in other patients' stories, so that instead of helping me, they made me feel sad or I started having unnecessary symptoms. Besides, one of the strongest reasons for having individual therapy was the possibility of a weekly visit to my therapist, and seeing her so strong and healthy twenty years after she had been diagnosed with cancer.

During this time, I was able to combine my therapy with the work I had done at the drama workshops with my friend Giselle Barret.*

---

* Gisélle Barret. "Myth of the Amazons Workshop" given at Cadir, Mexico. (personal notes) October 1995.

Working through dreams and the interpretation of the symbols and archetypes hidden in these dreams enabled me to rediscover myself through the myth of the Amazons. By this means, and with the help of visualization techniques and guided fantasies, I understood mastectomy as a powerful ritual which made me live like a warrior prepared to fight.

It is necessary to understand that illnesses do not happen on their own, nor are they a divine punishment. They are part of each person's life and process of growth.

Cancer in particular tends to be an excessively "personalized" illness. In other words, each tumor is slightly different from the next. A tumor is like a finger-print which has only one owner and therefore, only the owner can find the key to his cure or healing by using the knowledge of doctors, healers and therapists, among others.

For this reason, in my view the patient should ask himself how he has contributed to disabling his immune system and the formation of a tumor.

Generally speaking, two or three years before the diagnosis, critical or difficult events have occurred in the lives of most patients, events associated with losses such as divorce, grief, and unemployment, among others. Even in child patients, significant stress levels or depressive states are recorded prior to diagnosis.

If we examine where and how this process began, this will give us clues as to how to reverse it. Indeed, it is as a result of this constant exercise of SELF-OBSERVATION that we are able to decipher which places, people, food, music, colors, plants, etc. give us energy and which make use feel strong, happy, self-confident, and vital. We can also OBSERVE the opposite, i.e. which situations, people, food, etc. weaken us, or make us sad, exhausted or tired. It is on the basis of this exercise that we can determine what we need to find the way to a cure and health and thus choose our personal route. Focusing on the positive things that happen to us is the most effective means of preventing depression. It is like finding a wave-length or vibrating at a frequency that helps us to produce endomorphines and other substances that fortify us naturally.

We don't seem to be in control of our genes or our destiny, but if we examine the obscure part of our psyche through our behavior, we will find the start of the somatizations and processes of illness that are sometimes produced in childhood, to shed light on this dark part and make ourselves conscious of these events.

This process of exploring the dark part so that we can transform it deliberately and consciously and thereby make decisions and carry them out in the form of concrete actions is what we call healing.

Healing our inner lives may sometimes result in the healing of our bodies.

In my journey through the healers' surgeries, I would have liked to have been given more specific guidance regarding SELF-KNOWLEDGE through the use of the techniques proposed by each of the trends of thought.

In most cases, what I encountered was an area of implicit struggle and competition to prove that such and such a technique or substance was best and also a sort of non-declared but ongoing war between traditional medicine and alternative resources. I found healers who would help me if I gave up chemotherapy and doctors who were prepared to help if I gave up Ginseng or Shark tablets or whatever.

Finally, after traveling along this tiring VIA DOLOROSA in my search, I finally came across healers who work with doctors AS A TEAM and vice versa. With their help, I began to achieve my aim.

I now wonder whether a degree of understanding and integration between both positions might not be possible, which would help the patient to choose without having to give up in the face of doubt. Isn't it enough to be ill? Why does one have to feel parceled out between doctors, healers, lawyers and therapists who have already proved successful in every case but our own? Why are we only accepted as patients if we make an absurd pledge to accept only the panaceas that each has to offer? Why should we doubt whatever path is suggested?

I honestly believe that the solution to each illness lies in the hands of the patient, the affection he receives from his loved ones and the team of people who counsel and accompany him in the search to recover the balance and harmony involved in healing and health.

I think that the work of counselors, whether doctors or otherwise, lies in discovering this important degree of ethics, harmony and balance in themselves to be able to transmit it and accompany their patients with the best disposition, while being aware of the limitations of human beings, without creating doubt, guilt or contradictions in the patient, or, worse still, unrealistic expectations.

When I talk about harmony for both the patient and those who accompany him, I mean the harmonic level of the body's cells as well as each person' subtle levels of emotional and spiritual life. I am referring to the cultivation of a significant degree of Humble WILLINGNESS TO LEARN, the acceptance of the paradigm of science and also the unconditional acceptance of the mystic paradigm, since any scientific truth has its origin in mysticism from the moment the researcher posits a hypothesis or a question that does not arise from knowledge but from what is understood through intuition, perception and other sources. It is no doubt because of this that the TALMUD says that when a person asks a question, he already knows half the answer.

Accepting scientific truth as the ultimate level of truth is only possible if we have accepted the origin of this ultimate truth, in other words "OTHER FORCES OF THE UNIVERSE" as proposed by the new QUANTUM PHYSICS.

"A form of Intelligence exists that is superior to our own. This intelligence is at least partly responsible for what happens in the Universe. It is possible to communicate with this intelligence and the more ethical we are, the easier communication becomes."†

This continuous process of communication between ourselves and the information-filled QUANTUM UNIVERSE in which we live is acknowledged under different names, Baraka, Prana, Nirvana, CHI, Shejina, while some of us call it GOD.

I should like to end my participation by expressing once again my thanks to the organizers of events such as this, who permit communication and the unification of efforts, transcending all political and social borders, and distances. I should also like to invite you to take part in continuous, sustained team-work, work performed in unity by us all.

Illness is expressed as an interruption in the flow of love's energy which everyone needs to live. Keeping in constant touch with this energy is the best means of preventing or curing illness.

We can find love's energy in our degree of SELF-ESTEEM, our partners and children, our relatives, friends and fellow-workers, and our pets. We also subtly receive it from the Quantum Universe, and it is expressed through Nature and other languages such as music, painting and movement. Whether or not we keep in touch with this inexhaustible source is our choice. The Quantum Universe exists and it is up to people to make an effort to maintain this contact.

Contact is achieved through faith in ourselves, others and God.

Faith may be a permanent state of communication with these forces, or it may be a permanent state of availability to enable these forces to flow and function in and with us.

---

† Johon David García. "Creative Transformation." Noetic Press Whitmore Publishing Co. U.S.A.1991.

It may also be a clear, precise INTENTION on which we focus all our energy. Faith may be a permanent state of gratitude, or it may be all this and more.

We need to help each other to restore this energy of Love, which, in my view, is the path that must be taken.

Just as the problem involves everyone, so the solution is also to be found in everyone.

METHIONINE ENKEPHALIN

A New Cytokine with Antiviral and Antitumor Activities

Nicholas P. Plotnikoff

College of Pharmacy and College of Medicine
University of Illinois at Chicago
Chicago, Illinois 60612

Recently met-enkephalin has been identified to be derived from its prohormone, pro-enkephalin A in macrophages and T helper cells. Its immunological profile is similar to IL2 and gamma interferon. Met-enkephalin specifically increases numbers of cytotoxic cells (CD8 and NK) that are effective against viruses and tumors. Anti-viral activity was demonstrated against Herpes, AIDS, CMV, coronavirus, FluA, and Japanese Encephalitis either alone or in combination with known anti-virals. Anti-tumor activity was seen against melanoma, sarcoma, lung cancer, colon, leukemia, and neuroblastoma. In addition, clinical studies in AIDS and cancer patients are discussed.

Mind-body relationships probably involve interactions between hormones, neurotransmitters, cytokines and many other factors. Methionine enkephalin, which was originally identified to be an opioid peptide, neurotransmitter and hormone was more recently identified also to be a cytokine of the immune system (1) (2). The immunological effects of methionine enkephalin have been extensively reviewed and include increases of T cell subsets, NK cells, as well as macrophage activation (3). These effects were particularly striking in the treatment of AIDS and cancer patients. The presentation today, will focus on the broad spectrum of antiviral and antitumor activities of methionine enkephalin.

ANTIVIRAL ACTIVITIES OF METHIONINE ENKEPHALIN (MET ENKEPHALIN)

Influenza A Virus

Methionine enkephalin was found to increase natural killer and cytotoxic T lymphocyte activity in mice infected with influenza A/NNS/33 (H1 N1). Maximal increases in cytotoxic cells were seen when met-enkephalin was administered 72 hours after viral infection (4). It is interesting to speculate whether met-enkephalin enhances other cytoki-

Potentiating Health and the Crisis of the Immune System
edited by Mizrahi *et al.* Plenum Press, New York, 1997

nes, that are released from activated lymphocytes. Thus, met-enkephalin has been found to increase interleukin II levels and expression of receptors (5). This study provides a basis for considering treating hospitalized patients with "viral pneumonia". In a pilot study, AIDS patients with pneumonitis, were successfully treated with met-enkephalin (6). Since diagnosis of hospital viral pneumonias is difficult, it would not be unreasonable to consider combining met-enkephalin with antibiotic therapy. Precedence for this clinical approach was established with the cytokine, G-CSF and antibiotics in the treatment of immunosuppressed cancer patients (7).

Japanese and Tick-Borne Encephalitis

Peptides, derived from bone marrow, were found to have therapeutic effects against Japanese and tick-borne encephalitis. The principle components of these peptides were found to be enkephalins with opioid binding properties and were named myelopeptides. These peptides were found to stimulate anti-viral antibody production (2 to 5 fold) against Japanese encephalitis virus and also the langat virus of the tick-borne encephalitis TP-21 complex. Treatment with the peptides doubled the survival rate. These protective effects were seen both in acute and chronic infections. The maximal effects on antibody production was seen in activated cells at the peak of the immune response (8). This enhancement of activated cells suggests that there may be synergism with other cytokines such as interleukin II and gamma interferon (5)(9). No toxicity was observed with the use of these peptides.

Rauscher Virus (RLV)

AZT has shown potent activity against Rauscher murine leukemia virus complex (RLV) in vitro and in vivo. The pathology of RLV in BALB/c mice involves massive splenomegaly in two to three weeks post-inoculation (plus an increase in serum virus titer and serum reverse transcriptase activity). "Met-enkephalin prevented splenomegaly and reverse transcriptase activity slightly, but not significantly, at both 1.2 and 3.4 mg/kg/day doses. Also, the addition of 1.2 mg/kg/day of Met-enkephalin to an oral dose of 1.6 mg/kg/day of AZT produced a statistically significant prevention of splenomegaly and reverse transcriptase activity compared to untreated controls, while an oral dose of 1.6 mg/kg/day of AZT by itself did not". Combination therapy of antivirals and cytokines is now believed to be the best approach for the treatment of HIV (10).

Friend Leukemia Virus (FV) and Maids Virus (BM5)

FV induces a severe immunodeficiency and leukemia in susceptible BALB/c mice. AZT (azidothymidine) reduces morbidity and lengthens survival of infected mice. Combination of AZT and met-enkephalin decreases morbidity and increases survival beyond that seen with AZT alone. Both agents were administered to infected mice beginning 3 days p.i. and administered over four weeks. All mice surviving at 70 days p.i. were killed and their spleen weighed. Approximately fifty percent (14/30) of the mice survived at 70 days using the combination of AZT and met-enkephalin (11). Ciolli et al reported that administration of IL-1 and IL-2 resulted in increased survival of infected mice (12). Since met-enkephalin has been reported to increase IL-1 and IL-2 production, it is possible that the three cytokines may be acting together synergistically in protecting FV infected mice (13) (5). In the case of the MAIDS virus, AZT/met-enkephalin combination markedly reduced splenomegaly.

Herpes Simplex Virus Type 2

Methionine-enkephalin significantly increased the short-term survival of mice after they were challenged with an LD80 dose of HSV-2. The treatment was conducted over a period of 24 days. In a separate study met-enkephalin was added to culture media of VERO cells infected with HSV-2. No direct antiviral effects were seen. However, met-enkephalin was demonstrated to significantly increase NK activity of splenic lymphocytes (14). In addition, clinical studies in AIDS patients showed that met-enkephalin treatment markedly reduced herpes lesions (15).

Coronavirus MHV-JHM

The neurotropic strain of the mouse hepatitis virus produces encephalitis and a paralytic-demyelinating disease in rodents. It has been used as a model of multiple selenosis. B-endorphin was found to reduce the paralytic demyelinating disease in mice by 40–50%, compared to controls. Virus replication in the brain was significantly reduced 3 days post-infection (16).

Cytomegalovirus (CMV)

Only slight protection was seen against murine CMV when mice were treated with ganciclovir plus met-enkephalin (at low threshold doses). The combination treatment appeared to diminish mortality and increase the median day of death (17).

AIDS Patients

In a serious of clinical studies in ARC and AIDS patients met-enkephalin treatment was found to increase T cell subsets (CD3, CD4, and CD8), NK cells, IL2, and blastogenic responses to PHA (3). Also, met-enkephalin and reduced viral load, as measured by p24 levels (15). Finally, met-enkephalin increased cell-mediated immunity (via-THI cells) as well as IL2 and gamma interferon production (18). Combination studies of met-enkephalin and the new anti-virals are currently in progress.

ANTITUMOR STUDIES OF METHIONINE ENKEPHALIN

Human Colon Cancer

Methionine enkephalin at dosages of 0.5, 5.0, or 25mg/kg diminished the development of human colon cancer HT-29 xenografts in nude mice. Seven weeks after cancer cell inoculation, 57% of the mice treated with methionine enkephalin did not develop a tumor. A new opioid receptor (zeta) is believed to regulate oncogenesis with met-enkephalin being the specific ligand (19).

Human Lung Cancer Cell Lines

Opioid receptors have been identified on lung cancer lines. Various opioid ligands, including enkephalin, were found to inhibit lung cancer growth in vitro. The opioid peptides are believed to be part of an endogenous tumor suppressor system (20).

B-16-BL6 Melanonia

Both methionine enkephalin as well as leucine enkephalin diminished the growth and number of spontaneous pulmonary metastases in C57BL6J mice implanted with B16-BL6 melanoma. At the same time, it was demonstrated that the enkephalins increase the natural killer cell activity of splenic lymphocytes (21).

PYB6 Fibrosarcoma

The maximum inhibition of tumor growth achieved with methionine enkephalin was 72% on day 15 post tumor transplantation. The enkephalin was found to increase the blastogenic response to the T-cell mitogen, concanavalin A. This would suggest that an increase in cytotoxic T Cells could explain the antitumor effect (22).

L1210 Murine Leukemia

Both methionine enkephalin and leucine enkephalin were found to increase the number of survivors, compared to controls, when mice were inoculated with L1210 leukemia. In addition, the enkephalins were reported to increase blastogenesis to PHA, which would support the mechanism of increasing cytotoxic T cells that control tumor growth (23).

Neuroblastoma

Zagon and McLaughlin identified a new opioid receptor, zeta, which appears to regulate oncogenesis. The most potent ligand found to date is methionine enkephalin. Thus, tumor appearance was delayed and survival of mice was prolonged when mice were inoculated with S2OY neuroblastoma and treated with met-enkephalin (24).

Cancer Patients

Several different types of cancer patients were treated with methionine enkephalin, including lung cancer, melanoma, kaposis sarcoma, hypernephroma, and pancreatic cancer. Antitumor activity of met-enkephalin was presumably expressed as increases in cytotoxic cells (CD4, CD8, NK) as well as levels of interleukin II. In addition, increased blastogenic responses to mitogens were also recorded. Several patients reported improvements in mood (antidepressant effect) after one to two weeks of treatment with met-enkephalin (25) (1) (2) (3).

DISCUSSION

Methionine enkephalin has now been identified to be a new cytokine that is derived from T helper cells and macrophages (3). The immunological effects of met-enkephalin are similar to those seen with IL2; increases of T cell subsets, NK cells, and blastogenic responses to mitogens (26). In this regard, met-enkephalin was found to be synergistic in LAK cell production (27). Since met-enkephalin increases cytotoxic cells (CD4, CD8, and NK), it is not surprising that antiviral effects against herpes, Flu A, Japanese encephalitis, Friend leukemia virus, MAIDS, Rauscher, and AIDS were seen. Probably the most striking therapeutic effects of met-enkephalin were seen in ARC-AIDS patients with herpes, pneumonitis, and HIV. Marked reductions in symptoms of herpes, pneumonitis, and AIDS were noted. Constitutional symptoms of weight loss, fevers, night-sweats, diarrhea, thrush, and herpes, were almost completely eliminated. In addition, the patients reported improvement in mood (possible antidepressant activity). Studies of combinations of antivirals and met-enkephalin are in progress (1) (2) (3) (15).

The anti-tumor activities of met-enkephalin are also attributable to the increases of cytotoxic cells as well as activation of macrophages (25) (3). Thus, antitumor effects were seen against L1210 leukemia, B16 melanoma, fibrosarcoma, neuroblastoma, ovarian cancer, pancreatic, colon cancer, and lung cancer. Clinical studies were carried out in melanoma, lung cancer, hypernephroma, and pancreatic cancer with met-enkephalin treatment. Immunological end-points all showed significant elevation. Tumor regression was most noted in Kaposis sarcoma patients (28).

Of special interest are the studies carried out by Faith et al where they showed that NK cell activity was increased by met-enkephalin in many diverse cancers: thyroid carcinoma, acute myelocytic leukemia, small cell carcinoma of the lung, breast cancer, ovarian cancer, gastric carcinoma, and lymphoma, Hodgkins disease, and chronic myelogenous leukemia. The latter study suggests that NK cells (as well as LAK cells) may be increased in vivo in a large spectrum of patients. In addition the increases seen in cytotoxic T cells (CD4, CD8) in patients on met-enkephalin treatment gives further hope for treating solid tumors.

In conclusion, methionine enkephalin has been identified to be a new cytokine which activates cytotoxic cells that are effective against a large spectrum of viruses and tumors (2) (30). The prohormone, proenkephalin A, is found in the central nervous, neuroendocrine, and immune systems. In times of stress the prohormone is processed to met-enkephalin and other peptides. Met-enkephalin undoubtedly contributes to mind-body relationships and interactions and may explain, in part, "spontaneous healing" (32) (33).

REFERENCES

1. Plotnikoff N. P., Faith R. E., Murgo A. F., Good R. (eds), Enkephalins and Endorphins Stress and the Immune System. Plenum press, N.Y. 1986.
2. Plotnikoff N.P., Murgo A. J., Faith R. E., and Wybran J.(Eds.) Stress and Immunity, CRC Press, Boca Raton, FL, 1991.
3. Plotnikoff N. P., Faith R. E., Murgo A. F., Herberman, R. B., and Good R.A. (1997) Methionine Enkephalin: A new Cytokine - Human Studies. Clin. Immun. And Immunopath., 82, 1, 1–9.
4. Burger R.A., Warren R.P., Huffman J.H., and Sidwell R.W 1995. Effect of Methionine Enkephalin on Natural Killer Cell and Cytotoxic T Lymphocyte Activity in Mice Infected with Influenza A Virus. Immunopharm. Immunotox. 17(2), 1995, 323–334
5. Wybran J. And Schandene L. Some immunological effects of methionine-enkephalin in man: Potential therapeutical use, In "Role of Leukocytes in Host Defense" (J. Oppenheim, Ed.) A.R. Ziss, New York, 1986. 205–212.
6. Wybran J., and Plotnikoff N. P. Methionine Enkephalin, A New Lymphokine for the treatment of ARC patients. Stress and Immunity (Plotnikoff N.P., Murgo A. J., Faith R.E., and Wybran J. Eds.) CRC Press, Boca Raton, FL, 1991 417–432.
7. Herrmann R., Schulz G., Wieser M., Kolbe K., Nicolay U., Noack M., Lindemann U., and Mertelsmann R 1990. Effect of Granulocyte-Macrophage Colony-Stimulating Factor on Neutropenia and Related Morbidity Induced by Myelotoxic Chemotherapy. Amer. J. Med. 88, 1990, 619–624.
8. Petrov R.V., Mikhailova A. A., Sergeev U.O., and Sorokin S.V 1987. Regulator Bone Marrow Peptides and Immunocorrection. Estratto dalla rivista EOS. 2, VII, 1987, 88–93.
9. Brown S. J., and Van Epps D.E.(1986) Opioid Peptides Modulate Production of Gamma Interferon by Human Mononuclear Cells. Cell. Immunol. 103, 19–24.

10. Selleseth D. (1994) Division of Virology, Burroughs Wellcome Co. Personal Communication.
11. Specter S., Plotnikoff N., Bradley W.G., and Goodfellow D. (1994) Methionine Enkephalin Combined with AZT Theraphy Reduce Murine Retrovirus-Induced disease. Int. J. Immuno- pharmac. 16, 11, 911–917.
12. Ciolli V., Gabriel L., Sestili P., Varano F., Proietti E., Gressor I., Testa U., Montesor E., Bulgarini D., Mariani G., Peschle C., and Belardelli F. (1991) Combined interleukin-1 /interleukin-2 therapy of mice injected with highly metastatic Friend leukemia cells. Host antitumor mechanisms and marked effects on established metastases. J. Exp. Med. 173, 313–322.
13. Youkilis E., Chapman J,. Woods E., and Plotnikoff N.P. (1985). In vivo immunostimulation and increased in vitro production of interluekin 1 (IL-1) activity by met-inkephalin. It. J. Immunopharmacol. 7 (3), 79.
14. Faith R. E., Murgo A. F., Clinkscales C. W., and Plotnikoff N.P. (1987) Enhancement of Host Resistance to Viral and Tumor Challenge by Treatment with Methionine - Enkephalin. Ann. N. Y. Acad. Sci. 496, 137–142.
15. Wybran .J., and Plotnikoff N.P. Methionine Enkephlin – a new Lymphokine for the treatment of ARC patients. In "Stress and immunity" (N. P. Plotnikoff, . J. Murgo, R. E. Faith and J. Wybran, Eds.), CRC Press, Boca Raton, FL, 1991, 417–432.
16. Gilmore W., and Moradzadeh D. L. (1993) B-Endorphin protects mice from neurological disease induced by the murine coronavirus MHV-JHM. J. Neuro-immun. 48, 81–90.
17. Kern E. (1994) University of Alabama, Personal Communication.
18. Street N.E., and Mosmann T.R. (1991) Functional diversity of T lymphocytes due to secretion of different patterns. FASEB J. 5, 2, 171–177.
19. Zagon J. S., Hytrek S. D., Lang C. M., Smith V. P., McGarrity T. J., Wu Y., and Mclaughlin P. J. (1996) Opioid Growth Factor ([Met5] enkephalin) prevents the incidence and retards the growth of human colon cancer. Am. J. Physiol. 271(Regultory Integrative Comp. Physiol. 40): R 780-R-786.
20. Maneckjee R., and Minna J. D. (1990) Opioid and Nicotine receptors affect growth regulation of human lung cancer cell lines. Proc. Natl. Acad. Sci. 87, 3294–3298.
21. Murgo A. J., Faith R. E., and Plotnikoff N.P. Enhancement of tumor resistance in mice by enkephalins. In stress and Immunity (eds. N. P. Plotnikoff, A. J. Murgo, R. E. Faith and J. Wybran), CRC Press, Boca Raton, Fl, 1991, 357–372.
22. Srisuchart B., Fuchs B. A., Sikorshi E.E., Munson A. E., and Loveless, S. E. (1989) Antitumor activity of enkephalin analogues in inhibiting PYB6 tumor growth in mice and immunological effects of methionine enkephalinamide. Int. J. Immunopharm. 11, 487–500.
23. Plotnikoff N.P., and Miller G. C. (1983) Enkephalins as Immuno-modulators. Int. J. Immunopharm. 5, 437–441.
24. Zagon I.S., and McLaughlin, P. J. The role of endogenous opioids and opioid receptors in human and animal cancer. In stress and Immunity (eds. N. P. Plotnikoff, A. J. Murgo, R.E. Faith and J. Wybran) CRC Press, Boca Raton, FL 1991 43–55
25. Plotnikoff N. P., and R. E. Faith. Modulation of anti-tumor immunity by enkephlins in Psycho-immunology of Cancer (Eds. C. E. Lewis, C. O'Sulivan, and V. Barraclough) Oxford University Press, New york, 1994, 373–384.
26. Smith K. A. (1988) Interleukin-2: Inception, Impact and Implications. Science, 240, 1169- 1176.
27. Wybran J., and N. P. Plotnikoff (1989) Enhancement of Immunological Mechanisms, including LAK Induction, By Methioine Enkephalin. 7th Int. Congr. Immun. Berlin.
28. Plotnikoff N. P., Miller G. C., Nimeh N., Faith R. E., Murgo A. V., and Wybran J. (1987)Enkephalins and T-cell enhancement in normal volunteers and cancer patients. Ann. N. Y. Acad. Sci., 496, 608–619.
29. Faith R. E., Liang H. J., Plotnikoff N. P., Murgo A. J., and Nimeh, N. F. (1987) Neuroimmuno-modulation with enkephalins: in vitro enhancement of natural killer cell activity in peripheral blood lymphocytes from cancer patients. Nat. Immun. Cell growth regul., 6, 88–98.
30. Plotnikoff N. P., and Wybran J. Methionine enkephalin: Activation of cytotoxic cells in cancer and AIDS patients. In "Psychoneuroimmunology" (H. J. Schmoll, U. Tewes, and N. P. Plotnikoff, eds.), Hogrefe and Huber, Toronto, 1992 251–259.
31. Jankovic B.D. (1994) Neuro-immunomodulation. Ann. N.Y. Acad. Sci. 1–38.
32. Cousins. N. The healing heart. Norton, N.Y. 1983.
33. Weil A. (1995) Spontaneous Healing, Alfred A. Kopf, N. Y.

VG-1000 – A THERAPEUTIC VACCINE FOR CANCER

Harris L. Coulter,[1*] John Clement,[2] Valentin Govallo,[3] and Frank Wiewel[4†]

[1]Empirical Therapies, Inc.
4221 45th Street, N.W.
Washington, D.C. 20016
[2]IAT Clinic
Lucaya Medical Center East
Sunrise Highway, P.O. Box F-40827
Freeport, Commonwealth of the Bahamas
[3]Moscow
Russian Federation
[4]People Against Cancer
Box 10
Otho, Iowa 50569

ABSTRACT

This paper discusses the treatment of carcinoma with a vaccine made from human placenta. The Russian physician and immunologist, Valentin I. Govallo, concluded twenty years ago that cancer does not reflect immunodepression of the host organism. However, the tumor is capable of selectively "turning off" host defenses. In this it resembles the fetus in utero, which is also capable of "turning off" or "blocking" the action of the maternal immune system. Govallo further concluded that cancer is best treated not by enhancing the host immune system but by suppressing the defense mechanisms of the malignant tumor. This is done by injecting an extract made from the human placenta collected after a normal live birth; the extract apparently acts as a "vaccine" which, perhaps through production of antibodies and immune complexes, neutralizes the tumor defenses ("blockers"). We discuss the theory underlying the use of the Govallo extract (known as VG-1000) and present the results of cases treated by us in Moscow in 1994 and in the Bahamas in 1995 and 1996. Reference is made to the Law of Cure which is, in many cases, exemplified by the patient histories given below.

* Tel. 202–362–3185 and 202–364–0898 (voice-mail); Fax: 202–362–3407; E-mail: hlcoulter@msn.com. Internet website: http://home.earthlink.net/~emptherapies/
† Tel. 515–972–4444.

Potentiating Health and the Crisis of the Immune System
edited by Mizrahi *et al.* Plenum Press, New York, 1997

INTRODUCTION

This paper discusses some initial results obtained in treating carcinomas with a vaccine made from human placenta.

The vaccine was discovered by coauthor Valentin Govallo, a physician and immunologist living and working in Moscow. He has treated about 250 patients with results which seem substantially better than those achieved by other cancer treatments. Coauthors John Clement, Frank Wiewel, and Harris L. Coulter have commenced treatment with VG-1000 in the IAT Clinic, Freeport, Commonwealth of the Bahamas, and the Max Gerson Memorial Cancer Center in Tijuana, Mexico. Some results of treatment are given below.

1. IMMUNOLOGY OF MISCARRIAGE

Dr. Govallo's understanding of cancer, and his therapeutic recommendations, reflect his early work on immunologic factors in miscarriage.

The beginning of his career in the early 1960s coincided with the onset of an epidemic of miscarriage (spontaneous abortion) not only in the Soviet Union but in all industrialized countries. For reasons which seem to be immunologic in origin and are probably related to environmental contamination, since the 1960s women in many countries have experienced increasing difficulty bringing babies to term; more and more pregnancies have been ending in a spontaneous abortion in the second trimester.

Dr. Govallo reasoned that these environmental stresses were affecting the pregnancy by undermining the fetal immune system, and that this was the reason for the epidemic of miscarriage.

Pregnancy has long been a puzzle to immunologists in the sense that the fetus is composed, in part, of genetic material from the father, hence incompatible with the maternal immune system. And yet, in the great majority of cases, the mother does not react to this "non-self" tissue.

The reason, in Dr. Govallo's view, is that the fetal immune system is usually strong enough to withstand the inherent impulse of the maternal immune system to expel the fetus as "non-self." But environmental and other stresses can undermine the fetal immune system to such an extent that it can no longer be protective.

1.1. Treatment of Miscarriage

Dr. Govallo treated chronic miscarriage due to immunologic weakness by subcutaneous injections of a suspension of the father's lymphocytes—the purpose being to stimulate placental production of the factors ("blockers" or "blocking proteins") which strengthen the fetus and protect it from the maternal immune system.

Using this technique he treated more than 600 women who had had multiple miscarriages, with a normal birth ensuing in 553 of them (91.3%).

2. IMMUNOLOGY OF CANCER

In the 1970s Dr. Govallo became interested in the problem of cancer and took the mother-fetus relationship as his model. For, like the fetus in utero, the tumor possesses its own immune system which protects it from the immune system of the host.

Dr. Govallo noticed that the tumor from the beginning is capable of selectively "blocking" or "turning off" host defenses, just as a thief (he says) will dismantle the alarm system before burglarizing a home. And he saw the physician's intervention as being to counter this ability of the tumor. If it were possible in this way to breach the tumor's "immunologic shield," the organism's immune defenses could then cope adequately with the tumor.

Dr. Govallo was probably the first immunologist to conclude that cancer does not reflect a generally depressed host immune system. The immune depression often noted in cancer patients is the consequence of radiation and chemotherapy which are known to have an immunosuppressive effect.

Especially in the early stages of cancer, the patient's immunity is normal, except that the anti-tumor defenses have been selectively "blocked" or suppressed.

2.1. Immunologic Treatment of Cancer

To counter the tumor "blocking mechanism" Dr. Govallo eventually settled upon an aqueous extract of human placenta (VG-1000). Placentas after abortions are never used. While it is not entirely clear how the placental extract operates to neutralize tumor "blockers," Dr. Govallo thinks that it is through the principle of "similars," with the placental extract acting as a vaccine.

For optimal results he urges that the tumor be removed surgically to the extent possible—leaving less tumor mass for the body's immune system to cope with.

The mechanism of action of placental extract may be as follows. The extract contains the "blockers" which the fetus has needed for its own protection. When these blockers are injected into the cancer patient, they generate antibodies ("antiblockers") which combine with the tumor blockers to form immune complexes which are then removed from the organism.

Once deprived of its "shield," the tumor is vulnerable to the action of the immune system, specifically T-lymphocytes, natural killer cells, lymphokine-activated cells, and macrophages. Under this assault it becomes soft and proceeds to dissolve.

3. TREATMENT OF CANCER PATIENTS WITH VG-1000

The following accounts, selected from 40 patients treated in Moscow in 1994 and at the IAT Clinic, Freeport, Bahamas, in 1995 and 1996, illustrate how, in Dr. Govallo's view, the vaccine is designed to operate. It would be premature to conclude that VG-1000 does or does not "cure" cancer. Some of the patients have, in fact, died, and those who are still alive and in apparent good health have not survived long enough to permit conclusions to be drawn. Furthermore, many of these patients were extremely ill (in stages III or IV, often with metastases) when first presenting for treatment. However, in virtually every case administration of the vaccine evoked a notable curative response. Reactions took the form of somnolence, tachycardia, headache, depression, irregular heart beat, fever, sweat, chills, pain and/or feeling of warmth at sites of metastases, pain at sites of previous surgical operations, etc.

We found that, as was to be expected, patients in stages I or II at the time of treatment are still alive today and largely cancer free, while those in Stages III or IV have in some cases died, but still they showed remarkable improvement for several months after taking VG-1000.

The cases described below are not always "pure" in the sense that patients often try other kinds of therapy before, after, and during treatment with VG-1000. Many are on special diets, taking vitamins, dietary supplements, and the like. While the possibility of interference from these adjuvant therapies cannot be ruled out, one does not have the right to ask patients whose lives are hanging in the balance to reject any mode of treatment which seems to them promising, and compliance would be impossible to ensure. In any case, the fact of tumor "melting" seems peculiar to, and characteristic of, VG-1000.

3.1. The Law of Cure

We were particularly impressed to note that in many of the cases the curative response exemplified the operation of the Law of Cure.

The Law of Cure is a concept elaborated by the nineteenth-century pioneers of homoeopathy, Samuel Hahnemann (1755–1843) and Constantine Hering (1800–1880). Today it is a therapeutic guide employed in homoeopathic medicine and also in chiropractic, osteopathy, acupuncture, and several other healing modalities (not, however, in conventional allopathic medicine).

It sets forth the conditions which must be met if the patient is to be truly "cured" of his or her disease: namely, that in a true and lasting cure the symptoms and pathology must move: 1) from the inside of the body toward the outside, 2) from the more vital organs toward the less vital organs, and 3) from the upper part of the body toward the lower. Finally, they should disappear in the reverse order of their appearance.

The Law of Cure contrasts vividly with the conventional allopathic approach to the treatment of cancer which aims essentially to shrink the tumor through application of one or another toxic chemical. But oncologists do not know if this is really an appropriate "end point." It is merely an easily measurable parameter.

In any case, when patients are treated with VG-1000, their reactions generally follow the Law of Cure, signifying that the changes observed are truly in the direction of cure. In the best cases the tumor becomes soft and starts to dissolve.

3.2. From Inside to Outside; from More Important Organ to Less Important

The movement of symptoms and pathology from the inside of the body to the outside is probably the most striking phenomenon noted after treatment with VG-1000. For instance, two patients were treated for lung cancer of long standing and so advanced that, in both case, the tumor occupied one whole side of the lung. In both cases, a few days after injection of VG-1000, the tumors started to dissolve and to "melt" to such a degree that every morning the patient had to cough up and spit out necrotic tumor cells. At the same time, tumor also exited the body through the urine, and the patients noted thick deposits of sediment in their urine.

Patients with colo-rectal and prostate tumors noted the same phenomenon: thick deposits of sediment in the urine and intense diarrhea with voiding of cancerous tissue in the form of a thick ropy mucus.

A typical instance of pathology moving from the inside of the body to the outside, and also from a more important organ to a less important one, was a patient with a colorectal tumor in whom the tumor disappeared from the colon and then emerged on the skin of the perineum;thereafter it commenced to disappear spontaneously from the skin (without further treatment), promising a permanent cure.

3.3. Disappearance of Symptoms and Pathology in Reverse Order of Appearance

In several cases the most recent symptoms and pathology have disappeared before older ones. A patient with non-Hodgkin's lymphoma, who had a number of tumors on his skin found that the most recent tumors disappeared first, followed by the earlier ones. By the same token, in a case of advanced prostate cancer with metastases to the bones of the legs, the most recent metastases seemed to heal first (at least, they ceased being painful) while the older ones took longer.

In a woman who had been operated for breast cancer, several weeks after received VG-1000 the pain of an earlier operation returned in an intensified form.

We have not yet noted any instances in which symptoms moved from the upper part of the body to the lower, but, as we continue treating patients in Freeport and Tijuana, Mexico, we will certainly have many instances of this illustration of the Law of Cure as well.

3.4. Relief of Pain; Extension of Life

A remarkable feature of VG-1000 is that it seems to confer some benefit—in the sense of relief of pain, improved emotional state, and extension of life (several months)—in even the most advanced cases. One patient with multiple myeloma who died nine months after being treated in Moscow, was playing the banjo most of the last day of his life and expired peacefully in his sleep. These effects of treatment should not be disregarded.

4. THE BURTON IMMUNOCOMPETENCE INDEX

We have made frequent use of a test of immunocompetence developed by Dr. Lawrence Burton, founder of the IAT Clinic in Freeport, Bahamas. While the details of the test are commercially protected information, it involves a comparison between "normal" and "abnormal" prealbumins in the patient's blood. The test result is given as a number between 0 and 2.00: the range 1.05–2.00 representing a healthy immune system, 0.91 to 1.04 reflecting a weak or precancerous immune system (or a case of incipient cancer), and the range 0.30 to 0.99 reflecting the immunosuppression typical of the cancer patient.

While developed to monitor the progress of cancer patients under other types of treatment, this test has been found eminently applicable to monitoring patients treated with VG-1000. It closely mirrors the patient's actual status and has excellent prognostic value as well.

Treatment with VG-1000 causes the immunocompetence index to rise by ten or fifteen 15 points, usually within 2–3 days. Sometimes this rise is preceded by a drop in the index which lasts several days before rising, similar to the "latent period" or "negative phase" well known to practitioners of vaccine therapy in the 1920s and 1930s and indicating a momentary lowering of the patient's resistance.

If the patient can maintain a high immunocompetence index for several months after receiving VG-1000, the prognosis for recovery is excellent. But if the index starts to decline a few weeks treatment, the prognosis is less good. We have given booster shots to some patients two or three months after treatment, but they are less effective than the initial treatment. We do not yet possess sufficient experience to reach firm conclusions about the benefit of booster shots.

5. COUNTERINDICATIONS TO VG-1000

VG-1000 is better adapted to carcinomas and melanomas, less indicated in sarcomas. It is counterindicated in liver cancer or liver metastases, since the liver cannot then process the toxic matter of tumor breakdown.

It is probably not indicated in cases with advanced metastases, especially to the bone.

Preliminary surgery, to minimize tumor bulk, is always recommended, since otherwise the burden on the liver is too great. For this reason we do not treat patients with liver metastases. And patients who have had recent chemotherapy or radiation do not respond well to VG-1000. A waiting period of several months is recommended.

6. CONCLUSIONS; FUTURE RESEARCH

Our primary aim is to cure cancer when this is possible, otherwise to extend life and reduce the patient's suffering. Dr. Govallo has many cases of ten-year survival among his patients, and we are convinced that we will have the same results—especially if patients are treated in the early stages.

A number of problems remain to be addressed, among them: (1) is there any reason to prefer one blood type over another in preparing VG-1000; while we do not think that the blood type of the placenta has to be matched with that of the patient, it may be that one or another blood type provides a more effective medicine. (2) how often can the patient be retreated? (3) could VG-1000 be improved, for instance, by isolating an "active principle." (4) in which types of cancer is it more, or less, effective? (5) are there other counterindications besides the presence of liver and bone metastases? (6) how long must the patient wait, after chemotherapy or radiation, before VG-1000 can be administered with benefit?

SUMMARY

We discuss a new therapeutic vaccine for cancer, made from human placentas gathered after a normal live birth, which has been named VG-1000. The vaccine, which was developed by a Russian immunologist in Moscow, Dr. Valentin I. Govallo, has been brought to the West and introduced into clinics in the Bahamas and Mexico through the efforts of Dr. Harris L. Coulter, Dr. John Clement, and Mr. Frank Wiewel of the patients' advocacy organization, People Against Cancer.

About forty patients, with varying kinds and degrees of cancer, have been treated with VG-1000 during the past two years. It is still too early to draw definitive conclusions, but VG-1000 has clearly benefited most of them to some degree. Many are today symptom-free and apparently in good health.

The patients' reaction to the medicine usually follows the Law of Cure, with symptoms and pathology disappearing from the inside to the outside, from the more important organs to the less important, and in the reverse order of their appearance.

We have been able to monitor the patients' status and progress using an immunocompetence index developed at the IAT Clinic in the Bahamas. Vaccination of the cancer patient with VG-1000 invariably brings about an improvement in this index, and, if ability to maintain an elevated immunocompetence index is an excellent prognostic sign.

Some unresolved issues are: (1) which blood type makes the best vaccine? (2) how often patients can be retreated? (3) can VG-1000 be improved by isolating an "active principle"? (4) in which types of cancer it is more, or less, effective? (5) what are the counter-indications? (6) how long must the patient wait after chemotherapy or radiation before VG-1000 can be administered with benefit?

It is hoped that within the next few years we will be able to provide answers to these questions.

BIBLIOGRAPHY

Valentin I. Govallo, Immunology of Pregnancy and Cancer. Commack, NY: Nova Science Publishers, Inc., 1993. Address of publisher: 6080 Jericho Turnpike, Suite 207. Commack, New York 11725.

"FEMALE RELATED SURGERIES" – PRE- AND POST-SURGICAL MASSAGE AS A KEY TO SUCCESSFUL HEALING

Roseann Gould*

Reidman International Center for Complementary Medicine
109 Hayarkon Street
Tel Aviv 63571, Israel

ABSTRACT

How many of us don't know at least one woman who has undergone an hysterectomy or other "female related" surgery (cesarean section, mastectomy, breast reduction/enlargement/reconstruction, etc.). Common to all of these procedures is the risk of adhesions keloid scarring and reduced range of motion (ROM). All of these conditions begin to occur even as the surgery is in process. Deep tissue massage helps to prevent this "gluing" of the tissue layers. The atmosphere in the pre-surgical massage sessions encourages the client to express her fears and talk about them and as a result she can go into surgery feeling relaxed, self assured and empowered. Post-surgical massage will motivate and assist the woman to actively participate in her own healing process. Post surgical massage combined with ROM stretches, in addition to the psychological benefits, can assist in speeding up the recovery process, create greater elasticity and softening of tissue, help prevent adhesions and keloiding as well as create more comfort in the body and a more aesthetic surgical repair site. Specific techniques which have been developed will be discussed.

INTRODUCTION

For the past 25 years I have been integrating massage, range of motion stretches (ROM), body/mind techniques, relaxation, visualization and guided imagery techniques to empower my clients, assist in their healing and increase their self esteem and acceptance

* You can contact Roseann Gould in Tel Aviv, Israel at 972-3-560-5114 or write c/o Reidman Center at the above address. Please leave an Email address if you have one.

Potentiating Health and the Crisis of the Immune System
edited by Mizrahi *et al.* Plenum Press, New York, 1997

of self. I am constantly amazed at how quickly these techniques work to release pain and increase flexibility and body comfort (often during the very first session).

TEXT

How many of us don't know at least one woman who has undergone an hysterectomy or other "female related" surgery (cesarean section, mastectomy, breast reconstruction, facelift or liposuction). Common to all of these procedures can be any or all of the following: risk of adhesions, keloid scarring, reduced range of motion (ROM) and edema. According to Dr. Vicki Hufnagel "All of these conditions begin to occur even as the surgery is in process. Deep tissue massage helps to prevent this 'gluing' of the tissue layers".

Pre surgical massage encourages the client to express her fears and talk about them so that she can go into surgery feeling relaxed, self assured and empowered. Post surgical massage will assist the woman to actively participate in her own healing process and prevent or greatly reduce many of the above mentioned conditions.

PRE-SURGICAL MASSAGE

Benefits

- Creates an environment where the woman can express fears and ask questions.
- Empowers the woman.
- Allows the client/patient to take personal responsibility for her healing process.
- Massage therapist can act as liaison between doctor and patient.
- Patient learns progressive relaxation and visualization techniques to relieve stress and accelerate healing.

I'll venture a guess that every person in this room has either experienced some type of surgical procedure or knows someone who has. Therefore we can all identify with the fears and stresses that a patient feels as her date draws near. Whether or not the surgery is elective or non-elective is not important; surgery triggers an emotional response!

According to Marian Williams, R.N. and coordinator of California Pacific Medical Center, San Francisco Massage Therapy Program and Internship "when patients receive massage, it makes them better prepared to heal". She further states, "The purpose of massage therapists is not to 'cure' disease, but rather to facilitate healing".

At this time pre-surgical massage can be of great benefit. The positive, relaxed environment of the massage clinic, combined with the gentle touch of the massage therapist nurtures the trusting relationship the client needs to allow her to express unspoken fears and concerns and to ask questions she may not have had time to address with her doctor. These can then be conveyed to the doctor by the massage therapist. Ideally we will be working as a team-doctor, anesthesiologist (if necessary), nurse, psychologist, massage therapist, etc. I often use guided imagery, visualization and breath work to assist the client to express held back emotions. In this manner she goes into surgery feeling clear and better prepared for the experience. I use this approach because I have found over the years that it is important for the client to *consciously experience* her *feelings* of fear and discomfort, fear of loosing control or fear of asking others for support. The process is analogous to talking about riding a bicycle versus getting on the bicycle and *feeling* what it is like to ride it. They are very different! It is the *allowing* of oneself to feel that is transformative.

Going into surgery can leave one feeling alone. Even though we have loved ones to talk with and to support us, when you go through those doors you are alone on the table. Through the gentle touch of massage the client begins to relax and feel safe and trusting. It is the *touch* that enables her to give voice to her feelings at an emotional level as well as a verbal level.

POST-SURGICAL MASSAGE

Benefits

- Reduce or alleviate pain.
- Reduce edema.
- Alleviate fears of damaging self through touch.
- Teach client to work on herself.
- Stimulate circulation and lymphatic flow.
- Accelerate healing and promote healthy, flexible tissue.
- Discourage adhesions and keloiding
- Mobilize site for better ROM
- Heal the *"whole woman"* – physical, emotional and spiritual.

According to Drs. Field and Miller as stated in American Family Physician, 1992, "Gentle and frequent massage of the scar reduces thickening, stimulates revascularization and promotes mobility and elasticity of the skin". Further, Marci Javril, the massage therapist who co-developed a post surgical massage program with Dr. Vicki Hufnage, Ob/Gyn in Los Angeles, CA, succinctly states the benefits of post surgical massage for women who will have an hysterectomy (and of course this is extended to mastectomy and other procedures that invade the body and alter the self image). She states, "This is an empowerment process, one that reaches very deeply into the self image and spiritual concept of health and life. It is important for her to know that she can help herself, that she need not be a victim or a passive subject in the consented penetration of the internal organs; rather, she can be a participant in the release of trauma, and the acceptance of tender care after this physical, emotional and psychic confrontation (she) needs to reassured that it is O.K. to touch the surgical site, even though she may have a bit of apprehension about pain upon being touch. My participation in her healing process is to show her how to massage herself safely and firmly, adding to the blood flow and lymphatic drainage in the surrounding area". This has been my experience over the years also - that the woman is frequently concerned and fearful that she will inflict damage to herself by touching or moving the tissue around the surgical site. It is very reassuring to her to first experience someone else (the massage therapist) working on her and seeing that not only has no damage been done, but that she, in fact, feels much better. Now she is ready and eager to learn how to work on herself!

Just as doctors have discovered in the past that getting a patient up and walking the day of surgery to stimulate circulation, prevent edema and adhesions, bowel problems or other possible complications, so to are doctors and other health practitioners finding that the use of massage can assist in and speed up the recovery process. Massage combined with ROM stretches helps to prevent adhesions and keloiding and creates greater elasticity and softening of the tissue as well as a more comfortable body with a more beautiful surgical repair site—which from the point of view of the patient can be as important as the surgery itself. As a result of this groundwork, the client will want to actively participate in her post operative healing process.

The key to successful healing of our patients or clients lies in the willingness of the doctor, massage therapist, family members and others to work together as a team which has the best interests of the patient at heart.

TREATMENT APPROACH

Phase I

Weeks one and two. In the first days or weeks after surgery I use very gentle techniques that allow the woman to relax and realize that it is safe to touch the surgical site. I use a method of lymphatic massage that was developed by Lauren Berry, an American physical therapist and body therapist. This approach helps to reestablish peristalsis and reduce sensitivity. Also, according to the well known Dr.Vodder, lymphatic massage not only helps to reestablish circulation but assists the lymph passageways to flush proteins and other by products of stress out of the system. I work on the abdomen creating a soft, continuous undulating motion that gently stimulates the descending, transverse and ascending colon. This is usually very pleasant and relaxing for the patient. Next the sigmoid flexure is gently straightened, the area under the diaphragm-the spleen on the left and the liver and gall bladder and ileocecal valve on the right are stimulated. This work, besides gently stimulating peristalsis, assists in releasing gas which not only causes discomfort and pain in the abdominal cavity, but can be experienced as pain in the levator area because of the pressure exerted on the diaphragm. In the breast area, at this stage, I often utilize gentle somatic work (like rocking and shaking) to release the tissues and let the woman know it is safe to be touched.

Phase II

After two to three weeks, when the tenderness, edema, bruising and other trauma have lessened or disappeared, I start to work directly on the scar tissue. We are now starting to work with the connective tissues which can be likened to a stocking which has been wrapped around every fiber, every bundle of fibers, every muscle, joint, ligament and under the skin. If you can imagine this and then notice that if one part of this continuous tissue is pulled or shortened, it will effect all other parts and create some type of stress and/or misalignment. Deep tissue massage is especially effective in reestablishing balance at all levels in the body.

The scar tissue and surrounding tissue is stretched and rolled to foster better circulation, drain the lymph, prevent adhesions, keep the scar tissue soft and pliable and to discourage keloiding. The use of skin rolling and plucking (pulling up on the tissues) seem to stimulate the nerves and help to more quickly replace numbness with normal sensation. This method also seems to help create new lymph pathways.

> "Two years ago I had a facelift. Afterwards I experienced severe edema. Since then my head has never returned to its original shape and big parts of my head and neck feel numb and swollen. Headaches bothered me day and night. As a result of this I became very depressed. I was afraid the pain would never go away...... I read about Roseann and her method appealed to me....I arrived hoping to get some relief from the pain and feeling of dead skin (numbness-

> author) in my face and neck. The first session was painful but indicated to me that there was hope. After the second treatment progress was faster. After the third session (one week apart-author) I realized many hours had passed without feeling headaches. Now after my fourth session the parts that are numb become smaller and smaller. The headaches are rare. I never expected to feel such a change so fast". C.W., Tel Aviv

This client was very depressed when she first came to me. The changes both physical and emotional have been amazing to me. It is experiences like the above that excite me about this work and have kept me in the field for so long. To be present at another persons healing process is an indescribable experience.

This work is also extremely effective after liposuction to reduce edema and quickly decrease pain and discomfort in the area and soften the tissue.

> "Following a liposuction procedure, I had a series of massages with Roseann Gould. Only twice before in my 53 years have I had any kind of massage, and those experiences were not at all like the ones with Roseann. The massage was deep, but not painful, and I was surprised to feel an immediate positive change in the swollen, bruised and stiff areas of my body. I am convinced the massage has contributed significantly to my post-procedure healing." K.S., Tel Aviv

With facial surgery I, of course, work in a gentler fashion, but I still incorporate skin rolling after four to six weeks to stimulate nerve sensation. Numbness is a common complaint after face surgery. Clients tell me that it changes their self image—they look in the mirror and see 'their face' but what they feel as 'face' is very different. It can be very disturbing to only feel one side of your face, lips or chin or not have any sensation on top of our head where there may be a line of sutures. My experience has been that the sooner massage is started the quicker these feelings diminish.

In another few weeks we can go into a deeper phase using deep skin rolling and plucking (pulling up on the scar tissue to break up adhesions) and scrubbing motions to further release and prevent adhesions and keep the scar tissue soft. I also teach the client to work on herself on a daily basis as this brings the best long term results.

EMOTIONS

During massage sessions women will often release emotions they have been holding onto-possibly for a long time. With any surgery we have to go through a grieving process. Whether it is saying good bye to a uterus or a breast or saying farewell to the "old look", this is an important part of the healing process. After the farewell, and of course after the surgery, the woman can, if she desires, welcome the new self and even celebrate the new self, either in our session or with family and friends.

REFERENCES

Cyriax, James, "Deep Massage". Physiotherapy, 1977, vol 63, no 2.

Ebner, Maria. "Connective Tissue Massage". Physiotherapy, July 1978, Vol 64, no 7, pp. 208–210.

Field, Dean Arden, Miller, Sandra. "Cosmetic Breast Surgery." American Family Physician. Feb. 1992, vol. 45, no 2. pp. 711–719.

Frazer, F.W. "Persistent post-sympathetic pain Treated by Connective Tissue Massage". Physiotherapy, July 1978, vol 64, no. 7, pp. 211–212.

Gilliam, Lydia. "Lymphoedema and Physiotherapists: Control Not Cure". Physiotherapy, Dec. 1994, vol 80. no 12, pp. 835–843.

Hendricks, Tom "The Effects of Immobilization on Connective Tissue". The Journal of Manual and Manipulative Therapy. 1995, vol 3, no. 3, pp. 98–103.

Javril, Marci, with Hufnagel Vicki. "Post Surgical Massage for Female Disorders". Massage Magazine. issue 17, Dec/Jan 1988–89, pp. 20–23.

Pasini, Katy. "Hospital Based Massage Programs: One California Hospital Leads The Way". Massage Magazine, Issue 42, March/April 1993, pp. 44–50.

Wittlinger, H. and G. Manual of Dr. Vodder's Manual Lymph Drainage, vol 1: Basic course, 2nd revised edition 1982, Haug Publishers.

PARASYMPATHETIC STIMULATION THROUGH THE "VODDER" ORIGINAL MANUAL LYMPHATIC DRAINAGE

Virginia Cool

School of Parasympathetics
66300 Llauro
Pyrénées Orientales, France
Tél: 33. 04. 68. 38. 85. 92.; Fax: 33. 04. 68. 38. 88. 38.

OUR BODILY AND MENTAL BALANCE DEPENDS ON THE PARASYMPATHETIC SYSTEM, WHICH IS STIMULATED BY THE "VODDER" ORIGINAL MANUAL LYMPHATIC DRAINAGE

A note on the autonomic nervous system. Our smooth muscles (viscera, vessels, and our whole lymphatic system) are stimulated in their automatism by the autonomic nervous system which consists of two complementary and antagonistic parts: the orthosympathetic and the parasympathetic, together with the hypothalamus and the vagus nerve.

The balance between these two parts ensures our good health at all levels.

The autonomic nervous system, just as the hormonal system, is related to our emotions through neurotransmitters, which accounts for its precarious balance. Hence, because of the norms of living which we now inflict upon ourselves, **the orthosympathetic system is far more called upon**, leading to "stress".

Moreover, the reactions thus prompted on the hormonal level can result in less and less bodily responses and remain painfully imprinted, finding expression in tissue suffering and pollution of our inner *milieu*. Our acido-basic balance is therefore perturbed.

This awareness of the predominance of the orthosympathetic is crucial in order to give the parasympathetic a fair chance, so that our vital balance can be restored. Thanks to the regular working of the parasympathetic system, a healthy body organizes its own defence system and its faculties for self-recovery. The hypothalamus working in close collaboration with the vagus nerve, going through the cerebral trunk, conveys the parasympathetic influence to the whole body.

The parasympathetic chain equals to psycho-neuro-endocrino-lympho immunity.

"Vodder" Original Manual Lymphatic Drainage checks the autonomic neuro system through inhibitory orthosympathetic effects and stimulatory parasympathetic effects. Its "basic" technique (jugular chain, abdomen, mobilization of the diaphragm), immediately

Potentiating Health and the Crisis of the Immune System
edited by Mizrahi *et al.* Plenum Press. New York, 1997

sets its action on the vagus nerve. It requires much precision in the "Vodder" gesture: its rhythm is adapted to each individual, its intensity suits the tissue, its spiralling movement inspires a pulsation and follows the waves of our inner fluid. It demands a lot of rigour from the specialized practicioner; he/she also acts in the parasympathetic, so that his/her awareness and attention are at their best throughout the treatment. Tuned into the parasympathetic, healer and patient alike become aware of the fluid and dense parts of the body. Observing together, from both the inside and the outside, they establish a wonderful non-verbal dialogue. They give a free hand to the parasympathetic, enabling it to drain and purify body and mind. **In short, and as a result, being in the parasympathetic restores confidence and well-being.**

Our bodily and mental balance depends on the parasympathetic system, which can be stimulated by the "Vodder" Original Manual Drainage.

I thank Professor Mizrahi and the organizers of the 3nd Dead Sea Conference "Potentiating Health and the Crisis of the Immune System" for giving me the opportunity of presenting my experiences with Manual Drainage according to Dr Vodder — and about the possibilities this method offers us, for global well-being, through the stimulation of the Parasympathetic system, with the result of purifiing the complex field of our internal water-environment – the "psycho-neuro-endocrino-lympho-immunity" chain.

First, a short introduction on *evolution to global medecine* and the *"humoral approach"* in the treatment of our *internal environment (milieu)* in general.

The relationship between his quality and *cell-regeneration,* with the vital role of the lymphatic system.

Lymphatic system, which in his "vaso-motricity" depends on the autonomic nervous system, with predominant "Parasympathetic".

Then briefly, a look at the *Autonomic Nervous System.*

The method I propose is the *MANUAL LYMPHATIC DRAINAGE according to Dr VODDER,* which before starting the specific Drainage action, places a large importance to the specific manoeuvers, to stimulate the *"Parasymphathetic."*

I will finish with *experiments* we did in my school, on this subject.

EVOLUTION FROM AND TOWARDS GLOBAL TREATMENT

Even though Hippocrates remains the symbolic father of modern western medicine, his theory on humoral therapy had fallen into oblivion for almost 2000 years.

This is normal evolution, following the astronomical 26.000 years cyclical rotation of our planet earth, which alternated every 13,000 years the influence into a period off introversion and a period off extroversion.

During the period of introversion, duality appears (Plato underlined the duality between Corpus and Anima) and evolution tends towards individuality. The study of the material, visible and palpable, with an analitical physical approach becomes the most important The invisible world was hidden or could only be approached through religion. This results in a total separation, between physical and spiritual matters.

The global approach to the human-being was forgotten. We had to rediscover the psycho- somatic, and the influence of the environment and vis-versa.

The "Humoral" Medicine

It was only at the beginning of this century, when there was a great impetus towards healing the "total" human being, that, at last, some remembered the fact that the human body is composed of 2/3 of liquid, similar to sea water, 2/3 the same percentage as our blue planet Earth.

The quality of these fluids is vital and results in an acido-basic equilibrium, necessary for cell-regeneration.

Alexis Carrel received the Nobel Prize in 1912 for his experiments, which involved keeping living cells alive, solely by means of pure water, being continually renewed.

Pure water has always been considered "the life source for the human being, like a whole in a environment."

Some Reflections on Water

- WATER, element from which eveything begins.
- Life on planet Earth began in the mother "ocean"– Amniotic fluid.
- WATER, 2/3 of the planet and 2/3 of the human body.
- Our inner environment (milieu) still bears a strange resemblance to the ocean.
- WATER, insipid, odourless, colourless liquid, no possible differentiation, we are (of) WATER.
- How do we resolve the pollution WATER - earth - without first being aware of the WATER that we are?
- The quality of our inner sphere (milieu) is also measured by pH (acido-basdic equilibrium).
- WATER reflects the image and the colour of the sky, the movements, the flight of clouds, WATER sparkles.
- WATER connects us with our sources and environment.
- The quality of our WATER is determinant in relation "human being- environment".
- WATER is an intermediate element between matter and more subtle energies.
- On the Earth, everything resounds with the atmosphere WATER.
- WATER is a vehicle of energy.
- The molecule of WATER adjusts in the body the electromagnetic differentiation necessary for the cell metabolism.
- The neuro-transmitters permeate our WATER with our emotions, what determines its quality.
- WATER expresses rhythm, ebb and flow, constant change, spirals, waves.
- WATER runs, flexible element, and forges the shape of hard elements.
- WATER behaves in the same way inside and outside the body: rising and downward motion, stagnation, attraction by some elements, transport element, purifying element with a very intensive bacterial life, and so on.
- The rising impulse given by the WATER flood, is the base of energy that restores confidence by sweeping along the forces of life towards the sun, towards the joy.
- It is the same for the LYMPH: it rises like the sap, sweeps along and purges heavy elements that are stocked in it.
- The care of WATER, with WATER, is like a "key that opens". It is the starting care above all else, in the image of our origin.
- The rising impulse of WATER, of LYMPH, helps the patient to stand up straight again, to purify himself, to revive his senses in order to communicate.

It is in the joy of the reunion with confidence that the well-"being" can express itself. What Emil Vodder had already understood in 1930.

VODDER'S HUMORAL DRAINAGE, CELL-REGENERATION AND THE VITAL ROLE OF LYMPHATIC SYSTEM

Emil Vodder, Danish Biologist (Paris) and Doctor of Philosophy (Brussels),"drains" the marshes of the body in 1932 with a new massage technique, which was the concretisation of the idea, "the global approach of the human being through its own liquids." He was a Biologist, and of course, what interests a Biologist ? — the cell and it's regeneration. — And *a cell, in order to live, needs a good environment*—a good environment of liquids—and these liquids must circulate Emil Vodder was very aware of the fact, that it is the quality of the interstitial fluid, in association with a good blood circulation and lymph system , upon which the life of the cell depends. It is there that the fight for survival takes place between bacteria, wasteproducts, nutritious matter, lymphocytes, natural killers, etc. For the cell to regenerate itself, the surrounding liquid must have the right PH and of course be in motion. The stagnation here, of wasteproducts and of the larger protein molecules, (who attract water) is responsible for oedemas and tissue deterioration, leading to cellullar degeneration on all levels. The role of the lymphatic system is the most important in this cleaning up action.

Contrary to the circulation of blood, which receives its impulse from the heart, the *lymphatic system,* a one way system, which is not connected to the heartpump, begins in the tissues and is the only issue for the evacuation of macro-molecules. Lymph, a heavy liquid, rich in proteins and also in wasteproducts, is confronted by the Lymphocytes in the Lymphnodes. This during its journey, against gravity, in the lymphvessels, going up, from the extremities, towards the jugular fossa .

The progression of lymphflow, depends on the "vasomotricity" of the lymphangions. Because they are not connected to the heartpump, they need the help of the whole environment (muscle-contractions, peristalsis of the organs, movement of the diaphragm ect.) but especcially, like all smooth muscles, a good balance of the autonomic nervous system.

Lymphatic Drainage is a normal and essential function of a healthy body but it can be affected for example, by an accident, illness, radiotherapy, or simply by stress - the body only allows itself to be drained if it is relaxed, that is to say, in "parasympathetic". By inhibiting the "ortosympathetic", the "parasympathetic" is able to fulfill its function in the autonomous myogenic automatism of the lymphatic vessels (angions). The original Manual Lymphatic Drainage ad modum "Vodder" (MLDV) stimulates first of all the "parasympathetic".

ABOUT THE AUTONOMIC NERVOUS SYSTEM, ANS

The automatism of our *smooth muscles* (viscera, vessels and lymphatic system) is stimulated by the ANS.

The ANS is involved in the functioning of all the automatic mechanisms,

- the pulse, blood pressure and lymph- flow - heart beat,
- vaso -constriction and -dilation of blood and lymph-vessels,
- body and skin temperature – perspiration,
- respiration,
- digestive system (stomach, intestines, pancreas) - liver, kidneys,
- hormonal and sweat glands, regulation of temperature,

- supply of glucose and oxygen to the active muscles,
- secretory and execretory systems (intestines, kidneys, skin, mucus)
- healing wounds and injuries
- recovery and recuperation of the body through rest and sleep
- storage of digestive products (glucides, lipides, proteins etc.)
- adaptation of the body to its environment,
- elimination of waste,
- storage of glycogen in the liver, transformation of glycogen (starch) into sugar,
- hormonal production,— etc.

The center for the regulation and integration of the ANS, is in the hypothalamus which works in combination with the central nervous and endocrine systems. The hypophysis, stimulated by the hypothalamus, produces along with other endocrine glands, more than 30 hormones, which in turn stimulate the hypophysis and the hypothalamus.

Our emotions have a direct influence on the *autonomic neurvous and hormonal systems through the action of neuro-transmitters which explains their delicate equilibrium.*

The balance between the two complementary and antagonistic parts of the ANS, the (ortho)sympathetic and the parasympathetic with its principal nerve the vagus, is what maintains our total well-being.

THE ROLE OF THE SYMPATHETIC AND THE PARASYMPATHETIC (Table 1)

Because of our present way of life, the orthosympathetic system is far more in demand, resulting in "stress" which expresses itself in these 3 phases:

1. alarm
2. resistance
3. exhaustion

In addition to this, the reactions provoked at the hormonal level are less and less able to achieve the physical reactions corresponding to phase 2, and pass directly from phase 1 to phase 3, leaving a painful imprint. (Selye) This results in tissue damage, pollutes our inner environment and disturbs the acido-basic balance.

It is therefore essential to be aware of the predominance of the "ORTHO" in order to restore the balance indispensable to our well-being through recognition of the "PARA". Thanks to the regular working of the parasympathetic system, a healthy body organises its own defence system and has the ability to heal itself. It is the hypothalamus working in close collaboration with the nervus vagus and passing through the cerebral trunk which conveys the parasympathetic influence to the whole body.

This "parasympathetic" chain == the "psycho-neuro-endocrino-lympho-immunity".

The psychosomatic notion becomes a reality, according to Prof. LABORIT and the classification of the influence of stress by Dr. SELYE, who also describes the effect on the lymphatic system. Already, in '85, Dr. Steven Locke wrote his book "The Foundation of Psycho-neuro-immunology". The medical faculty of the University of Massachussets, encouraged by Jon Kabat-Zin and Saki Santorelli uses this research in its "stress reducer" clinic.

Prof. Mislin has already demonstrated that the ANS has a direct influence on the lymphatic system through the vasomotricity ot its angions. Prof. Hutschenreuter of Ulm University submitted a paper to the congress for "Stroke Management" (organised by the

Table 1.

(ortho)SYMPATHETIC	PARASYMPATHETIC
"Survival"	"Recuperation"
When the human body, often stimulated by *fear or ambition*, has to protect itself, fight flee, or be at its best.	When the human body has to be restored to health by *confidence*, rest, sleep, relaxation, enjoyment, nutrition, elimination of waste products ...
The sympathetic (catecholamines) STIMULATES the ACTIVITY of certain organs or mechanisms:	The parasympathetic (acetylcholine) STIMULATES the ACTIVITY of, (for example, the pancreas secretes insulin to activate):
a. The adrenal glands secrete adrenalin to activate the muscles, the heart, respiration, the sweat glands.	• mental activity • stomach, intestines, kidneys
b. The liver transforms glycogen into sugar to feed the muscles.	• superficial blood circulation and that of the non-active zones of the organism
Consequences: • enlarged pupils • muscles contracted • heart beat accelerated to feed muscles and active organs • increased blood pressure • enlarged bronchia • deeper, faster respiration, etc.	Consequences: • superficial(skin) blood circulation improved • deep blood circulation normalised • resumption of gastric secretions and peristalsis • elimination of waste etc.
At the same time the sympathetic, under the influence of adrenalin, REDUCES the ACTIVITY of certains organs: • restricted mental capacity • cessation of stomach and intestinal activity • constricted throat • paralysed vocal chords • slower circulation in the non-active zones, etc.	At the same time the parasympathetic under the influence of insulin, REDUCES the ACTIVITY of certain organs: • heart activity reduced • liver no longer makes sugar • respiration regulated etc. BUT ALSO • increased mental capacity • digestion restored
Consequences: • the gastric juices are no longer secreted, etc. • blood drained to active zones diminished superficial circulation • vessels contracted in the non-activated zones • lymphatic blockage.	Consequences: • blood pressure regulated • relaxation of muscles • heart beat slowed down • vessels contracted in the zones which were active • vessels enlarged in the zones which were blocked etc. • lymphatic stimulation.

W.C.P.T.E. in Copenhagen in '95) showing how a purely neurological problem affects the lymphatic vasomotricity and is the primary cause of oedema of the hand, frequent in cases of cerebral vascular accidents. Endocrinologists are increasingly confronted by cases involving "water retention" following neuro-vegetative imbalance (Congress of Beauty and Fitness, Uruguay'90).

The direct link between the ANS and auto-immune and auto-deficient illness is now recognised in Immunology. At the University of Utrecht Psychologists, Neurologists, Lymphologists, Physiotherapists specialising in DLMV, and others are working together to study this relationship.

Because of its automatic functionning, medical science has long considered it difficult to exert influence over the ANS which, in any case, generally works well. This may explain why some eminent lymphologists continue to deny this relationship and exclude the ANS from their research. Nevertheless it is at the origin of many diseases affecting the lymphatic system. The facts demonstrate that a majority of patients suffer from neurovegatative imbalance with a disorder in the area of the neuro-transmitters, leading to serious lymphatic problems.

In addition to this, it has now been shown that it is possible to influence the ANS- which explains the trend for research into relaxation techniques (often involving the extensive cooperation of the patient, who collaborates whole-heartedly from the start, unfortunately at the risk of producing a new ortho stimulation!!)

THE "VODDER" TECHNIQUE REGULATES THE ANS

This results in orthosympathetic inhibition and parasympathetic stimulation.

Its BASIC technique i.e.*"jugular chain – abdomen – mobilization of the diaphragm"* immediately locates its action on the vagus nerve. The *nervus vagus* coming out of the cerebral trunk, being much in evidence in the area of the neck and even around the fossa jugularis, will be in direct connection with the diaphragm and the sacral region. Thus, stimulating the nervus vagus by the movements in the area of the jugular chain also repercutes in the sacral region.

In addition, *the rhythm* is primordial in the "Vodder" gesture. It adapts to each person's rhythm whilst leading it to a deceleration. It respects the slow rhythm of the lymphatic vasomotricity (as studied by Mislin, and also in a study described by Prof. Michael Foldl in 1971 in his book "Erkrankungen des Lymphsystems", comparing "slow" MLD and so-called "normal" MLD.)

The very light pressure, complementary to the hydro-static presure of the tissue, also has a parasympathetic effect (study of Prof. Eberhard Künckhe, Bonn.)

The pumping movement, however light it may be, provokes a discharge of histamine under the skin which results in a hyperaemia (to be avoided in drainage) (Dr. Johannes Asdonck, Feldbergklinik)

The spiral movement activates the pulsation of angions (Mislin) and inhibits the orthosympathetic system. (Prof. Hutzchenreuter, Ulm.)

The experience of the practitioner means that he can tune himself into the "parasympathetic" with his patient in order to be as receptive as possible during the treatment.

Practitioner and patient together in "parasympathetic" become aware of the fluid and dense areas of the body through MLDVodder°

They allow the "Para" to do its work, draining the lymph and purifying the body.

EXPERIMENTS, CARRIED OUT, AT THE VIRGINIA COOL SCHOOL OF PARASYMPATHETICS, THROUGH MANUAL LYMPHATIC DRAINAGE "VODDER" ABOUT "THE RELATIONSHIP BETWEEN MLDV AND ANS"

1// 27 years of practical experience in MLDV with its specificity concentrating on the jugular chain and slow, rhythmic spiral movements.

The testimonies of 20 years of "case reports" from specialised student physiotherapists.

2// *Electrospectography* (KIRLIAN) (before and after treatment and cours,) with the physician Herve Moskovakis, (Paris) and Interpreted by Dr.Med. Andre Banos (Bordeaux /F):

Evelyn Janke-Brione, physiotherapist, Paris treated 2 patients, with 15 minute MLDV, (basic, jugular chain and abdomen) The feet reappear on the photo, immediately after- and remain the following 8 days.

12 MLDV practitioners at the Virginia Cool School,drain each other, the amount of time they desired, only 2 improved- for the remaining 10, the stress symptoms are visible on the photos. They were in an "ortho" state concerning the success of the experiment.

During a basic 3-day MLDV Course (electospectography by Tony De Beuckelaere, Antwerpen and Phillippe De Paepe, Brugge, Belgium) out of a group of 30 students, the first day of the course, everyone showed a predominance regarding the hands to the detriment of the feet. However by day 3 of the course, they all show a neuro-vegetative rebalancing which is revealed by the longer "streamers" and a re-establishment of energy in the feet.

The teacher, following the three day course, also shows the same improvement.

The same group 1 month later: there is a clear difference between 15 physiotherapists who have drained regularly (an average of half an hour daily) and 15 who have only done classic physiotherapy. The first group continued to improve, the second group regressed to their original situation. This has been the subject of a study by Dr. Banos in Bordeaux, "*the MLDV practitioners themselves improved.*"

3// Measures of subtle energy with the "*Rathera*" instrument of Drs. Voll and Schimmel, with 50 mll. Amp. intensity, taken by the physiotherapist A. van der Burgh (Rotterdam.) Basic MLDV on 10 patients. Immediate energy improvement for everyone at all levels, with a clear predominance of rebalancing on the point D. of Voll i.e.meridian nerve between the index and second finger, and in the area of the meridian of the heart.

The lymph system is stimulated, it is measured on the external side of the thumb.

4// The *biological evaluations* (with Basic and Abdomen MLDV.)

Test carried out by Dr. Med. Josiane Bru, (Millau), Proteinic profiles and Lymphocitic sorting.
13 students (5 Day Course MLDV.) 10 patients (10 sessions in one month.)

The interpretation of Dr.Biologist André Burckel (Paris): "too little time for a Proteinic Profile (6 months), however in terms of cellular immunity, everyone experienced a clear immuno-regulatory effect."

Test carried out on the neuro-transmitters by Dr. Corinne Scoruptka, and Dr. André Burckel, on 10 patients:10 X MLDV by Evelyne Janke-Brione.

All evolve towards the norm, and even the evaluations of the patients tending to auto-immunity (dry syndrome) regularised.

It is clear that one cannot claim a "scientific" value for these experiments, but they were made with the single aim of reassuring us in our work. They show in another light the improvement we acknowledge in our patients after practising the "basic" treatment specific to MLDV. It is precisely these "basics" which certain scientists judge superfluous. We are practitioners who witness from our experience, our work in interaction with the patient, and not with our dossier.

In addition to this, there seems to be a reluctance on the part of the scientific community to tackle the subject of ANS. It is certain that until recently it was difficult to measure. Now, however, we have the possibility of objective measures thanks to biological evaluations which have evolved with the tests of neuro-transmitters. It is a pity that electrospectography, despite its objective image, continues to give rise to much scepticism.

TO SUMMARISE

The physical and mental health of "human-being-as-a-whole-in-an-environment" depends on the fluidity and adaptability, of his inner shere (milieu).

The action proposed is simple and manual. "The Vodder Technique" acts directly on the pulse, motion, flow, biological rhythm, quality of our internal water (milieu), by its spêcifical rhythm, its spiral wave motion, and its gentleness.

Specific manoeuvres act immediately on the Nervus Vagus, provoking stimulation of the Parasympathetic system, great restorer of the organismus, by its direct action on the entire lymphatic system.

Physical relaxation and mental awarness result from this stimulation.

The PSYCHO-NEURO-ENDICRINO LYMPHO IMMUNITY chain is engaged, procuring confidence and well being, both for the practicioner and the patient.

BIOLOGICAL RESONANCE AND THE STATE OF THE ORGANISM

Functional Electrodynamical Testing

Gábor Lednyiczky and József Nieberl

Hippocampus Research Facilities
Nánási út 67, H-1031 Budapest, Hungary
Tel: +36-1-430.8343; Fax: +36-1-430.8342
e-mail: hippocam@mail.datanet.hu

> To discover the true causes of everything that happens to a being, we must always look to the possibilities inherent in the very nature of the being itself.
>
> —*René Guenon*

ABSTRACT

A brief survey of electromagnetic field interactions in living systems leads into the theoretical foundations for biofeedback processes. The continuous information exchange between a living system and its environment and within the system, is shown to occur via electromagnetic field (EMF) interactions. The quantification of this continuous information exchange is employed in a device (the Cerebellum Multifunction Medical Instrument, CMMI) which allows a substance-specific monitoring of the ongoing regulative processes of the body. With this device, complex adaptation processes in an organism can be tested. Such processes are based on the information exchange between a living system and its environment, so that the procedure is actually a functional electrodynamic testing (FEDT). With an FEDT instrument, a physician can determine the electromagnetic state of a patient, and from this can make a diagnosis without the necessity of invasive methods. In order to further avoid an invasive character, extremely-low-intensity EMF signals are used in the CMMI. Due to the fact that the patient is exposed to the informational character of the homeopathically prepared body-specific constituents, the diagnostic procedure is also a kind of treatment itself.

Potentiating Health and the Crisis of the Immune System
edited by Mizrahi *et al.* Plenum Press, New York, 1997

INTRODUCTION

Just as a society cannot be reduced to a sum of individuals, a living organism and its functions cannot be reduced to a set of chemical reactions, even if it were possible to account for all of them. Yet, the dominant paradigm thinking suggests just this, that the dynamics of the whole could be understood from the parts. The list of medical disasters that have occurred due to the shortcomings of this kind of thinking is long, and will continue to grow unless a shift in paradigms takes place. One of the essential points of a new paradigm is that the properties of the parts can be understood only from the dynamics of the whole [1 and references therein].

The Cerebellum Multifunction Medical Instrument (CMMI) (Hippocampus Institute, Hungary) was designed and developed to be able to estimate this dynamic relationship and hence make a diagnosis. A perturbance at the cellular, tissue, organ, or system level will effect the functioning and performance of all other levels in an organism. Functional Electrodynamical Testing (FEDT) with the CMMI is a very effective tool for estimating the condition of the body, as the entire body's conditions are measured and hence can be treated as a whole.

Trying to understand an organism, or the health of an organism, by only considering the static elements (as with the old paradigm) can be compared to trying to learn a foreign language by only reading the dictionary. The underlying guidelines may be understood, but the dynamics of the living language will remain unknown and hence the language would still be unintelligible. In order to understand the wisdom of an organism, we have to be able to 'speak its language'.

The CMMI is a unique synthesis of modern thinking and modern technology with ancient wisdom about the body and its relationship to the environment. The body's reactions are measured as a part of its dynamic relationship with the environment and a diagnosis can be made which will ensure a higher degree of complementarity with the environment.

Recent studies (by ourselves and many others [2–6]) have shown the crucial role of continuous information exchange within living matter up to the most subtle levels of biological functioning. This has made it possible, even obligatory from the clinical point of view, to construct a device which will allow a substance-specific monitoring of the ongoing regulative processes of the organism. With such a device, the human body is exposed to the information (stored in the endogenous electromagnetic signals) of homeopathically prepared body-specific constituents and is then free to respond (also by altering its endogenous electromagnetic field pattern) to any piece of this information according to the organism's own choice. This way, the organs functional states in the endogenous information processing are measurable and classifiable. As a result, the vital reserves of the body as a whole are reflected through the vital reserves of the actually tested sub-system. In this way, one can test nearly all aspects of the body (ammino acids, enzymes, fatty acids, hormones, minerals, the impact of viruses and vitamins, etc.).

In order to understand the functional aspects of the CMMI, it is necessary to establish a theoretical understanding of the electromagnetic attributes of all organisms and their interrelationship.

ELECTROMAGNETIC (EM) BIO-COMMUNICATION

All organisms radiate a very low intensity endogenous electromagnetic field (EEMF) in the range below 1 Hz up to 10^{15} Hz, as a result of biological processes. Electrolytes moving in an organism (e.g., via the circulatory system or within cells) create an electromagnetic field. Low frequency fields are generated in cells from the alteration of protein configurations, changes in the amount of lipids, and across cell membranes due to the migration of ions. High frequency fields are generated by enzymatic peroxidation, ATP production, the Krebbs cycle, and natural luminophores in nucleic acids and proteins. These high frequency radiations from cells are what Fritz-Albert Popp has termed "biophotons" in order to emphasize their endogenous origin and substantial role in biological communication as well as the optical aspect (visible and ultraviolet light) of the electromagnetic spectrum. Biophotons are considered to be the mechanism by which intra- and intercellular communication takes place. The weak light emission ("dark luminescence", "ultraweak" photon emission) from biological objects has been measured by photon counting techniques and has an intensity of a few up to some hundred photons per cm^2 surface area per second [2, 3, 7].

The theory which inspires many current studies is Alexander Gavrilovich Gurvich's notion of the "vectorial biological field", or "morphogenetic field" [8]. Gurvich introduced the notion of the morphogenetic field to account for a wide range of biological phenomena from metabolic processes to the psychic sphere. Many investigators, most notably Fritz-Albert Popp and Mae-Wan Ho, have established a link between EMF and the morphogenetic field. The morphogenetic field contains the information of the whole from a part; each cell in a system is a reflection of the surrounding cell's architecture (spatial arrangement) and each cell makes a contribution to the architecture as a whole. Popp introduced the biophoton as a contributing factor in the morphogenetic field effect. The biophoton is a high frequency photon (in the UV and visible light range) emitted by a biological object which is then received by another cell, thus facilitating intercellular communication. Multiple studies of biophoton emission from unicellular organisms up to primates demonstrate that cells are influenced by the biophotons from other cells and "respond" with their own (endogenous) biophoton emissions [2, 9]. Ho has demonstrated (with *Drosophila Melaongaster* embryos) that the biophoton flux may have biological significance in the synchronization of development to external light [10].

COMPLEX ADAPTATION PROCESSES: INFORMATION EXCHANGE BETWEEN A LIVING SYSTEM AND ITS ENVIRONMENT

Frank Brown challenged the paradigm that circadian rhythms in organisms are linked to either sunlight exposure or tidal activity. In his experiment, oysters in an aquarium with constant light, temperature, and water levels opened and closed their shells in synchronicity with their compatriots who remained on a Connecticut beach. Then, the oysters were moved (by Brown) 1000 miles west into a light-proof box in Illinois where they were placed into an aquarium. Initially, the oysters remained on Connecticut time, but in a few weeks shifted to the would-be Illinois tidal pattern [11].

Rutger Weaver designed an experiment in which several hundred males lived in underground rooms for up to two months in an environment cut off from light, time, sound, and temperature which were initially presumed to be the normal cues of circadian rhythms. The two rooms were identical except that one was shielded from electromagnetic fields; various parameters such as sleep-wake cycles, body temperature, and urine content were charted for both groups, and both groups soon developed irregular rhythms. Those living in the shielded room became thoroughly desynchronized, while those still in EM contact with the Earth's fields held a rhythm close to 24 hours. Next, Weaver introduced various EM fields into the shielded room, none of which had any effect save a 10 Hz, 0.025 V/cm field, which restored most of the parameters to normal [12]. Arthur Pilla and Michael Manning conducted bone density's conservation circuitry experiments which NASA uses in the Space Shuttle currently [13].

These examples evidence the essential role of electromagnetic interactions in the information exchange with the environment that provides the complex adaptability of organisms.

ENDOGENOUS AND EXOGENOUS CONTROL MECHANISMS

Metabolic activity depends on the electric properties of membrane potential and environmental EM conditions [14–34]. Continuous adaptation to changing conditions, hence continuous readjustment of the parameters of the biochemical reactions inside the body, is characteristic for living matter. Any change or adjustment (with a rate exceeding a certain threshold determined by an organism's adaptability) is considered a perturbance of the system, irrespective of whether this change is intended to cause or prevent illness. Illness in general generates this communication breakdown within the organism's functional network. Since living beings are highly integrated open dynamic systems, wholeness in general is maintained by a permanent mass, energy and information exchange. The dynamics of communication are thusly vital for organisms.

When an organism is treated on a more general level of its functional dynamic hierarchy, it is easier to restore the physiological communication pathways within it and thus activate the endogenous healing processes. Alterations in the biophysical parameters, primarily electrophysical, occur at general levels of the organism's functional hierarchy. Therefore, they are responsible for the very subtle intimate mechanisms of an organism's self-regulation and interlevel communication through resonance (tissue coupling) interactions.

Every level of an organism's hierarchy possesses a characteristic spectrum of endogenous electromagnetic oscillations originating from various processes. Intra- and interlevel resonances should occur to maintain wholeness, more or less providing correlations between these processes. From this point of view a pathology, which may be born at any level, will perturb all oscillations via wave interactions, irrespective of the origin of such waves. The distorted interference pattern of the endogenous waves of a sick organism is a reflection of its improper biochemical processes.

Electromagnetic resonance interactions between the endogenous electromagnetic oscillations of organisms are suggested to occur in living systems; however, an attempt to detect them is a rather complicated problem [35]. Nevertheless, the still growing number of therapeutic devices, which use such kinds of interactions, is elaborated on in [36–39]. For example, the more than 15 years of 'devices for bioresonance treatment' utilization in various European clinics evidence their efficacy in the treatment of many diseases [38–41]. They are designed to use resonance interactions between endogenous electromagnetic oscillations.

Numerous positive experiences in the application of electromagnetic therapy devices makes it possible to assume that device-induced restoration of the interference pattern will renovate physiological order in a sick organism. The problem is to isolate basic processes (and the frequencies which correspond to their time scales) which are common to all levels of an organism's hierarchy in certain frequency ranges and can thus open pathways of interlevel signal transduction.

COMMUNICATION PATHWAYS AND RESONANT INTERACTIONS

The meridian system, according to Taoist description, is the communication system of the body which rises to the skin at certain points (acupuncture points) [42–46]. Becker, Voll and many others, have measured the relative resistance of the skin at the proposed acupuncture points and found that resistivity drops exactly at the points suggested by ancient Taoists [47–50].

At the microscopic level, numerous attempts to elucidate the extremely-low-frequency (ELF) signal transduction pathways of the interactions with cell membranes and subcellular components have been made by measuring various cellular and subcellular characteristics while exposing the studied systems to experimentally generated external ELF fields [22–30, 51]. Alternating electromagnetic field treatment induces observable responses in biological systems. Many processes turn out to be frequency dependent with thresholds or some peculiarities at certain values of external fields [25, 52–55].

The information obtained thus far is still insufficient to offer a reasonable mechanism for EMF interaction with biological tissues. Nevertheless, we would like to emphasize some general features of such kinds of interactions, primarily: the "windowing" of frequency and power in tissue interactions with weak EMF as revealed by William Ross Adey. He studies the behavioral and neurophysiological effects of ELF and modulated radiofrequency (RF) fields as well as the responses of calcium ion binding in tissues to ELF and RF fields [53–55]. The occurrence of amplitude and frequency "windows" in a biological object's response to external stimuli is essential for an understanding of how biological systems can show a high sensitivity to external ELI fields, yet remain stable under intrinsic fields several orders of magnitude larger [50, 56]. The natural dynamic complementarity of inherent and environmental electromagnetic signals (frequency, amplitude, phase and the composition of complex signals), ensures a very fine selectivity of the available information from electromagnetic noise as well as preventing a "dissolving" in environmental electromagnetic fields.

Bearing this in mind, various electromagnetic therapy techniques - especially, those employing low-frequency and low-intensity signals - are promising with respect to initiation of healing processes. Because of their low-intensity and non-locality, such signals come into play at very general levels and may be involved in the inherent mechanisms of the non-specific defense of the organism [34, 35, 57–59].

Resonance is an elusive property which is responsible for synchronizing pendula as well as destroying bridges and buildings. Resonant frequencies establish a kind of sympathetic relationship between objects which have the same period of oscillation, such that the motion of one object will influence the motion of another without any physical connection.

Bio-resonance works on this same principle. It is a method of communication between objects, and apparently, biological objects have the capability of "tuning" their resonant frequencies to match that of another object. Since this is a non-local event, it is difficult to determine the cause and effect of resonance between two or more biological objects. The argument has been made that in order for inter and intra cellular communication to be successful, that is, no disturbances or distortions in the message, the cells need to be in resonance.

To the degree that one can use phrases to re-enforce a notion, our common usage of "being on the same wavelength" to describe an understanding of what another is saying might convey human's more intuitive notion of resonance. This may also account for the good or bad 'feeling' that one gets when first encountering someone else. Is this other person capable of resonating with you?

The notion of canceling out certain undesired frequencies, as in Smith's allergy experiments [60], suggests that an organism would have to be able to attune itself to the frequency emitted by a certain substance. In order to cancel out a signal, or to receive information from a signal, the receiver and sender must be synchronized. Otherwise, the signal is just data, which will be unintelligible to a receiver which is not in resonance. Below, the evidence for this kind of bio-resonance is listed.

KNOWN FROM FORMER RESEARCH

The previous results of Hippocampus Institute's studies show that:

- The integrating regulatory role of cellular level endogenous oscillations is explicitly evidenced by their influence on malignant (human lymphoma and melanoma B16) cell cultures. Extremely-low-intensity endogenous electromagnetic fields (EEMFs) induce a regulative halt in cell division which is also proved by a comparative study of the cytotoxic and cytostatic effects of EEMFs of embryonic and malignant cell cultures [62, 63].
- The regulative function of the EEMFs in naturally occurring cell populations involved in a complex metabolic process is proved by the study of the EEMF effects on the phagocyting activity of human blood [63].
- EEMFs are shown to influence the functioning of the immune system and correct a condition of immune deficiency provoked by the exposure to continuous, low-intensity radiation (in mice kept in Chernobyl) [64].
- The EEMF feedback of women suffering with mastopathy is shown to improve the patients' state [65].
- EEMFs are shown to influence the viability of heat shocked *Drosophila melanogaster* chrysalises [66].
- At the subcellular level, modulated endogenous EM oscillations are shown to affect kinetic and thermodynamic characteristics (as well as structural dynamics) in water and biological solutions (human blood serum and nucleoproteid complexes) [67, 68].
- Human EEMFs are shown to influence the processes of self-regulation in chaotic chemical oscillations [69].

A LIVING SYSTEM CAN ONLY BE REGARDED AS A PART OF ITS ENVIRONMENT

Andrew Weil notes that most synthetic pharmacological substances are only semi-synthetic in that they are nearly always based on a natural compound (the 'active ingredient' of a plant or part of a plant) with a slight shift in the structure of the molecule. This is because humans do not generally respond well to purely synthetic substances, and in the case of psychoactive substances, the brain lacks the appropriate neurotransmitter to distinguish the substance and respond [70].

It is unlikely that this is mere coincidence. The dominant paradigm of evolution erroneously suggests that organisms accidentally respond to their environment (as it is shown in [71, 72 and references therein]), thus tending to suggest that an organism is not an aspect of its environment. And though this idea runs counter to intuition, it is what is generally accepted, and taught in schools. Unfortunately, the paradigm has lead to a feeling of being separated from nature, and many pharmacological disasters have occurred over the last century due to the failure of pharmaceutical companies to recognize that all organisms are derived from, and are an integral part of their environment.

Until recently, electricity was not considered as part of the environment due to the lack of evidence of EMFs having an effect on biological objects. There are still substantial disagreements as to the extent and effect of EMF on living systems, but many developed nations have adopted certain guidelines for maximum exposure limits to certain field strengths [73]. Westerners have considered electricity and magnetism to be vital to life processes for at lest 200 years (e.g., by Mesmer and the 18th-century vitalists) however, at the time it was largely discredited. Modern possibilities of detecting endogenous AC electrical oscillations in cells [13–15, 50, 56, 73 and references therein] reveal that the endogenous fields are strongest when cell metabolism is most active. No signs of AC oscillations are found in dead or heavily poisoned cells [14]. These measurements make it possible for Herbert Pohl to assume that endogenous oscillations must accompany cellular reproduction, and vice versa - reproductive processes cannot proceed without endogenous AC oscillations [16]. This testifies to the key role endogenous EMFs play in the dynamical maintenance of an organism's stability.

Many thousands of people with heart troubles have a "pacemaker" inserted in their body to keep the heart beating at a steady rate. This can be considered an external pacer as it has replaced the body's own (evidently faulty) internal pacing system. The aforementioned works by Weaver and Brown [11, 12] demonstrate that all organisms rely at least partially on certain pacers from the environment (external pacers). NASA has also realized the need for these environmental pacers and now equip manned space voyages with an artificial Earth field so that the astronauts can maintain their normal bodily rhythms.

The fundamental self-awareness of living matter has been based on continuous communication between living subjects and the environment, during the whole process of development from protozoa to primates. This development occurred in the natural electromagnetic field of a broad frequency range (at least, from the order of 10^{15} Hz (ultraviolet light) to the order of units of Hz (Schumann earth/ionosphere cavity resonance [74]). This supports the occurrence of the endogenous mechanisms of electromagnetic signal modulations in a broad range of frequencies. Internal electrodynamic field coherence is evidenced to be the instrument of biological organization at many levels starting from the observations of the embryonic field in eggs [72].

After his lifetime's work Brown concluded, "No clear boundary exists between the organism's metabolically maintained electromagnetic fields and those of its geophysical environment" [11]. Unfortunately, information obtained thus far is still insufficient to offer a reasonable mechanism for EMF interaction with biological tissue. This is not entirely surprising since the theoretical exploration of EMF interactions with organisms only began about 30 years ago when Fröhlich began to apply his theory of dielectrics to biological systems to describe the propagation of EM signals in a given system [75].

Even if we try to do our best in order to appear self-ruling, we actually find ourselves highly dependent on the weather and the mood of encounters, as well as ones own attitude of how "things are going". Speaking more scientifically, the human organism is essentially incorporated into the entire network of natural interrelations via a permanent balancing of the organism's personal integrity with the integrity of nature as a whole.

Self-regulation of biochemical oscillatory cycles defines the natural selectivity. As it has been already mentioned, such self-regulation is maintained through a permanent information exchange within living matter. Louis-Marie Vincent has recently proposed a new approach to information by providing a conceptual tool adapted to biology [76, 77]. According to this concept, a message (transmitted by a means of communication) does not carry any information, only data. It is the receiver which makes an identification by recognizing the forms. In developing a theory for the mechanism of homeopathy, Del Giudice suggests that the EM information of a substance can be transduced into the surrounding water molecules (for instance, by affecting the field produced by large clusters of molecules), so that when the substance is diluted out, the information is retained in the water. Del Giudice writes, "the homeopathic remedy works only if it is meaningful to the array of previously existing signals in the organism; otherwise it is washed out" [78].

WHERE IS THE BORDERLINE BETWEEN THE POSSIBILITY OF RECOVERING BY NATURAL SELF-REGULATIVE MECHANISMS?

The Biodiagnosis and Biofeedback functions of the CMMI provide a fresh perspective on the "health" of a patient. As physicians well know, as many parameters as possible of a patient's state should be taken into consideration when making a diagnosis. However, even the most extensive questioning of a patient may not yield proper results, as the patient may fail to answer questions correctly, forget certain details, over- or underestimate their condition, or, conceivably, deliberately attempt to deceive the physician. The most common analogy of determining the health of a patient based on external factors alone is that of trying to determine the structure of an iceberg based on an examination of the above-water structure. The physician is often forced to use invasive techniques, which by definition of their being invasive, necessarily alter the patient's condition, further complicating the diagnosis (as well as often introducing unexpected side-effects). Extremely-low-intensity electromagnetic signals are used in the CMMI in order to avoid a possible invasive character of the diagnosis. Extremely-low-intensity signals make it possible to identify endogenous regulatory processes rather than to list all kinds of functional and morphological disorders based on deviations from normal values set by linear statistics. The latter is definitely not sufficient to estimate the extent of treatment intervention that will be adequate to the body's state. To be successful, the treatment should resonate with the pace of the changes - either positive (curative) or negative - in the body's state. The pace of such changes (adaptive changes during homeostasis), in turn, depend on the vital reserves of the organism. These vital reserves are tested by using the "adaptation test" option of the CMMI with respect to different information stressors about which are available as extremely-low-intensity electromagnetic signals of the corresponding substances implanted in the device. Needed information facilitates immediate regulatory activities (a triggering of phase changes in processes being in meta-stable states), the characteristics of the alterations of bio-electric parameters give an abundance of functional diagnostic information [60].

FEDT = FUNCTIONAL ELECTRODYNAMIC TESTING WITH THE CMMI

With the CMMI, a physician can determine the electromagnetic state of a patient, and from this can make a diagnosis without the necessity of invasive methods.

During the Biodiagnosis procedure, the patient is briefly (40ms, though adjustable) exposed to the magnetic fields of over 2000 homeopathic substances (not simultaneously, of course). Then, the patient's reaction to that substance is measured as value of the voltage changes (e.g., between the two wrist electrodes) and is translated into the reactivity (an average of the voltage fluctuations). Once all substances have been tested, the program establishes an overview of the patients integrity through the "Statistics" function.

As is mentioned above, every organism is in a constant feedback system with its environment. The statistics function measures to what degree a patient is capable of adapting (participating) to the environment. There may be a variety of reasons (both physiological and psychological) for this over- or under-adaptation, which naturally, the physician will have to determine. The uniqueness of this feature is that what a physician previously had to estimate (based on experience and intuition), can now be quantified, and even shown to the patient.

Each time the patient is tested, the substance list and statistics will be different. This may be alarming at first, as one of the most seemingly basic functions of a diagnostic device should be to deliver the same diagnosis given the same conditions. And here lies the key: the organism never experiences the same conditions from one moment to the next, so necessarily, the device will not provide exactly the same diagnosis from one minute to the next. In our age of massive toxic, nuclear, and electromagnetic pollution, as well as rapid transportation, an individual's fluctuations are more and more dynamic, they change every moment.

The device is a dynamic measurement system, and, as such, its power is in its ability to determine these very changes that happen from one moment to the next. Specifically, the physician, once having established the overall adaptability of a patient, can then go the repeat test function.

If the patient has either too narrow or too wide a statistics function the physician might consider terminating the test at that point and recommending to the patient possible ways of normalizing the overall reaction. In any case, the general adaptability of the patient must be considered for all repeat tests as all measured values are relative to the initial measurements.

The physician's initial assessment of the patient, combined with the statistics results, and the list of items on the "Substance List" can be used to then isolate the substances which are the most crucial to the contribution of the patient's condition. The adjusted values of +120 to -120 can be considered as informational values in that a strong reaction (far from 0) indicates the organism's missing or needed information which it is no longer getting from the environment. And here again, the dynamics of the reaction are crucial to a proper diagnosis.

Due to the fact that the patient is exposed to the informational character of the substances, the diagnostic procedure is a kind of treatment itself. This means that these initial reactions serve primarily as markers. It may be that the patient only needed the very brief exposure to the information, and reacted strongly initially, but the physician should not immediately assume that values which initially appeared with high values are what the patient's body needs.

Through repeated applications of the "Adaptation Test" (Transmit Button), the physician can determine to what extent the information is needed, or over saturated with. A patient's adaptability to a particular substance is indicated by the subsequent "Reactivity", "Rise", and "Fall" values as well as the character of the line. The adaptability to the information of a certain substance reveals how much that information is needed, as well as indicating possible pathologies from the inability to adapt to that substance.

In this way, the physician can simulate a homeopathic treatment and subsequently determine the expected and possible therapeutic effects regarding the tested substance.

DIAGNOSTIC PROCEDURE FOR THE CEREBELLUM MULTIFUNCTION MEDICAL INSTRUMENT

1. Anamnesis (Taking Medical History from the Patient and Creating the Null Hypothesis)

In order to understand the patient's state, we have to be informed about his or her situation on all levels of existence included physical, social, psychological, and emotional. Without understanding the context, the data cannot be converted into information. First of all, it's very important to get enough information from the patient in order to conceive the primary hypothesis (the null hypothesis). All kinds of data can provide important information concerning symptoms and their development (pathogenesis). In the holistic approach, all things and processes are considered to be interlinked. Despite the body's ability to express (by the measured electrical parameters) its reactions with a finite speed, time sets certain limitations to our measurements (that is, patients usually don't have an entire day for a diagnosis, also electrophysiological parameter are no longer 'natural' after one sits still for more than an hour). The null diagnostic hypothesis creates a kind of base line from which the outcome of the electrical diagnosis can be compared. This can provide a justification by the measured results or an adjustment where you can include pathogenic components or skip some of the ones you included in your null hypothesis.

2. Automatic Measurement

It is practical to start with a general screening, so choose "all substances" from the substance list. There are always substances (aspects) which may be forgotten while conceiving the null hypothesis, here you will have a chance to identify them on the screen. A reaffirmation of the null hypothesis is the other goal of this step of the diagnostics process. The differences between your null hypothesis and automatic measurement results remain for further testing. The permanent comparison of the results of the "Automatic Measurement" with the information from the anamnesis shall lead you on through further testing.

3. Statistics

The shape of the distribution diagram (number of substances plotted against the measured value of the electrophysical parameter under consideration) represents the ability of the organism to react (participate) to its environment, e.g. to the changes in it. This correlates to organism's reactions to the tested substances. On the x-axis, values from 0 to 4095 correspond to an adjustable voltage measurement of +3mV (4095) to -3mV (0), therefore, 2047 and 2048 signify no measurable reaction (0 Volts, in microvolt resolution). The number of substances is plotted on the y-axis. Usually, the distribution of the measured values is non-Gaussian. In all distribution diagrams (e.g., fig. 1), the Gaussian distribution is shown as a reference only, and does not correspond to the patient's actual distribution diagram. If the wings of the distribution of the plot are broad, it means that the organism reacted very intensely to many substances (fig. 1). An extremely high peak at the middle and diminished wings correspond to low reactivity (i.e., with the measurement

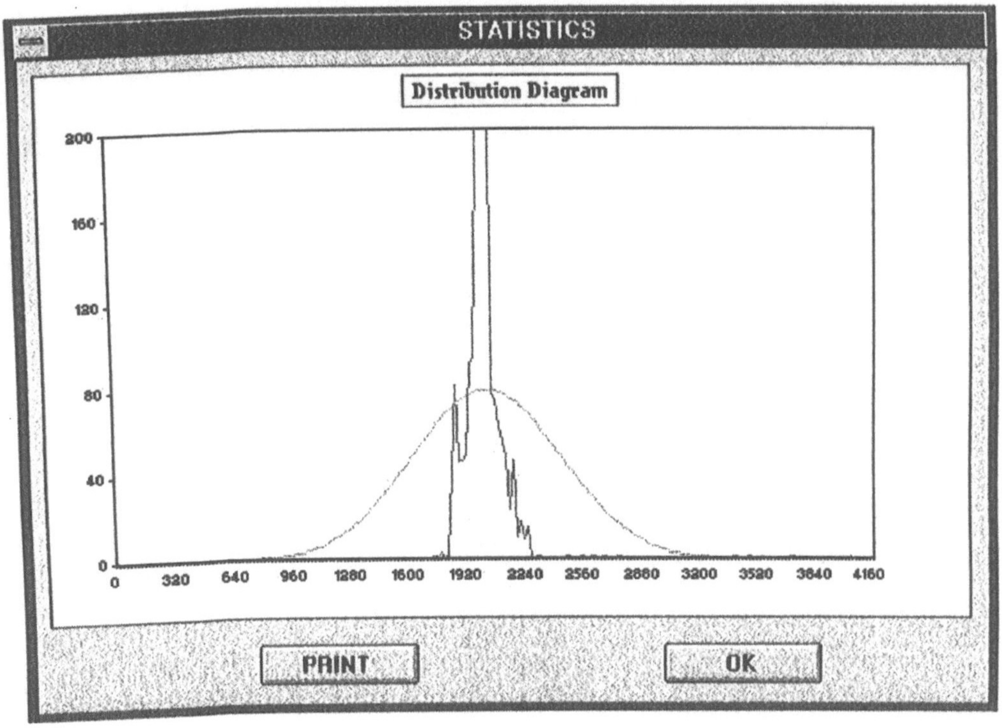

Figure 1. An example of a commonly found distribution diagram of a patient's reaction.

parameters applied (fig. 2)), and means that a fast, general test (short exposure time) cannot be performed on the patient. The distance between the maximum and minimum values (on the x-axis) measured from a particular patient is linearly subdivided into 240 units regardless of the number of measured substances. The newly obtained scale is labeled starts from -120 (at the minimum initial value) and goes up to +120 (maximum initial value). From this, the patient's individual distribution list is generated to serve as the base line for "Repeat test" measurements. We are interested in the amplitude of the changes due to adaptation which should be considered within the individual range of fluctuation of the electrophysiological parameters.

Broad wings signify a decreased integrity of the organism and an over-reactiveness to environmental changes without effective compensation. This distribution is mostly found in multiple-allergic patients and by those who are exposed to extreme environmental pollution.

A lack of wings indicates that the body doesn't want to show individual reactions (doesn't want to participate) and that it is keeping its integrity with all of its forces. This is a sign of an over-compensated state (called masking in allergy studies), when neural and informational systems are no longer responding because the organism does not want to risk its stability (integrity). This is found mostly in patients with a severe chronic overload coupled with strong determination, like many high position bank managers, and noticeable emotional control (stoic patients), and also in some introverted, psychologically unstable patients. Patients with this type of distribution need certain preparatory treatments before becoming electrically responsive. This is possible with the treatment unit (the treatment

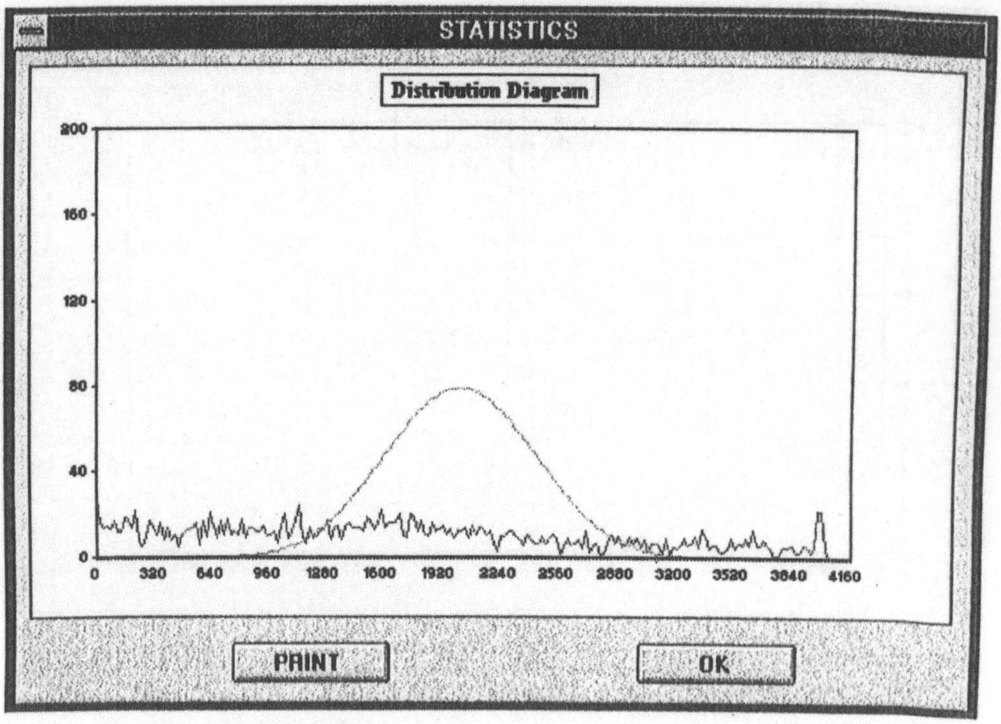

Figure 2. Distribution diagram of a hypererg patient's reaction during the automatic test.

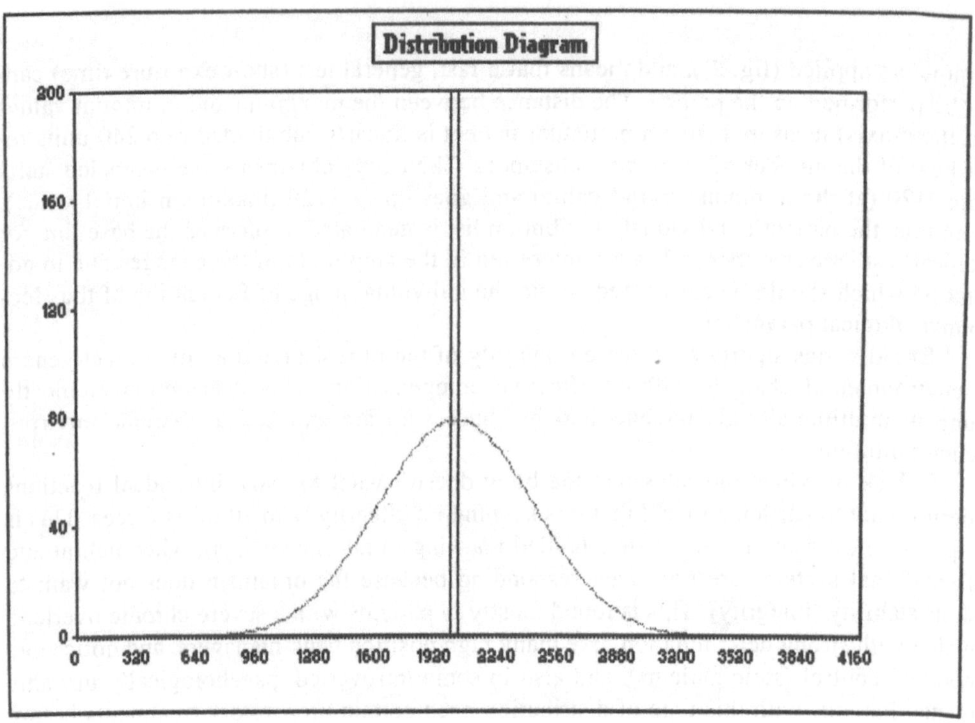

Figure 3. Distribution diagram with extremely diminished wings.

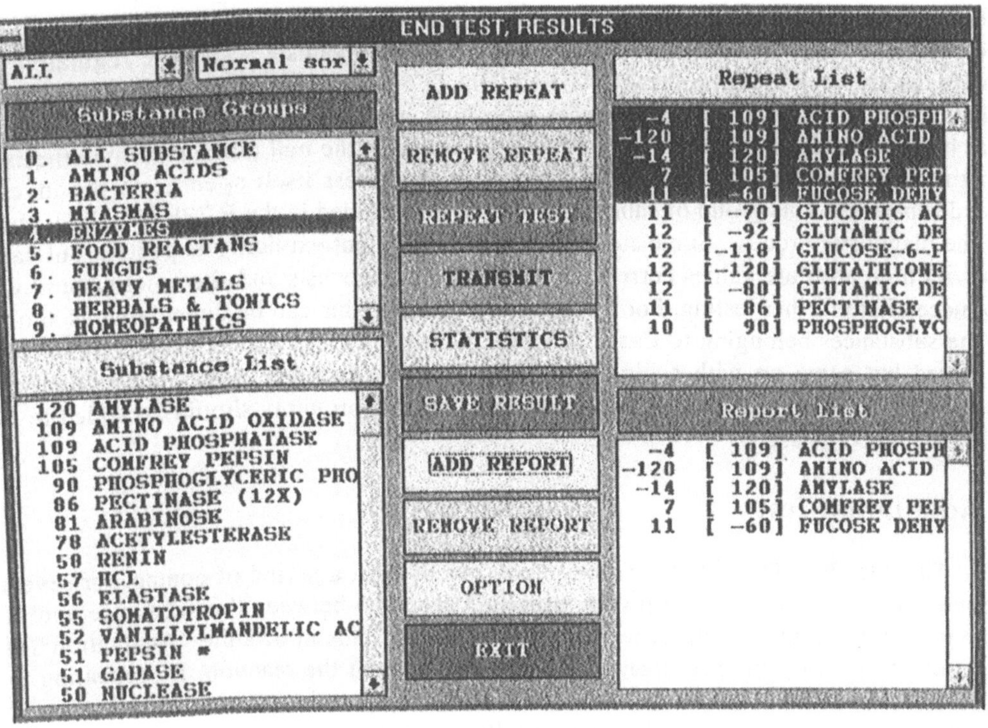

Figure 4. End test result window.

unit of the CMMI has specifically designed programs for this), or by applying longer "transmit" triggers of this diagnostic unit.

4. Result List

When the SUBSTANCE LIST window (located at the lower left corner of the screen, Fig. 4) appears after the calculations are finished, it begins by listing ALL SUBSTANCES. The x-coordinate of any substance (in the 240-point scale defined above) is listed starting with 120 (extreme right position in the distribution diagram) down to -120 (extreme left position). If you want to examine the results within a specific category of substances, you can click on a group name and open that group. The substances which compose the group will appear (in "Normal Sort" mode) listed by their x-coordinate in descending order. One should keep in mind that all values are measured on a relative scale, all values are relative to the patient in the moment that the measurements were taken. This serves the purpose of obtaining the dynamics of the patient's reactions to the tested substances during the REPEAT TEST period. However, a good starting point is to assume that values above 60 or below -60 indicate a "high reaction". Always relate those values to the shape of the "Distribution Diagram". Now we can compare the differences between the null hypothesis and those values measured by the functional electro-diagnostic testing (FEDT) in the automatic mode.

There will always be only a partial overlap between the lists of the substances with expected high values and actually measured high absolute values. Those which confirm the null diagnostic hypothesis need not be "Repeat tested". Yet, if you want to estimate the pathogenic priority of these substances and test the organism's specific regulative capacity toward each of them, the ADAPTATION TEST can be carried out. For substances which were expected but do not appear with high absolute numbers, the "Repeat Test" will help to decide whether we had a false assumption in the null diagnostic hypothesis or may indicate that the organism needs more time to express itself specific to the investigated substance. This group of substances should be included in the REPEAT TEST phase of the diagnostic process. Some substances which were not explicitly expected, but came up with high absolute values correlate well to the pathogenesis and the epicrisis and will not necessitate further testing. For further investigations, this can be considered as similar to the substances belonging to Case "A". Still more substances which were not explicitly expected but came up with a high absolute value will correlate either questionably or poorly. These substances need to be investigated more thoroughly should be added to the REPEAT TEST phase.

5. Adaptation Test

The ADAPTATION TEST function consists of first, a period of contact for several seconds (adjustable from the software, usually 5 seconds) between the substance and the patient, then the usual testing of the patient's reactions. This allows the "Adaptation Test Diagram" to provide an exact insight into the dynamics of the reaction. By repeating the ADAPTATION TEST, we can get the most detailed diagnostic information. "Reactivity" indicates the average of the 16 channels' results during the entire period which was monitored. "Rise" indicates the highest deviation upwards from the average. "Fall" indicates the largest deviation downwards from the average. In the "Adaptation Test Diagram" on the vertical axis, we can see the amplitude of the signal measured in the same units as the x-axis of the STATISTICS TEST. On the horizontal axis, time is indicated in milliseconds. The actual measured values are depicted in red, "Reactivity" is represented by the flat green line corresponding to the overall average during the actual measurement period. (In later models, the vertical amplitude axis is represents only the actual interval according to the actually measured intensities during the "Test Measure Time" .) Normally, we expect "Reactivity" to settle down around 2048, and the "Rise" and "Fall" values should be lower than 40. In order to understand the dynamics of the ADAPTATION TEST, the following tendencies are listed.

First of all, we follow the changes in the "Rise" and "Fall" parameters. The dynamics of the patient's adaptation to the field of a therapeutically diluted substance is very significantly characterized by the fluctuations of its electric parameters in both frequency and amplitude. The maximum amplitude of fluctuation is characterized by the "Rise" and "Fall" values. The frequency of the phase changes (where the slope of the line changes from positive to negative, or vice versa) are also observed as well as the average value over the entire measurement time (this is the reactivity value which is displayed by the device).

Usually, we include in the ADAPTATION TEST only substances which seem to play a key role in the epicrisis presented by the patient and by the former automatic measurement. The dynamics of the changes of reactivity in the adaptation test will allow classification of the meta-stable states. In the adaptation test (and in the whole diagnostic process) first of all, we want to collect information about which aspect of the pathology should be treated first. We are looking for not only the primary causes but the information which facilitates necessary self-regulative activities, and also the principle stressors, the elimination of which brings about the opportunity for normal self-regulation of the organism. The ADAPTATION TEST also makes it possible to identify the conditions under which the body's self-regulative activities seem to be insufficient and substitution therapy or other more invasive treatments may be required. The self-limitation of the method (stimulation of self-regulative activities) is also revealed by the method itself, so we can identify the sub-systems which are unable to communicate in a self-sustainable way.

In the case of a fast normalization of the "Rise", "Fall", and "Reactivity" values (e.g., within two ADAPTATION TESTS), we can conclude that a cell communication enhancement treatment (such as bioresonance therapy) would give the necessary first push and the organism could accomplish the rest of the recovery.

If the "Rise" and "Fall" values are only partially normalized (e.g., a stepwise function coming down from a higher value to a significantly lower value) we should consider a temporal administration of home medication in addition to cell communication adjustment treatments.

If "Rise" and "Fall" values don't tend to normalize even after 3–4 ADAPTATION TESTS, then the organism is incapable of any detectable degree of self-correction concerning the tested substances. In this case, the revealed principle pathogen agent cannot be treated by a local enhancement of the damaged endogenous control mechanisms. Thus, the identified substance is a significant contributor to the patient's pathology, and a longer preparation phase in the treatment is required before causal therapy can be applied (since at the moment, the organism has no vital reserves to react to a specific treatment regarding this substance); or, if irreversible tissue damage is threatening, we find an indication for immediate invasive intervention.

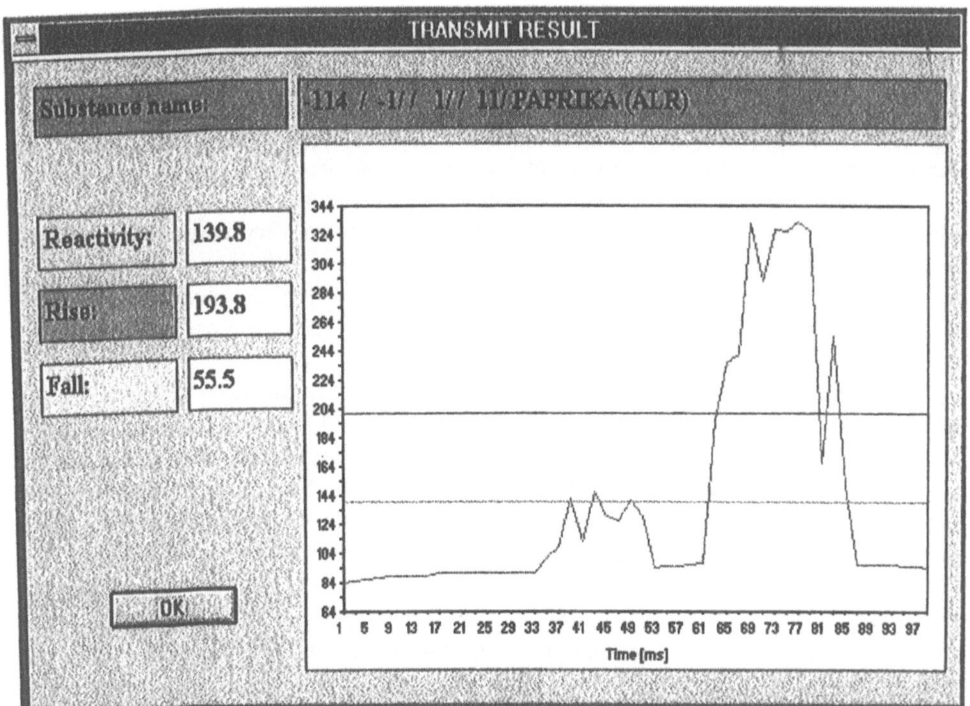

Figure 5. Adaptation Test result window.

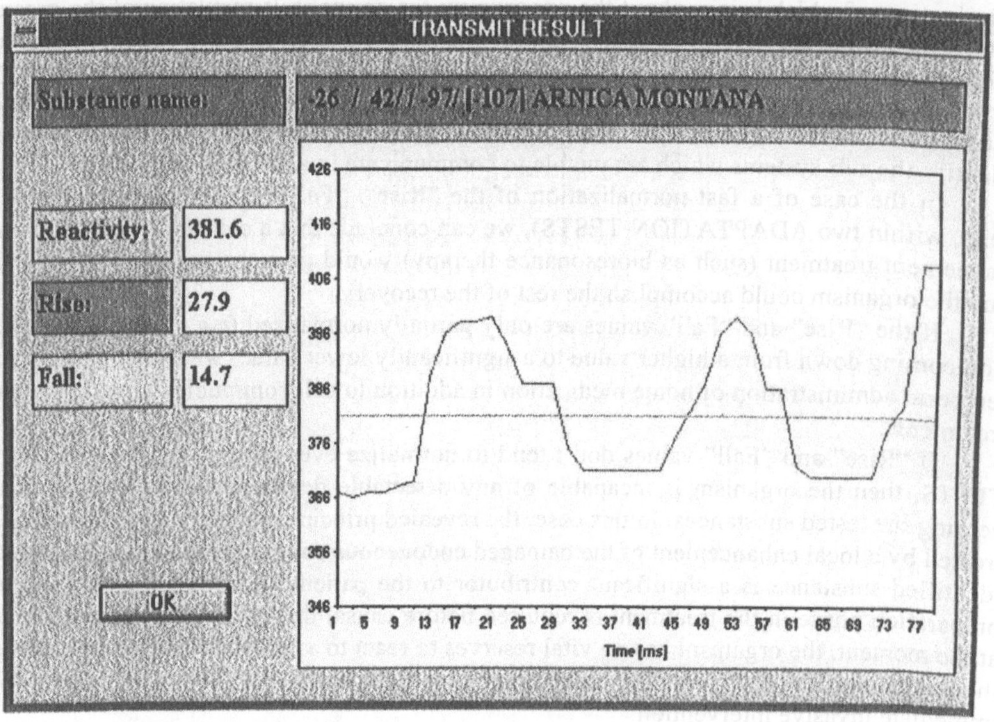

Figure 6. An example of a patient's reaction to a prolonged exposure (transmit test).

In some few cases, we find a strange kind of extreme fluctuation of the "Rise" and "Fall" values (from several hundred to almost zero and back repeated several times). This indicates a very severe situation, potentially due to a lack of vital reserves. The body is trying to decide whether to interrupt reflex pathways and ignore this substance or adapt to its field, and confusion ensues. This situation must be treated very carefully, and it requires that the body's vital reserves are built up first because if the body is immediately treated with regards to this function or substance, the patient may develop hypererg reactions. Therefore, don't include this substance in the treatment until the patient's substance or function specific vital reserves are at a sufficient level.

As Fig. 7 shows, all measurement characteristics can be adjusted in the software. We usually connect the patient and the test-substance for 40 ms and start measurement with the smallest possible delay time of 1 ms. One measurement block usually consists of 110 ms. With the usual measurement time of 80 ms this provides a 70 ms relaxation time for the body after connection with each subsequent test substance. Since immediate adaptational activity is the object of interest, these parameters are convenient in clinical practice. However, in the case of an extreme mesenchymal block [59] (a solid disturbance in the mesenchymal matrix which makes endogenous information transfer, i.e., electric connection, very poor), the measurement and relaxation times should be elongated.

For carrying out the adaptation test, a longer period of connection between the test-substance and the patient (the *Transmit* time) is applied, usually 5–7 seconds. These are the dynamics of the tendency for the substance-specific therapeutic signal to be saturated which is monitored by the changes in the dynamics of the reactivity.

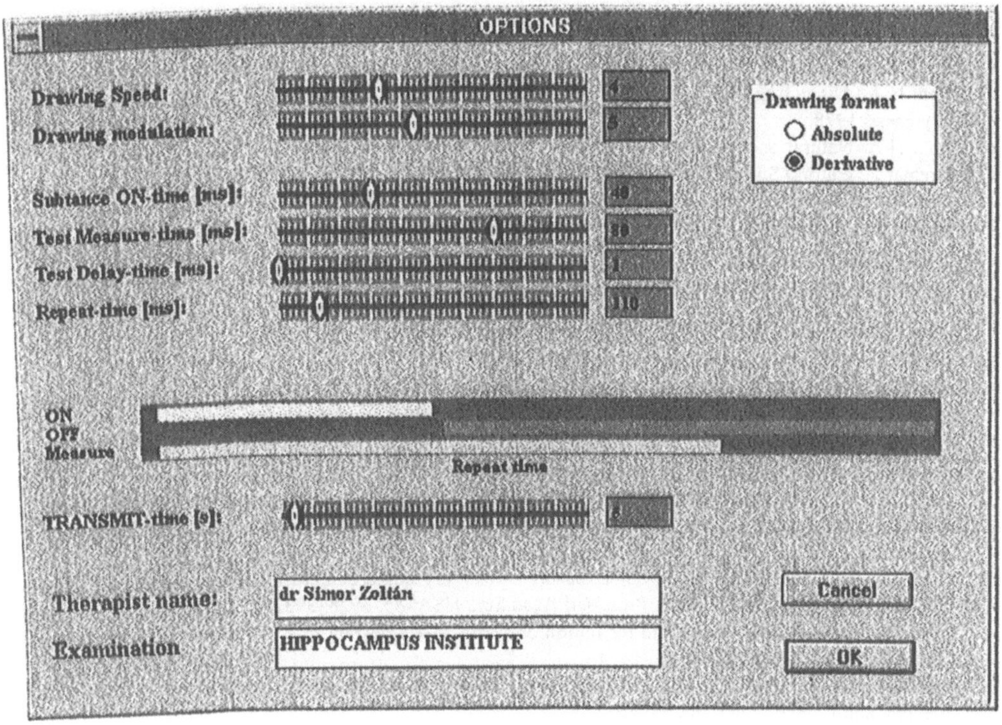

Figure 7. Adjustable parameters of the device.

Ultimately, the human being, the doctor, in the context of all the data through a holistic synthesis conceives the diagnosis by converting all the data into information using personal experience and intuition.

REFERENCES

1. Peat F.D. Synchronicity: The Bridge Between Mind and Matter. Bantam Books N.Y. 1987.
2. Recent Advances in Biophoton Research and its Applications (Popp F.-A., Li K.H. and Gu Q. Editors) World Scientific Singapore. 1992.
3. Bischof M: Biophotonen. Das Licht in unseren Zellen. Zweitausendeins Frankfurt-am-Main. 1995.
4. Zhalko-Tytarenko O., Lednyiczky G. (1997) Endogenous Electromagnetic Oscillations in the Consciousness Field Pattern Formation. World Futures. (in press)
5. Homöopathie - Bioresonanztherapie (Endler P.C., Schulte J. Editors) Maudrich Wien, München, Bern. 1996.
6. Zhalko-Tytarenko O., Lednyiczky G., Topping S. (1997) A Review of Endogenous Electromagnetic Fields and Potential Links to Life and Healing Processes. Alternative Therapies. (in press)
7. Popp F.-A. Biologie des Lichts: Grundlagen der ultraschwachen Zellstrahlung. Verlag Paul Parey Berlin, Hamburg. 1984.
8. Gurvich A: Selected Works. Meditsina Moscow. 1977.
9. Warnke U. Influence of Light on Cellular Respiration in: Electromagnetic Bio-Information (F.-A. Popp Editor.) Urban & Schwarzenberg Munchen. 1989 pp.213–220.
10. Ho M.-W., Xu X. Ross S., Saunders P.T. Light Emission and Rescattering in Synchronously Developing Populations of Early Drosophila Embryos - Evidence for Coherence of the Embryonic Field and Long Range Cooperativity in: Recent Advances in Biophotons Applications (Popp F.-A., Li K.H. and Gu Q. Editors) World Scientific Singapore. 1992 pp.287–306.

11. Brown F.A. (1954) Persistent activity rhythms in the oyster. American Scientist. *178*, 510.
12. Weaver R. (1970) The effect of electric fields on circadian rhythms in men. Life Sci. Space Res. *8*, 177.
13. Pilla A., Manning M., Personal communication with G. Lednyiczky.
14. Pohl H.A.: Dielectrophoresis: The Behavior of Matter in Non-uniform Electric Fields. Cambridge University Press. 1978.
15. Pohl H.A. Natural oscillating fields of cells in: Coherent excitations in Biological Systems. (Frölich H., Kremer F., editors) Springer Verlag Berlin. 1983 pp.199–210.
16. Pohl H. A. (1983) Natural ac electric fields in and about cells. Phenomena. *5*, 87–103.
17. Hölzel R., Lamprecht I. (1984) Electromagnetic fields around biological cells. Neural Network World, *3*, 327–337.
18. Gow N.A.R. (1984) Transhyphal electrical currents in fingi. J. Gen. Microbiology. *130*, 3313–3318.
19. Gow N.A.R., Kropf D.I., Harold F.M. (1984) Growing hyphae of Achlya bisexualis generate a longitudinal pH gradient in the surrounding medium. J. Gen. Microbiology. *130*, 2967–2974.
20. Pohl H.A. (1981) Natural electrical RF oscillation from cells. J. Bioenerg. Biomembr. *13*, 149–169.
21. Toko K., Hayashi K., Yamafuji K. (1986) Spatio-temporal organization of electricity in biological growth. Trans. IECE Japan. *69*, 485–487.
22. Yamaguchi H., Hosokawa K., Soda A., Mizamoto H., Kinouchi Y. (1993) Effects of seven months exposure to a static 0.2 T magnetic field on growth and glycolytic activity of human gingival fibroblasts, Biochem. Biophys. Acta, *1156*, 302–306.
23. Azadniv M., Miller M.W., Cox C., Valentine F. (1993) On the mechanism of a 60-Hz electric field induced growth reduction of mammalian cells *in vitro*. Rad. Environ. Biophys. *32*, 73–83.
24. Goodman R., Henderson A.S. (1988) Exposure of salivary gland cells to low frequency electromagnetic fields alters polypeptide synthesis. Proc.Natl.Acad.Sci., USA. *85*, 3928–3932.
25. Blank M., Soo L. (1992) Threshold for Inhibition of Na, K-ATPase by ELF Alternating Currents. Bioelectromagnetics. *13*, 329–333.
26. Blank M., Soo L. (1993) The Na, K-ATPase as a model for electromagnetic field effects on cells. Bioelectrochem. and Bioenergetics. *30*, 85–92.
27. Blank M., Soo L., Lin H., Henderson A.S., Goodman R. (1992) Changes in transcription in HL-60 cells following exposure to alternating currents from electric fields. Bioelectrochemistry and Bioenergetics. *28*, 301–309.
28. Blank M., Khorkova O., Goodman R. (1994) Changes in polypeptide disribution stimulated by different levels of electromagnetic and thermal stress. Bioelectrochemistry and Bioenergetics. *33*, 109–114.
29. Serpersu E.H., Tsong T.Y. (1984) Activation of electrogenic Rb+ transport of Na,K-ATPase by an electric field. J.Biol.Chem. *259*, 7155–7162.
30. Mevissen M., Stamm A., Buntenkötter S., Zwingelberg R., Wahnschaffe U., Löscher W. (1993) Effects of Magnetic Fields on Mammary Tumor Development Induced by 7,12-Dimethylbenz(a)anthracene in Rats. Bioelectromagnetics. *14*, 131–143.
31. Biological Effects and Dosimetry of Static and ELF Electromagnetic Fields, (Grandolfo M., Michaelson S.M., and Rindi A. editors) Plenum Press New York, London. 1985.
32. Abstract Book of the 17th Annual Meeting of BEMS, Boston, MA, June 18–22, 1995, The Bioelectromagnetics Society, Frederick, MD. 1995.
33. Blank M. (1992) Na, K-ATPase function in alternating electric fields. FASEB Journal. *6*, 2434–2438.
34. Goodman R., Blank M., Lin H., Dai R., Khorkova O., Soo L., Weisbrot D., Henderson A. (1994) Increased levels of hsp70 transcripts induced when cells are exposed to low frequency electromagnetic fields. Bioelectrochemistry and Bioenergetics. *33*, 115–120.
35. Popp F.-A. Coherent photon storage in biological systems in: Electromagnetic Bioinformation (Popp F.-A. Editor) Urban & Schwarzenberg München-Wien-Baltimore. 1989 pp.144–167.
36. Ludwig H.W. (1988) Die Debatte um die Magnetfeldtherapie aus der Sicht der Biophysik. Erfahrungsheilkunde. Acta medica empirica. *12*, 735- 739.
37. Schumacher P: Biophysikalische Therapie der Allergien, Sonntag Verlag Stuttgart. 1994.
38. Brügemann H: Bioresonance and Multiresonance Therapy (BRT). Haug International Brussels. 1993.
39. Ludwig W: SIT- System-Informations-Therapie. Spitta Balingen, 1994.
40. Proceedings of the Annual Meetings of the International Medical Society of BRT and International Therapeutic Society of BRT, RTI Heft I-XVII. (Brügemann Inst. ed.), Lochhamer Schlag 5, 82166, Gräfelfing (Germany).
41. Lehmann H. (1993) Elfolgreiche Behandlung primärer Dysmenorrhoe - fast ohne Therapieversager. Der Freie Arzt. No. 4.
42. Schmidt H: Konstitutionelle Akupunktur. Hippokrates Stuttgart. 1988.
43. Acupuncture (O'Connor J., Bensky D. Editors) Eastland Press Seattle. 1981.

44. Bensoussan A: The Vital Meridian. Churchill Livingstone Melbourne, Edinburgh, London, N.Y. 1991.
45. Prkert M., Hempen C.-H: Systematische Akupunktur. Urban & Schwarzenberg München, Wien, Baltimore. 1985.
46. Heine H: Lehrbuch der Biologischen Medizin. Hippokrates Stuttgart. 1991.
47. Rossmann H: Organometrie nach Voll. Haug Heidelberg. 1988.
48. Vill H: Vom Impuls- zum Decoder-Dermogramm. Haug Heidelberg. 1982.
49. Langreder W: Von der Biologischen zur Biophysikalischen Medizin. Haug Heidelberg. 1991.
50. Becker R. O., Selden G: The Body Electric. Morrow New York. 1985.
51. Bawin S.M., Sheppard A.R., Mahoney M.D., Adey W.R. (1984) Influences of Sinusoidal Electric Fields on Excitability in the Rat Hippocampal Slice. Brain Research. *323*, 227–237.
52. Adey W. R. (1980) Frequency and Power Windowing in Tissue Interactions with Weak Electromagnetic Fields. Proceedings of the IEEE. *63*, No. 1. 119.
53. Lin-Liu S., Adey W.R. (1982) Low frequency Amplitude Modulated Microwave Fields Change Calcium Efflux Rates From Synaptosomes. Bioelectromagnetics. *3*, 309–322.
54. Adey W.R. (1981) Tissue interactions with nonionizing electromagnetic fields. Physiol. Rev. *61*, 435.
55. Interaction Mechanisms of Low-Level Electromagnetic Fields in Living Systems (Adey W.R. Editor) Oxford University Press New York. 1992 pp.47–77.
56. Smith C.W., Best S: Electromagnetic Man. J.M. Dent & Sons, Ltd. London. 1990.
57. Pischinger A: Matrix and Matrix Regulation. Haug Brussels. 1991.
58. Perger F: Kompendium der Regulationspathologie und -Therapie. Sonntag München. 1990.
59. Normal Matrix and Pathological Conditions (Heine H., Anastasiadis P. editors) Gustav Fischer Stuttgart, Jena, New York. 1992.
60. Choy R., Monro J.A., Smith C.W. (1987) Electrical Sensitivities in Allergy Patients. Clinical Ecology. *4*, 93–102.
61. Lednyiczky G. (1993) BICOM In-vitro-Modulation der Sphaeroidformation. Kolloquium des Internazionalen Medizinoschen Arbeitskreises BRT (IMA). Fulda, 1–3 Oktober, 1993. Brügemann Institut Gauting RTI. Heft *13*, pp.152–154.
62. Lednyiczky G. (1993) BICOM In-vitro-Modulation der Tumorzellen-Entwicklung. Kolloquium des Internazionalen Medizinoschen Arbeitskreises BRT (IMA), Fulda, 1–3 Oktober, 1993. Brügemann Institut Gauting RTI. Heft *13*, pp.146–150.
63. Lednyiczky G., Osadcha O. (1994) In-vitro-Modulation der Phagozytose durch die Bicom-Technologie. Acta medica empirica. *43*, 3a, pp.184–188.
64. Lednyiczky G., Savtsova Z., Sakharov D. (1995) Endogenous Electromagnetic Field Corrects the Immunodeficiency of Chernobyl Mice. Abst. book of The 17th Annual Meeting of the Bioelectromagnetics Society, 18–22 June, Boston, Massachusetts, p.210.
65. Nosa P.P., Penezina O.P., Brener I.P., Lednyiczky G., Tcheshuk V.E., Lakotosh V.P., Serbynenko V.G. (1995) Treatment of Women Suffering from Mastopathy Employing Homoeopathic and Bioresonance Methods. Abstr. in: Forschende Komplementarmedizin. *2*, no. 6, p.335.
66. Lednyiczky G., Waiserman A., Sakharov D., Koshel N. Geschädigte Drosophilalarven und Information von nicht Geschädigten Tieren in: Homöopathie - Bioresonanztherapie (Endler P.C., Schulte J. Editors) Maudrich Wien, München, Bern. 1996. pp.181–188.
67. Zhalko-Tytarenko O., Lednyiczky G. (1994) Bioresonance-induced tunneling in serum albumin. XXV Congress of the Internat. Soc. Hematology. La Revista de Investigacion Clinica, Suplemento, Abril, p. 303.
68. Zhalko-Tytarenko O., Liventsov V., Lednyiczky G. (1996) Endogenous electromagnetic field influence on the free energy of hydrogen bond formation in water. Proceedings of the Second Annual Advanced Water Sciences Symposium, 4–6 October, Dallas Texas, part 7, pp.23–27.
69. Savostyanova A., Zhalko-Tytarenko O., Lednyiczky G. (1996) The influence of the human endogenous electromagnetic oscillations on non-regular chemical oscillations. Inorganic Chemistry (in Russian) (submitted).
70. Weil A., Rosen W: From Chocolate to Morphine. Houghton Mifflin Co. Boston, N.Y. 1993.
71. Evolutionary Processes and Metaphors (Ho M.-W., Fox W. Editors) John Wiley & Sons Chichester, N.Y., Brisbane, Toronto, Singapore. 1988.
72. Ho. M.-W. Coherent Excitations and the Physical Foundations of Life in: Theoretical Biology: Epigenetic and Evolutionary Order from Complex Systems. (Goodwin B. and Saunders P. Editors) Edinburgh University Press Edinburgh. 1989 pp. 162–176.
73. EMF in the Workplace. U.S. DOE, NIOSH, NIEHS Washington. 1996.
74. Schumann W. (1954) Uber die strahlunglosen Eigenschwingungen einer leitenden Kugel, die von einer Luftschicht und einer Ionosphärenhülle umgeben ist. Zeitcshrift für Naturforschung. pp.149–154.
75. Biological Effects and Dosimetry of Static and ELF Electromagnetic Fields (Grandolfo M., Michaelson S.M., Rindi A. Editors) Plenum Press N.Y., London. 1983.

76. Vincent L.-M. (1993) Theory of Data Transferal. Principles of a New Approach to the Information Concept. Acta Biotheoretica. *41*, pp.139–145.
77. Tamba-Mecz I: La sémantique. Presses Universitaires de France Paris. 1988.
78. Ultra High Dilution: Physiology and Physics (Endler P.C. and Schulte J. Editors) Kluwer Dordrecht, Boston, London, 1994.

APITHERAPY (BEE VENOM THERAPY)

Literature Review

Christopher M.-H. Kim

International Pain Institute, Inc.
Red Bank, New Jersey 07701

ABSTRACT

Apitherapy is the medicinal use of various products of Apis mellifera (the common honeybee) including raw honey, pollen, propolis, beeswax, royal jelly and venom. Various studies attribute antibacterial, antifungal, anti-inflammatory, antiproliferative and anti-cancer potentiating properties to honey.(1)

In China, for example, raw honey is applied to burns as an antiseptic and a pain killer.(2–3) Recently, propolis has been identified as containing substances called caffeic esters that inhibit the development of precancerous changes in the colon of rats given a known carcinogen.(4) Preparations from pieces of honeycomb containing pollen are reported to be successful for treating allergies and bee pollen is touted as an excellent food. This review focuses on related research materials about bee venom to treat chronic inflammatory painful illness.

1. INTRODUCTION

Apitherapy (bee venom therapy) has been used since ancient times. Ancient writers as diverse as Hesiod (800 BC), Aristophanes (450–388 BC), Varro (166–27 BC) and Columella (1st century AD) all wrote on the cultivation of the hive. Hippocrates (460–377 BC), the Father of Medicine, used it and call it Arcanum—a very mysterious remedy. Galan (131–201 AD), the Father of Experimental Physiology, mentioned it in his 500 treatises on medicine. Charlemagne (742–814 AD) is said to have had himself treated with bee stings. The Koran (XVI:71) refers to bee venom in the following terms: "There proceeded from their bellies a liquor wherein is a medicine for men." (5) For apitherapy and the scientific understanding of bees, real progress began about 100 years ago when physician Phillip Terc of Austria advocated the deliberate use of bee stings in his work: Report about a Peculiar Connection Between the Beestings and Rheumatism.(6)

Potentiating Health and the Crisis of the Immune System
edited by Mizrahi *et al.* Plenum Press, New York, 1997

Today's proponents of apitherapy cite the benefits of bee venom for alleviating chronic pain and for treating many ailments including various rheumatic diseases involving inflammation and degeneration of connective tissue (e.g., several types of arthritis), neurological disease (migraine, peripheral neuritis, chronic low back pain), autoimmune disease (multiple sclerosis, lupus) and dermatological conditions (eczema, psoriasis, herpes virus infections).

In contrast, interest in bees has been sporadic in conventional medicine, focusing mainly on two areas unrelated to the therapeutic uses proposed above. These areas are: (i) the danger of hypersensitivity reactions, including anaphylactic shock, from the sting of insects of the genus Apis; (ii) the use of bee venom itself as immunotherapy for allergic reaction to such stings, especially to prevent life-threatening anaphylactic reactions in adults.

Apitherapy is still being widely used today, especially in China, Korea, Russia, Eastern Europe and South America, but its use in the U.S. is still considered to be controversial. It must, however be acknowledged that, in spite of this controversy, apitherapy has been the subject of many studies in animals as well as in human subjects.

2. METHODS

A Medline literature search was carried out for the period 1960 to 1994. In addition, the references given in relevant publications were further examined. More than 3,000 pages of documentations were reviewed. All of the papers were analyzed and special attention was paid to discussions and conclusions of the different authors.

3. BIOCHEMICAL COMPOSITIONS

The scientific interest in the bee venom's composition was first reported late in the nineteenth century.(7) These somewhat crude studies indicated that the venom is a complex mixture. More definitive results date back only 40 years, when Neumann and coworkers(8) showed that it is the venom's protein components that are associated with the biological activity. The introducton of an ingenious method of allowing the collection of high grade venom initiated an active biological research to define the structure and pharmaco/biochemical properties of its individual components. There are many studies identifying various biological properties for semi-purified fractions of bee venom and for more purified products to help explain the curative properties attributed to this venom. Table 1, adapted from Kim(9), summarizes these components.

Bee venom consists of a large number of more or less complex molecules. Certain components, especially melittin, apamin and peptide 401, have been aggressively researched as to their physiological effects, probably due to the fact that they are contained in great proportions in bee venom and due to the compatibility of their effects upon different anti-inflammatory mechanisms.

4. PATHOPHYSIOLOGY AND MECHANISM

As remarked by Pincus and Callahan(26), between 1958 and 1987 the standards of classification of the American Rheumatic Association were applied in all clinical studies. These standards were revised in 1987. As these standards have been designated to be

Table 1. Biochemical compositions of the bee venom (Source: Adapted from Kim, 1992)

Description	Component	Mol. Wt.	% (Dry Venom)	Reference
Peptides	Melittin	2,840	40-50	Neumann et al. 1952(8)
	Apamin	2,036	2-3	Habermann et al. 1965(10)
	MCD-Peptide (Peptide 401)	2,588	2-3	Fredholm 1966(11)
	Adolapin	11,500	1.0	Shkenderov 1982(12)
	Protease Inhibitor	9,000	<0.8	Shkenderov 1973(13)
	Secarpin		0.5	Gauldie et al. 1976(14)
	Tertiapin		0.1	Gauldie et al. 1976(14)
	Melittin F		0.01	Gauldie et al. 1976(14)
	Procamine A, B		1.4	Nelson and O'Connor 1968(15)
	Minimine	6,000	2-3	Lowy et al. 1971(16)
	Cardiopep?		<0.7	Vick et al. 1979(17)
Enzymes	Hyaluronidase	38,000	1.5-2.0	Habermann & Neumann 1957(18)
	Phospholipase A2	19,000	10-12	Habermann & Neumann 1957(19)
	@-Glucosidase	170,000	0.6	Shkenderov et al. 1979(20)
	Acid Phosphomonoesterase	55,000	1.0	Shkenderov et al. 1979(20)
	Lysophospholipase	22,000	1.0	Ivanova et al. 1982(21)
Physiologically Active amines	Histamine		0.6-1.6	Reinert 1936(22)
	Dopamine		0.13-1.0	Owen 1971(23)
	Norepinephrine		0.1-0.7	Owen 1982(24)
Nonpeptide Components	Carbohydrates:			
	Glucose and Fructose		<2.0	O'Connor et al. 1967(25)
	Lipids:			
	6 Phospholipids		4-5	O'Connor et al. 1967(25)
	Amino Acids:			
	r-Aminobutyric Acid		<0.5	Nelson and O'Connor 1968(15)
	b-Aminoisobutyric Acid		<0.01	Nelson and O'Connor 1968(15)

"classifications" and not "diagnosis", it is reasonable to presume that these standards allow the identification individuals who are relatively homogeneous with regard to the pathogenesis and prognosis. According to the authors, individuals that meet these criteria are particularly heterogeneous.

They explain that many facts suggest that patients who meet these standards of rheumatoid arthritics in the general population, differ substantially from patients identified as afflicted with rheumatoid arthritis in a clinical surrounding. The observations suggest that the 1958 standards of the American Rheumatism Association identify at least three types of the pathogenic process. In imitation of pathophysiological factors involved in the sudden appearance and development of inflammatory illness, our knowledge of the mechanism of treatments of inflammations are very limited. The following section presents different mechanisms which have been proposed for bee venom, or certain of its constituents, as well as for other anti-inflammatory treatments.

4.1. Immune Modulations

The immune system provides the body's defense against invading pathogenic microorganisms and tumors, as well as being importantly involved in tissue repair after injury.

According to Artemov(27), bee venom attacks the vital system of an organism and in turn mobilizes the organisms' defense-forces. Shipman and Cole(28) found that the resistance of mice against X-irradiation increases greatly after injection of melittin.

According to Zurier et al.(29) it is conceivable that a subtle interference with the immunological function of the lymphocytes (independent of adrenal stimulation) would be invisible in this, or in the mechanism by which the bee venom suppresses induced arthritis in the rats.

According to the works cited by Chang and Bliven(30) the effects of anti-inflammatories and Immunosuppressants can be differentiated in using the model of induced arthritis. Immunosuppressants are the most efficient when used early during the period of sensitization (the first four days after completed inducement) and become ineffective when the illness is established. Given the fact that the venom is most efficient when administered before inducement or during the period of sensitization, this effect gradually diminished when the administration was delayed. The authors suggest that the inflammatory action plays a contributory primary role.

Chang and Bliven(30) conclude that the mechanism of the venom involved, above all, is the modification of immunological response through the competition of the antigens, and to a lesser degree, an anti-inflammatory action originating in the cortico-adrenal area or by another mechanism not as yet discovered.

Hyre and Smith(31) state that mice injected with bee venom prior to, and following, an injection of sheep red blood cells produced significantly more direct IgM plaques than the sham-injected group. The results of this study indicate that bee venom can effect both T and B lymphocyte functions.

Using the example of induced arthritis, Hadjipetrou-Kournunakis and Yiangou(32) have shown that the venom does not directly affect the T cells in vitro but seems to affect the macrophages which lessen their production of interleukin II. They suggest, in agreement with the works of Somerfield(33), that the anti-inflammatory action of venom (the development of induced arthritis) seems to be due to the inhibition of certain functions of the macrophages which inhibit directly the activation of T cells and B cells.

Concerning the mechanisms, Kim(9) mentions that an effect upon the immune system could be involved since animal studies have shown effects upon the lymphocytes T and B and in the response of the splenocytes in the mitogenes of the T cells.

Belliveau(34) reported that bee venom is effective in treating an animal model of colon cancer. Bee venom did not have a direct effect on cancer cells but it stimulated immune system so that it did have an effect indirectly on cancer cells.

4.2. Anti-Inflammatory Effects

4.2.1. Hypophysis and Adrenal Gland. Artemov(35) has shown in a controlled study with 248 rats (bee stings versus fine needle), that venom considerably lowers the level of ascorbic acid in the adrenal glands, evidence of strong stimulation of the activity of secretion of the corticoadrenals.

Artemov(36) has also shown in a controlled study of four groups of 10 rats, that the eosinopenia, provoked by the bee venom, is not different from that introduced by the administration of the adrenocorticotropic hormone (ACTH).

Hammeral and Pitchier(37) described the anti-inflammatory action of venom which produces the freeing of the corticosteroides endogenes.

Intravenous injection of bee venom significantly raises the plasmatic cortisone level in dogs.(37) This is also true of melittin and apamin.(38)

The anti-inflammatory effect of bee venom occurs through corticoglandular stimulation.(29,30,37–39)

Whole bee venom produces a high increase in the level of cortisol in dogs and monkeys, while hypophysectomy inhibits the stimulation of the adrenal glands through the venom in these animals.(38) A similar effect upon the ACTH could be the reason for the effect bee venom in rheumatoid arthritis.(39) Bee venom is mainly composed of melittin and, surprisingly enough, only a hundredth of a dose of melittin is sufficient to obtain the same elevation of plasmatic cortisol (cortisone) in dogs.(37) A possible explanation would be that the presence of the phospholipase A interferes with the activity of the corticoadrenals.(40)

According to Couch et al.(41), whole bee venom increases corticosterone in the blood vessels of rats, and one can expect an anti-inflammatory effect.

Raising of cortisone levels following an injection of certain components of bee venom in dogs, suggests general stimulation of the adrenal gland.(37)

In contrast to the intraperitoneal channel and subcutaneous injection of the individual components (melittin, apamin, phospholipase A), the whole venom, injected subcutaneously, produces a rapid augmentation and persistent rate of the corticosterone.(29)

The whole venom has no effect upon rats that have been adrenalectomises, which leads Zurier to propose, that it suppresses induced arthritis with the help of hypophysis and/or the adrenal glands.

Chang and Bliven(30) reject a mechanism through cortico-adrenal (cortisone-glandular) stimulation, since in the model of induced arthritis, the effect of venom is only temporary in arresting the progress of the illness, contrary to hydrocortisone. They conclude that the mechanism of the venom is by a modification of the immunological response by competition of the antigens, and to a lesser degree, an anti-inflammatory action of cortico-adrenal origin, or by another mechanism not yet discovered.

Hanson et al.(39) have shown that an adrenalectomy diminishes, but does not suppress, the anti-inflammatory action of peptide 401 and concludes that the change of secretion of the corticosteroids only play a minor role in the anti-inflammatory activity, contradicting the conclusions of Zurier(29) and those of Lorenzetti et al.(42)

Subcuteneous administration of venom raises the level of cortisol and motor activity in arthritic dogs.(43)

Shkenderov and Koburova(12) have shown that adolapin has a powerful anti-inflammatory action in the example of edema of the foot in rats, induced by carrageenan and that of the inhibition of the edema induced by prostaglandin E1. The edema induced to the bradykinin by histamine and serotonin have less influence, while they have been influenced by the adrenalectomy.

Knepel and Gerhards(44) demonstrated the effect of melittin on the release of adrenocorticotrophin (ACTH) and beta-endorphin from the corticotrophic cells of rat adenohypophysis. Melittin stimulated ACTH-Immunoreactivity and beta-endorphin-immunoreactivity release.

Among the possible mechanisms, Kim(9) discusses the effects of melittin and apamin upon plasmatic cortisone which contributes to the lessening of the inflammation. He cites, equally, the direct action of granulative peptide on the mastocytes which restores the vascular anergic endothelium to its phlogistic stimulations.

There are arguments against the effect of venom by means of the hypophysis and/or the adrenal gland, as the stimulation of the gland in rats is generally accompained by lymphocytosis(45) which is not being observed with bee venom.(29)

There are arguments against the effect of venom by means of the hypophysis and/or the adrenal gland because in rats treated with paramethasone, polyarthritis develops in less than a week after the treatment has been discontinued, but the pre-treatment with bee venom has remained without effect.(46)

There are arguments against the effect of venom by bias of the hypophysis and/or the adrenal gland, because the production of antibodies introduced by erythrocytes of the sheep are little reduced by treatment with bee venom. In tests of arthritis induced in rats, a more marked reduction of humoral response has been observed with prostaglandin E, and it has been suggested that the prostaglandin E affects only a small population of lymphocytes.(47)

4.2.2. Prostaglandins. Habermann and Reiz(10) have shown that the fraction Oa have analgesic properties and inhibit the activity of the cyclooxygenase (synthetize prostaglandin).

Peptide 401 is capable of reducing and suppressing a response to the intradermal injected substances such as histamine, 5-hydroxytryptamine, bradykinin and prostaglandins.(39)

Billingham et al.(48) have shown that peptide 401 is a powerful anti-inflammatory in tests of edema in the feet of rats. It reduced, considerably, the inflammation associated with primary and secondary lesions in established illness. If administered together with the peptide 401, it generally prevents the development of the illness.

It is possible that the anti-inflammatory action of the peptide 401 is related in a certain way to the distribution of the mastocytes (or other tissue damages) as the elevated dosage of prostaglandin E diminishes the severity of induced arthritis.(49)

According to Hanson et al.(39), upon interaction, the synthesis of the prostaglandins cannot be excluded. The existence of such a mechanism implies that cutaneous reactions, produced by histamine, 5-hydroxytryptamine, and bradykinin, depend upon the simultanous increase to the non-steroid anti-inflammatories to suppress all such responses in rats.(50)

Banks(51) reports on the prostaglandins which are implicated in the phase called secondary inflammation. The non-steroid anti-inflammatories, such as aspirin and indomethacine, are inhibitors of the synthesis of the prostaglandins in vivo and in vitro. Bank has shown that peptide 401 is also an inhibitor in the conversion of arachidonic acid to prostaglandin E in vitro. One believes that the prostaglandins are synthesized in the leukocytes at the site of the inflammation.

Billingham et al.(48) have shown that peptide 401 has a marked anti-inflammatory effect upon the rats, using induced edema in the carrageenan. The authors show that peptide 401 suppresses the deveopment of induced arthritis, even in the one rat in which it was already established.

Chang and Bliven(30) confirm that bee venom possesses anti-inflammatory properties because it suppresses edema, induced in the carrageenan, and there exists a relationship between the administered dosage and the percentage of inhibition of the edema. Their results confirm those of Billingham(48) and those of Hanson.(41)

Shkenderov and Koburova(12) have shown that adolapin has analgesic properties due to inhibiting of the cyclooxygenase of the microsomes of the spleen of cats and of rats. In the tests used, the efficient dosage to produce a 50% inhibition (DE50) is 13 and 16 microg./kg. On a molar basis, adolapin is an inhibitor of the cyclooxygenase of the brain, 70 times more forceful than indomethacine. The inhibiting effect is biphasic and merits a more profound investigation.

Shkenderov and Koburova(12) also propose that the inhibition of cyclooxygenase represents the fundamental mechanism of anti-inflammatory and analgesic effects of adolapin. These suggestions are based upon the well known relationship between the inhibition of prostaglandin synthesis and anti-inflammatory effects(52) and analgesic effects(53) of the non-steroid anti-inflammatories.

Shkenderov and Koburova(12) have shown that adolapin has an anti-inflammatory action, powerful in the tests of edema in the feet of rats, induced by carrageenan, and also in the inhibition of edema induced by prostaglandin E1. Edema induced by bradykinin, histamine, or serotonin are less influenced just as they have not been influenced by adrenalectomy.

Phospholipase A2 is the key enzyme which catalyzes the freeing of arachidonic acid, a substratum in the synthesis of the prostaglandins and the leukotrienes.(54)

Koburova(54) mentioned that adolapin, like aspirin and other similar components, has antipyretic properties, probably by inhibition of the biosynthesis of cerebral prostaglandins.

Koburova(54) goes so far as to propose that adolapin could be considered to represent a new class of compounds peptic inhibitors of the synthesis of prostaglandins and leukocytes which have noticeable analgesic and anti-inflammatory actions.

Shkenderov(55) has shown that carrageenan significantly increases the concentration of prostacycline and of thromboxane A2, but adolapin prevents this effect. The inhibition of an increased synthesis of the thromboxane A2 by adolapin could explain the aggregation of plaques and erythrocytes of the sedimentation which have previously been observed in vitro with the fraction Oa.

Skhenderov et al.(55) have shown that adolapin inhibits the primary reaction in the release in the arachidonic-cyclooxygenase. In contrast to indomethacine and other anti-inflammatory components similar to aspirin which inhibit in a irreversible way almost 100 percent of the action of cyclooxygenase, adolapin could interact in an allosteric way, only suppressing a maximum of 60–80 percent of the activity.(56)

Shkenderov(57) has also shown that the anti-inflammatory effect of bee venom's protease inhibitor on a model system of acute inflammatory edema in rats. Protease inhibitor suppressed about 40 percent model paw inflammations caused by carrageenan, prostaglandin E1, bradykinin and histamine.

Shkenderov et al.(57) have measured the plasmatic level of two important metabolites for release of arachidonic acid: the prostacycline (prostaglandin I2) and the thromboxane A2 in the inflammatory test with carrageenan after administration of adolapin. That adolapin has had no effect upon prostacycline or thromboxane A2 suggested that it does not affect the normal equilibrium between their opposed actions which serve to maintain cardiovascular homeostasis.(58) Thromboxane A2, produced principally in the thrombocytes and the lungs, produces bronchial constriction, vasodilation, and plaque-accumulation.(59) The prostacycline presented in the endothelium of the arteries, hinders the accumulation of plaque,(53) all being a powerful dilator of the arteries.(60) It is known that the gelatine (soluble collagen) s promotes the accumulation of plaque which results in an increase of thromboxane A2.(60)

Among the possible mechanisms, Kim(9) mentions peptide 401 as an inhibitor of the conversion of arachidonic acid in the prostaglandins. The inhibition of the cyclooxygenase by the adolapin could constitute the principal mechanism of anti-inflammatory and analgesic effects.

4.2.3. Implication of Free Radicals. When inflammatory cells are activated in a variety of ways, toxic oxygen radicals and metabolites are produced. These are thought to be beneficial in inflammation when such responses are host protective and appropriate, e.g., during bacterial invasion. However, during circumstances of chronic, inappropriate inflammation where, for example, immune complexes or sequestrated antigen causes chronic stimulation of the immune and inflammatory systems, then such toxic oxygen radicals and metabolites have been incriminated as contributors to inflammatory tissue damage, not only in rheumatic disease but also in cardiac, and central nervous system disease. In some experimental diseases, dismutation of superoxide radical following its production by leukocyte, can largely prevent the occurrence of the disease. Sommerfield(61) reported that whole bee venom inhibited superoxide production and also able to inhibit hydrogen peroxide production from normal human polymorphonuclear leukocytes in a dose-dependent fashion. This effect was found to be maxmal at 1.5 microg./ml, well below the toxic dose of >10 microg./ml.

Somerfield et al.(61) have shown that melittin is responsible for the effect of whole venom upon the production of superoxydal anion, which possesses similar effects to those of trifluoperazine upon the production of superoxide.(32, 62–64)

Somerfield(65) reports that whole bee venom is a powerful inhibitor of the production of superoxydal anion, and this effect is due to the melittin. The author also maintains that apamin, phospholipase A and peptide 401 (the peptide that destroys the mastocytes) have not shown this effect. The possibility that this inhibition is due to the presence of a very weak contamination of the existing melittin but it seems not very probable because of the combination of other minor components of bee venom that have not had an effect.

Somerfield et al.(65) reports that melittin diminishes the production of superoxydal anion, by a non-toxic and direct action upon the polymorphonuclear leukocytes, and this depends on the dosage. It is not a dismutation of an attack upon the free radicals, or a deceptive effect upon the indication of the milieu.

According to Somerfield et al.(65), melittin could interact with other components of the membranes, some of which could be important producers of superoxide.

Rekka et al.(66) have shown that at a dosage of 50 mg., bee venom inhibits almost totally the peroxydation of lipids and demonstrates a considerable predatorial activity upon hydroxy radicals. According to the authors, the traditional usage of bee venom as an anti-inflammatory and anti-arthritic agent could more or less explain its anti-oxidant activity.

Rekka(66) reports that certain aspects of radical oxygen are implicated in inflammatory reactions. The hydroxyl radicals (OH) are among the most active and are set free at the reduction of the peroxides generated in the release of lipoxygenase and cyclooxygenase. The authors's works has shown that bee venom inhibits significantly the non-enzymatic peroxydation of the lipids, and it is an excellent predator of the hydroxyl radicals. On the basis of the inhibition of the production of interleukin I, these results support the hypothesis that an anti-oxidant activity is involved in the anti-inflammatory effect of the bee venom.

4.2.4. Lysosomal Membranes and Phospholipids. The role of lysosomes in inflammation is one of the most interesting problems in human pathology. In an experimental test of inflammation, described by Chayen et al.(67), human skin maintained in a non-proliferative culture was treated with histamine. The anti-inflammatory action of the compound was evaluated upon the basis of its capacity to protect the lysosomal membranes from damages caused by histamine. Utilizing this test, Bitensky and Chayen, results not published but cited in Billingham(48), peptide 401 has been confirmed to be twice as efficient as the salicylates, weight for weight.

According to Dufourcq et al.(68), mellitin strongly disturbs the membranic structure of the liposomes into which it penetrates. It is not selective about the phospholipides, and it increases the permeability of all kind of cells, like the liposome, mastocytes and cytoplasmic bacterial membranes.

Shkenderov(69) has shown that the anti-inflammatory effects induced by low melittin dose (20g/kg). This became apparant as early as the second day of post-injection. This is probably due to the direct stabilizing effect of melittin on the lysosomal membrane.

Schwartz et al.(70) have reported that melittin associates itself with bilamellaric phospholipids. These properties could be considered as an adjuvant upon the specific antioxidant effect by bee venom.

4.2.5. Kinins and Complements. The mediators of inflammation (histamine, serotonin and prostaglandin) are set free at the time of the aggregation of the thrombocytes, while the fibrine produced is a hemotactical factor for the leukocytes. The kinins produce plasmin and kallikrein and act as typical mediators in inflammation, thus bringing about the edema, pain and migration of the leukocytes(71).

The inflammatory reaction causing damage to tissues and experimental edemas develops to a full extent only in the presence of an intact complement system(72). It is involved in the pathogenic mechanism of the so called immune complex diseases, to which belongs rheumatism. On the other hand, separate complement components or products of its activation alone exhibit anti-inflammatory activity. Gencheva et al.(73) reported that the low molecular fraction of bee venom injected into rats in a daily dose of 100microg/kg over a three week period reduced the complementary activity by 45 percent.

Apamin and protease inhibitor in a dose of 25microg/ml inactivated the complement in vitro; the former by 35 percent and the latter by 15 percent. Injection of rats with apamin, protease inhibitor, and melittin in daily dose of 20microg/kg and 100microg/kg for three weeks caused statistically significant drops in the complement levels. This effect was markedly more (over 60 perent inactivation) in the higher dose(73).

4.2.6. Calmodulin. Menander-Huber(74) have shown that melittin binds with calmodulin. By the way, this is the only peptide of bee venom which binds with calmodulin(75).

As explained in Somerfield(61), when the monocytes and the polymorphonuclear neutrophils are being stimulated by different particles and stimulants and are soluble, an explosion of the oxidative metabolism occurs. Due to the activation of NADPH-oxydase bound to membranes of subjacent cells, there occurs a liberation of superoxydal anions (O2) and the production of hydrogenic peroxide, hydroxyl radicals (OH), and a singlet of oxygen. The importance to superoxydal anions and to metabolites of the oxygen in the cytotoxic mechanisms is well documented, as is the role of the oxygen metabolism when induced by phagocytosis in the inflammatory processes of the tissues. The processes depend upon the calcium and can be inhibit by medicaments that bind with calmodulin, such as trifluoperazine and other phenothiazines.

The similarity of the results of Somerfield et al.(61) with the ones of several authors(62–64) are striking because of the inhibition of the production of superoxydal anion by the trifluoperazine, a medicament which binds easily with calmodulin. It has been demonstrated that NADPH-oxydase contained in the membranic fraction of the activated leukocytes (and which is responsible for the production of superoxydal anions) depends on the calmodulin(76). It could be that bee venom works via a similar mechanism. The principal fraction of the venom, melittin, shows a great affinity with calmodulin, and it is the only fraction of bee venom which has this characteristic.

Somerfield(65) has advanced the hypothesis that the effect of melittin could be produced thanks to this bonding to the inhibiting calmodulin and also to an oxydase on which it depends, in this case, the NADPH oxydase(76,77). Somerfield concluded that the peptides which have an effect on production of the superoxide and the oxydase NADPH, could control the inflammation in vivo. The new class of substances which bind with the calmodulin, could also have new therapeutic implications in inflammatory illnesses.

4.2.7. Counter-Irritant Effects. Chang and Bliven(30) propose that the venom could suppress induced arthritis by anti-inflammatory action, a counter-irritant effect and/or by an immunosuppressive action.

According to Chang and Bliven(30), counter-irritant action of the venom is not very probable since the irritation produced by bee venom is weak and of short duration; and a single dose of venom proves highly efficient in preventing the development of the illness, even if the injection is given 10 or 11 days before the inflammation. Also, the same single dosage has been established to have a much lesser effect when it was administrated during the development of the arthritis, at the stage of the illness most susceptible to the effect of counter-irritants.

Hanson(39) has shown that the anti-inflammatory activity of peptide 401 could not be attributed to the freeing of vasoactive amines, or to the modification of vasomotor activity, to the modification of the prostaglandin synthesis, or to their irritant inflammatory properties.

According to Banks et al.(51), peptide 401 does not act as a counter-irritant. While it is very forceful in tests of induced arthritis and in tests of vascular permeability, while other substances capable of destruction of the mastocytes (melittin and compound 40/80 for example), are not; in tests of permeability, its effect could be blocked by inhibitors of histamine or 5-hydoxytryptamine, without having an effect upon anti-inflammatories. The peptide in question is less efficient as an anti-inflammatory but is effective as a destroyer of the mastocytes.

Banks(78,79) has contradicted the works of several researchers and confirms that peptide 401 is not a real anti-inflammatory agent, but actually a counter-irritant.

4.3. Cytolysis

4.3.1. Destruction of Mastocytes. There are three components in bee venom that cause mastocytolytic action: phospholipase A2, melittin, and MCD-peptide. Of these, phospholipase A2 is an indirect mastocytolytic agent(80).

Breithaupt and Habermann(81) have shown that the anti-inflammatory effect of peptide 401 can perhaps be attributed to the freeing of vasoactives resulting in the destruction of the mastocytes while other destructors of the mastocytes, like the component 40/80, have not shown the anti-inflammatory effect in tests performed.

According to Hanson(39), peptide 401 and melittin increase the vascular permeability and lead to a destruction of the mastocytes in vivo and in vitro. While in these tests, melittin has, in effect, not shown anti-inflammatory properties, this suggest that the anti-inflammatory action of peptide 401 is not caused by its inflammation on destructive effects of the mastocytes.

Billingham et al.(48) have separated several peptides of venom of which they have shown peptide 401 to be identical to the peptide that destroys the mastocytes referred to by Breithaupt(81).

Zurier and Ballas(82) reported that it is possible that the anti-inflammatory action of peptide 401 is related in a certain way to the destruction of the mastocytes or to other tissue damages, since elevated doses of prostaglandin E diminished the severity of induced arthritis.

According to Banks(78), the anti-inflammatory action of peptide 401 and of the component 40/80 in the test of edema in the hind leg carrageenan show the destruction of the mastocytes in vivo.

Ziai et al.(83) have actually published a survey of the literature about the knowledge of the actual structure and qualities of peptide that destroys the mastocytes. The authors report that peptide is a powerful anti-inflammatroy agent, but in weak dosage it is a strong mediator for destruction of the mastocytes and for freeing histamine. The peptide is also a neurotoxic epileptoid, a blocker of the potassium channels, and can provoke significant hypertension in rats.

Among the possible mechanisms, Kim(9) mentions notably direct action of the peptide destroyer of the mastocytes which renders the vascular endothelium anergic to phlostic stimulations.

Other authors, however, reject the direct or indirect implication with the destruction of the mastocytes.

Hanson(39) has shown that the intradermal injection of peptide 401 diminishes considerably the increase of the vascular permeability as a result of the sub-planter injection of carrageenan or an intra-articular of terebenthine. Its effectiveness has proven superior to that of indomethacine, salicylate, and phenylbutazone.

Billingham et al.(48) concludes that the anti-inflammatory action of peptide 401 is not the result of its capacity of destructing the mastocytes.

According to Banks et al.(51), peptide 401 does not have a counter-irritating effect, which shows very strongly in tests of induced arthritis and in tests of vascular permeability, while other substances, capable of destroying the mastocytes (melittin and its components 40/80 for example) are not. In tests of permeability this effect can, perhaps, be blocked by inhibitors of histamine or 5-hydroxytryptamine without affecting the anti-inflammatroy effect. The pronounced peptide is much less effective as an anti-inflammatory, but is effective as far as destruction of the mastocytes is concerned.

4.3.2. Vasomotor Activity. Billingham et al.(48) have confirmed that peptide 401 possesses inflammatory properties based upon an increase in peripheral permeability and its abolishment by pretreatment with mepyramine and methysergide.

The standard global measure of extravasation of plasmatic proteins by Hanson(39), does not allow differentiation between the changes of pressure across the vascular wall, (caused, for example by the vasoconstriction or vasodilation of the capillaries) and by the alteration of the permeability itself. In spite of that, the persistence of anti-inflammatory properties of peptide 401 in rats treated with mepyramine or methysergide can be said to be associated with its incapability to react upon the absorption of 86Rubidium by the skin, and the absence of a vasoconstrictive action suggests that peptide 401 acts directly upon the vascular wall.

According to Foster(84), the pharmacological effects of bee venom are based upon all its components, in particular the effect upon the circulation of the blood, the permeability of biological membranes, and the organic exchanges. The venom has an vasodilation, accompained by an increase of the permeability, caused by hyaluronidase. He mentions also a hypophysical effect, demonstrated by the absence of an effect upon hypophysectomised animals. This effect upon the hypophysis brings about a discharge of adrenocortisone which, on its part, activates the formation of cortisone by the corticoadrenals.

There is also evidence that vascular effects are negligible or unrelated to the anti-inflammatory effects of bee venom.

Hanson et al.(39) have shown that anti-inflammatory activity of peptide 401, cannot be attributed to the freeing of vasoactive amine, resulting in the destruction of mastocytes. Other destructors of the mastocytes, like the compound 40/80, have not shown anti-inflammatory effects in those tests performed(81).

In spite of the fact that the adrenalectomy diminishes the effect of peptide 401, the denervation and pretreatment with phenoxybenzamine could have caused this effect and Hanson et al.(39) conclude that the increase of corticoadrenal activity or vasometric activity can not play but a minor role in the anti-inflammatory effect of the peptide 401.

4.3.3. Coagulation, Anticoagulation, and Hemolysis. Neumann and Hebermann(8) have shown that the direct hemolytic activity of bee venom was identified with melittin and the indirect hemolytic activity with phospholipase A. Heparin inhibits the hemolytic action of bee venom(85).

Hebermann(86) described antithromboplastic activity caused by phospholipase A and melittin. He postulated that melittin binds with basic groups to acidic groups of thromboplastin, thus preventing coagulation.

In contrast to anticoagulatory action with bee venom, Meszaros(87) mentioned that massive number of beestings leads to disseminated intravascular coagulation (DIC).

4.4. Neurotoxic Effects

It is obvious that bee venom could affect the nervous system either directly or indirectly by releasing neuroactive substance from, for example, mast cells. Hahn and Ostermayer(88) found the first evidence of neurotoxin in whole bee venom.

4.4.1. Postsynaptic Effect. Wellhoner(89) demonstrated that the apamin's activity on the spinal cord. It has been found that apamin augments polysynaptic reflexes and that is causes excitable polysynaptic pathways to become more effective than inhibitory polysynaptic mechanisms.

Vladimirova and Shuba(90) have shown that apamin abolishes the hyperpolarizing action of externally applied ATP on guinea pig stomach and taenia coli as well as the hyperpolizing (inhibitory) functional potential that follows stimulation of the noradrenergic inhibitory nerve, which is proposed to be purinergic. However, the inhibitory action of norepinephrine on intestinal smooth muscle is also blocked and even reversed(91–93).

The apamin is a specific blocker of Ca++-dependent K+ conductance when K+ flux studies show that apamin prevents the rise in potassium permeability in guinea pig hepatocytes and taenia coli(91,93) produced by ATP and norepinephrine. The apamin-sensitive Ca++-dependent K+ channel is only one of several types of Ca++-dependent K+ channels(94,95).

Jenkinson(96) concluded that apamin blocks inhibition by alpha and not by beta adrenoreceptors in intestinal muscle. Apamin also antagonizes the neurotensin-induced relaxation of guinea pig proximal colon(97).

Burgess et al.(98) reported apamin blocks the increase in K+ efflux that followed application of alpha adrenergic agonists to rabbit livers. The concurrent rises in Ca++ efflux and glucose release were unaffected. The authors also concluded that apamin must be able to block either the Ca++-dependent K+ channels present in the cell membrane, or the mechanism that controls these channels.

Myotonic muscular dystrophy is a dominantly inherited disease of muscle which arises from genetically induced alterations of the muscle membrane. Renaud et al.(99)

first showed that muscle membranes of patients with myotonic muscular dystrophy contain the receptor for apamin, a bee venom toxin known to be a specific and high-affinity blocker of one class of Ca++-activated K+ channels in mammalian muscle. The apamin receptor is completely absent in normal human muscle as well as in muscles of patients with spinal anterior horn disorders.

4.4.2. Central Effect. Hebermann and Reiz(10) have found that apamin is a long-acting centrally excitable toxin which is isolated from whole bee venom. Using rat brains, Hugues et el.(100) found that apamin binds highly specifically.

Habermann and Reiz(10) also reported that apamin is toxic to the central nervous system. The central neurotoxicity gives a much stronger response when injected intraventricularly or intralumbarly than the intravenous or subcutaneous route(101). The CNS effects of melittin were also reported by Vyatachannikov and Sinka(102).

Adolapin has an analgesic effect, in which central mechanism may also be involved. This is suggested by the fact that naloxone, a blocker of the opiate receptors, partly eliminates the analgesic effect of adolapin. This was discovered using the writhing test with mice(12).

Mourre et al.(103) have studied the autoradiographic localization of apamin-sensitive Ca++-dependent K+ channels in rat brains. Autoradiograms demonstrated a very heterogenous distribution of the apamin receptor throughout the brain.

Koburova(54) mentioned that adolapin, like aspirin and other similar components, has antipyretic properties, probably by inhibition of the biosynthesis of cerebral prostaglandins.

Koburova(54) concludes that the analgesic activity of the adolapin is being controlled by a part of the central nervous system since they have used the method of the "tail flick" characteristic of the central action of the analgesia as being partially suppressed by the administration of naloxone, an antagonist of morphine which blocks the morphine receptors of the brain.

4.4.3. Cardiovascular Effect. Vick et al.(104) have studied the cardiovascular effects of apamin. It would appear that apamin is a potent, non-toxic beta-adrenergic-like stimulant not entirely blocked by propranolol and possessing definite anti-arrhythmic properties. There are no significant changes in arterial blood pressure, central venous pressure or cortical activity.

Bee venom especially phospholipase A2, decreases the blood pressure and the rate of cardiac contractions. Phospholipase A2 decreased the arterial pressure and heart rate(105), and increased capillary permeability(106).

Vick et al.(17) reported a cardioactive compound that has been called cardiopep. This compound is relatively nontoxic. It produces a 50 percent increase in heart rate and a 150 percent increase in force of contraction, but no change in coronary vascular resistance. The compound is a potent beta-adrenergic-like stimulant not entirely blocked by propranolol and possesses antiarrythmic properties. Kim(9) mentions that the LD50 of cardiopep is equal to that of apamin(15mg/kg) and this peptide is probably an impure preparations obtained by an inadequate separation scheme.

4.5. Antibacterial, Antifungal, and Antiviral Effects

Arthritis originates from a deficiency in circulation of the blood and of the lymph in the tissues of the joints, and results in an accumulation of lactic acid which permits the bacterial increase. The vasodilating effect of bee venom increases local circulation, tends to correct the deficient circulation, and works against the spread of bacteria(107).

Schmidt-Lange(108) also refers to the antibacterial properties of bee venom, which was confirmed by Ortel et al.(109).

Fennell et al.(110) have shown that bee venom and a derived polypeptide fraction melittin have antibacterial activity against a penicillin-resistant strain of staphylococcus aureus (strain 80). Both whole bee venom and melittin were also able to inhibit the growth of 20 of 30 different bacterial organisms tested. More gram-positive organisms (86 percent) were sensitive to bee venom and to melittin than were gram-negative organisms (46 percent). Among the gram-positive organisms tested, the antibacterial effect of 1 mg. of melittin was equal to that 0.1–93 units of penicillin.

Dorman and Markey(111) also exhibited the antibacterial activity of melittin. Hadjipetrou-Kourounakis and Yiangou(32,112) reported bee venom treatment affects adjuvant induced disease development by inhibiting certain macrophage functions, and, thus, indirectly inhibits the activation of T and B cells, and possibly the activation of an endogenous virus which might be involved in adjuvant induced disease induction.

Kim(9) mentions that the risk of infection due to bee venom injection seems minimal and it seems that melittin has antibacterial and antifungal properties.

4.6. Radioprotection Effects

Shipman and Cole(28,113) found that the resistance of mice against X-irradiation increases greatly after bee venom injection. They found the main fraction of bee venom, melittin, was effective. Thirty days after irradiation 57 percent of treated group were still alive, while all the irradiated control group injected with saline had died.

Ginsburg et al.(114) repeated above study with the support of National Center for Radiological Health. They confirmed Shipman and Cole's results and found melittin was more effective compare to whole bee venom. Bee venom was highly effective against X-irradiation on mice and findings were statistically significant.

Kanno et al.(115) reported similar results on radioprotective action of bee venom.

4.7. Other Actions

Bee venom has histamine-releasing properties in animal tissues. Two principles are responsible for this effect(116,117): phospholipase A2, and a basic peptide(118). Melittin releases histamine from tissues by direct action. The action of phospholipase A2 is indirect.

Melittin binds to components of skin to create a tight complex formation(119). This complex formation appears to be at least somewhat specific and, as melittin does, will not form a complex with albumin.

Bee venom has an anti-alkylating activity and is effective against poisoning with bis(2-chloroethyl)- methylamine-HCl when given prophylactically(120). Melittin and apamin are not responsible for this action.

Blood glucose level is increased when bee venom is subcutaneously injected into rabbits. The venom may stimulate adrenaline secretion in the animals and thereby increase glyconeogenesis(121). But, blood glucose level is not increased when bee venom is injected intradermally into human subjects(122).

5. CLINICAL EFFECTIVENESS

In practice, the best results are obtained when there is a "good reaction" - considerable swelling and inflammation- at the site of injection. The optimal means of delivering venom is through a hypodermic needle administered by a licensed physician. However, practicing apitherapists use the original hypodermic needle developed by Mother Nature and the honeybee some 30 million years ago, i.e. the bee stinger. Procedures for obtaining and purifying venom have been developed. Physicians are now using a standard purified injectable bee venom(9).

The usual treatment involves injecting the patient at a specific site relative to the illness and repeating the injections over a period of time. For example, it is suggested that the venom be injected into arthritic patients at acupuncture points or trigger points in a twice weekly course of treatment that lasts six to eight weeks. Kim(5) indicates that there are typical patterns of responsiveness, depending on the ailment. A 50-year old patient with arthritis might note pain relief in two weeks, mobility in three weeks, and freedom from symptoms in six weeks.

It should be noted that results presented in this section were obtained upon clinical observations. The majority of the studies were not random assessments, or double blind studies, and therefore were not compared with a placebo or frame of reference. It must also be said that it seems there is a certain consensus, among authors, who have published about this subject, which supports the beneficial effects of bee venom in inflammatory illnesses.

5.1. Efficacy of Apitherapy

5.1.1. Favorable Results on Human Subjects. As early as 1888, Dr. Terc published a medical paper reporting on 173 cases of arthritis which had been successfully treated with bee stings. In 1912, more of Dr. Terc's work with bee venom was published(123). This publication included the results of 660 cases of rheumatics treated with bee venom. Of these, 544 (82.5 percent) were listed as cured, 99 (15.0 percent) improved and 17 (2.5 percent) were unimproved.

Becker(124) had studied 120 cases of arthritis and had obtained favorable results in 74 percent of these cases. The author enthusiastically recommended bee venom as a weapon against rheumatism. He found its usefulness as a treatment even more promising because of its relative safety and the absence of harmful side effects.

Burt(125) has analyzed the results of 50 patients, out of a total of 200, who had been treated with a solution of bee venom. He reports that 16 patients felt distinctly better (32 percent), 9 patients felt better (18 percent), 15 patients remained unchanged (30 percent), and 10 cases deteriorated (20 percent). The author concluded that the treatment with bee venom was definitely not effective in all cases, but definitely in certain cases.

Kroner et al.(126) treated 100 patients suffering from atrophic arthritis. 35 patients (35 percent) have manifested marked relief while 38 (38 percent) had moderate relief. The alleviation of pain was definite and of long duration, while others noted a reduction in the rate of sedimentation. No relief was manifested in 27 patients. The study concluded that bee venom tharapy had brought about impressive results, with marked alleviation of clinical symptoms (pain and swelling) in a significant number of patients.

Fellinger et al.(127) studied 34 cases of chronic polyarthritis using electrodermography in order to eliminate the subjective factor in his evaluation. The authors have shown objective signs of modification of the vegetative nervous system which indicates normalization and healing, corresponding with subjective improvement.

Fichkov(128) reports a 86 percent success rate in 110 cases of neuralgia, and 75 percent of 159 cases of lumbo-ischial radiculitis.

According to Artemov(35) the venom has been used with success in the treatment of inflammatory illnesses of the peripheral nerves like the radiculitis, neuritis, plexitis, and the neuromyositis. He particularly mentions the results of a study by Krivoloutskaya (1958–1960) of 50 patients, stricken with neuralgia of the trigeminal nerve. The author reports 33 with complete cures, 13 improvements and 4 cases did not respond.

Artemov(35) cites the works of numerous Russian authors, such as of Cherchevskaya (1949), Gaidar (1955), Zaitsev and Archangelsky (1956), Zaitsev and Poriadin (1958), Zinkov and Goldina (1957), Neverova (1958–1959), Lobatchev (1958), Chmeleva (1958), Krivoloutzkaya (1959–1960), Alesker (1959), Gorchkov (1960) and Poriadin (1960). He points out that certain of these works reached an elevated scientific values, and were executed in a modern clinical environment. All these authors have obtained good results and Artemov (1959) suggests that, therefore, the introduction of apitherapy in the medical practice would be highly probable. Artemov also mentions the work of Desmartis in France (1855–1859) and of Terc in Austria (1888) and those of Loubarsky in Russia (1887). All have obtained good results in the treatment of chronic illnesses of the joints, muscles, nerves, and the illnesses which are characterized by strong pain.

Zaitsev and Poriadin(129,130) reported on their experiments in treating surgical disease with bee venom. They found bee stings to be very effective in treating the following conditions: ankylosing spondylitis, polymyalgia rheumatica, endarteritis, atherosclerosis of peripheral vessels, thrombophlebitis, trophic ulcers and slowly granulating wounds. Successful results were obtained in 292 of 300 cases treated with bee stings applied every other day for thirty days. No side effects were noted and allergy occured in only three patients. Along with the alleviation of arthritic symptoms, some patients also experienced normalization of sleep, appetite and stool; improvement in general well-being; regularization of pulse and breathing; and lowering of arterial pressure in hypertension.

Kelman(131) treated a total of 140 patients having myalgia and myositis. The author had obtained good results in 105 cases (75 percent).

Zaitsev and Poriadin(132) reported that of 150 patients stricken with ankylosing spondylitis and polyarthritic deformity. Good results were obtained in 117 cases (78 percent), satisfactory results in 30 cases (20 percent), no change in two cases, and allergic reaction in one case.

A controlled clinical study was conducted by Steigerwaldt et al.(133) comparing 50 cases treated with a standardized preparation of bee venom against 11 cases treated with a placebo (injection of chloride of sodium). The result showed beneficial effect with 84 percent and 55 percent using a placebo. After elimination of the cases that showed only a weak improvement, the proportion of patients showing beneficial effects was 66 percent with those who had the venom and 27 percent with those who had been given a placebo.

Hurkov(134) has studied the use of venom with 180 patients stricken with osteoarthritis. Ninety-four had manifested a significant improvement and no adverse reaction was observed.

Nokolova(135) reported a success rate of 94 percent of his patients stricken with rhumatoid arthritis who had been treated without success by conventional treatments.

Zaitsev et al.(136) studied 415 patients: 77 were suffering with obliterating endarteritis, 138 had arteriosclerosis of the peripheral blood vessels, 65 Bachterow's disease, 50 spondylarthritis deformans, 85 polyarthritis deformans. He reported a success rate of 80 to 85 percent of the arterial diseases of the extremities and of the diseases of the spine and the joints.

Serban(137) has compared indomethacine (100mg/day) with purified bee venom (Forapin) in two groups of 50 patients. Each group contained 20 cases gonarthrose, 10 cases of coxarthrose, and 20 cases of spondylitis. After 24 days of treatment the results were in favor of bee venom.

Feldsher et al.(138) reported that bee venom therapy is highly effective for the treatment of chronic low back pain and lumbosacral radiculitis.

Mund-Hoym(139) reported the therapeutic results of 211 patients who suffers from mesenchymal diseases of the locomotion extremities. After treating for six weeks, average improvement was 70 percent. This results must be regarded as very satisfying.

Forestier and Palmer(140) report that among 1,600 cases that were treated with bee venom, an 80 percent success rate was obtained in the following cases: pain in the knee before the arthritis was not too developed; chronic periarthritis of the shoulder had resisted the injections of cortisone; epicondylitis after failure of local injections of cortisone; and relief of pain at the base of the toes. In the case of the rheumatoid polyarthritis, a certain effect noticed at the beginning, but it slowly lost effectiveness. There was minimal effect observed in the coxarthrose, ankylosing spondylitis, and osteoporosis at the vertebrae of menopausal women. They finally conclude that an increasing dose of bee venom could end the intense pain of severe rheumatism, which has existed for months or even years.

Lonauer et al.(141) has presented results of 30 patients stricken with rhumatoid arthritis in stages I-III. The usage of corticosteroid was forbidden, and most of the patients did not use any anti-inflammatory drugs (NSAIDs). A marked improvement of articular joint pain was observed in 74 percent of the cases, a moderate improvement in 9 percent of the cases and no improvement in 17 percent. Mobility of the joint was improved in 65 percent of the cases and swelling diminished in 56.5 percent of the cases. They could considerably lower the use of NSAIDs and the apitherapy constituted an efficient element and was well tolerated.

Kim(142) reports results of 53 patients with a long history of arthritis or neuralgia. While conventional treatments had failed with all of these patients, most of them benefited by the treatment with bee venom.

Kim(143) studied a total of 108 cases of polyarthritic rheumatism. The study did not include a placebo group and no statistics were presented. The author concludes, nevertheless, that the treatment with bee venom appears applicable, effective, and without serious side-effects in the absence of an allergy to bee venom.

Klinghardt(144) reported anecdotally that among 128 patients with a wide spectrum of illnesses, all but 11 appeared to improve (90 percent improvement). This report is typical of anecdotal apitherapy results that begin with stories of beekeepers recounting various health improvements after receiving accidental multiple stings from their bees. Klinghardt's patients had diagnoses of gout, rheumatoid arthritis, fibromyalgia, spinal strain or sprain, spinal disc injuries, postlaminectomy pain, bunion, postherpetic neuralgia, incomplete healing of a fractured bone, intractable pain from large burn wounds, osteoarthritis, ankylosing spondylitis, vertigo, and multiple sclerosis.

There are many articles which show scientifically sound evidence of effectiveness of bee venom against arthritis and painful inflammatory conditions(107,145–153).

In a recent study(154), a randomized, controlled trial was conducted comparing true honeybee venom therapy with a "sham" product for 180 patients suffering from chronic pain and inflammation who did not respond anymore to conventional therapies. Apitoxin (pure bee venom solution) was injected twice weekly for 6 weeks. Significant post-treatment reductions in pain and inflammation were recorded in the true bee venom therapy group and were maintained at 6-month followups.

The American Apitherapy Society (AAS) endeavors to coordinate information on bee venom research for the past 20 years. The society has now acquired more than 12,000 case reports on persons treated with bee venom(155). These 12,000 reports are the basis for the ongoing National Multicenter Apitherapy Study. Approximately 200 physicians and 300 beekeepers voluntarily contribute reports. The multicenter study has in its data base some 1,300 reports on patients with multiple sclerosis (subjectively reporting increased sensation and energy, and bowel and bladder control); 2,800 with rheumatiod arthritis and other groupings of data on such problems as gout, viral illnesses, and premenstrual syndrome - nearly 100 percent of 40 women being treated for premenstrual syndrome by apitherapy became symptom free.

5.1.2. Unfavorable Results on Human Subjects. There are only a few articles out of hundreds that show unfavorable results.

Nicholls(156) treated 27 patients with severe active rheumatoid arthritis with bee stings. The results were very disappointing. Five (18 percent) had to discontinue the bee stings because of severe reaction. Twenty patients continued for from three to eighteen months and only three (15 percent) showed any definite improvement.

Hollander(157) reported 42 cases of chronic arthritis treated by bee venom* injection. An improvement was only observed in 8 cases (19 percent). Seventeen patients, serving as control-group, were injected with Proteolac (foreign protein). Three cases (18 percent) of these patients also manifested an improvement. The author concluded that the incident of the degree of improvement was very discouraging.

5.1.3. Results in the Animal Models

5.1.3.1. Rat Models. Much of the controlled clinical research analyzing the effects of bee venom on arthritis has been done with arthritic animals (mice, rats, guinea pigs, dogs and horses). The adjuvant-induced arthritic rats is one of the most often used test models in the developmental research of new compounds which may be used to treat arthritis and other inflammatory conditions.

A study conducted by Lorenzetti et al.(42) showed that bee venom, when administered three times a week to adjuvant-induced arthritic rats, had a significant effect both in preventing the development of arthritis, and in reducing the severity of a pre-existing arthritic condition. However, these authors also found, that the anti-inflammatory effect of bee venom was much more pronounced when it was used in preventive treatments.

Weissman et al.(158) found that the effects of daily injections of three individual components (melittin, apamin, and phospholipase A), along with the effects of daily injections of whole bee venom, prevented adjuvant arthritis from developing in rats.

Chang and Bliven(30) examined the mode of action of bee venom on adjuvant-induced arthritis and concluded that at least two mechanisms were involved. They found that a single injection of bee venom administered on the day before, or the day of, the injection of adjuvant, effectively suppressed the development of arthritis in rats. Their results also suggested that this suppression was brought about not only through the anti-inflammatory action of the bee venom, but also through an alteration of the immune response.

A number of independent investigations utilizing rat models have been made of the anti-inflammatory properties of whole bee venom(30,42,159,160). These studies have demonstrated an anti-inflammatory effect of whole bee venom in the rat adjuvant induced arthritis and carrageenan footpad edema tests. Using these in vivo assays Chang and

* Hollander used Ven-apis in the first 24 cases and the Lyovac bee venom in the last 20. These two brands of bee venom were obtained by grinding of the venom sacks. These solutions cannot be considered to be a pure bee venom. -Author

Bliven(30) found that bee venom (0.5mg/rat subcutaneously) and cyclophosphamide (60mg/kg orally) behaved similarly in suppressing adjuvant induced arthritis in rats and showed a distinctly different temporal pattern of activity from steroid therapy, counting the argument that effects observed were due to stress induced steroid production. Their results also suggested that this suppression was brought about not only through the anti-inflammatory action of the bee venom, but also through an alteration of the immune response.

Eiseman et al.(159) studied the effects of bee venom on the course of adjuvant-induced arthritis and depression of drug metabolism in rats. The authors found the bee venom suppressed the primary and secondary inflammatory responses to the adjuvant hind paws. They also observed changes in heme metabolism elicited by the venom strongly suggestive of perturbations of the immune system causing alterations in hepatic microsomal enzymes.

Hadjipetrou-Kourounakis and Yiangou(32) have shown that Fisher rats treated with 0.5 mg/kg daily for 17 days did not develop adjuvant induced arthritis. Local therapy with 200 microg./kg in the hind paws three times weekly for three weeks inhibited development of adjuvant induced disease in the hind paws but not the non venom injected front paws.

5.1.3.2. Dogs. Dogs have also been used as experimental models in testing the effectiveness of bee venom as an antiarthritic compound. The US Army became interested in the possibility of using bee venom as a treatment for their guard and attack dogs which tend to develop severe arthritis in the hind legs. Vick et al.(161) began investigating the use of pure bee venom in arthritic dogs. In their study with sixteen beagles, bee venom therapy was found to be extremely effective in relieving the symptoms of the arthritic animals. Of the sixteen beagles, eight had been diagnosed by veterinarians as having osteoarthritis, while the other eight were free of any arthritis. Pain and stiffness in the arthritic animals kept them from moving around as much as the normal beagles, so pedometers were hung about the dogs' necks in order to measure improvement in terms of walking activity. Before treatment began, the normal dogs moved about an average of twelve miles a day in the kennels, but the arthritic dogs only moved an average four miles a day. After only three injections of bee venom (1mg. per injection), the arthritic dogs' average kennel activity approximated that of the normal dog population. Also, the venom appeared to have a long-lasting effect, as this increase in activity continued for sixty to ninety days after the last injection.

Short et al.(162) reported the results of treating seventeen dogs that had been diagnosed as arthritic. Fourteen of the seventeen dogs improved significantly following bee venom therapy, returning to normal or near-normal movement. Four of five dogs treated for joint complications (hip displasia and arthritic joints) showed improved movement; four of six dogs treated for poor surgical recovery responded well; and all dogs suffering from disc complications returned to normal or near-normal conditions after a series of bee venom injections administered at the sites of pain and stiffness. From the results of this study, authors concluded that bee venom therapy may be highly beneficial in alleviating certain arthritic conditions in dogs.

According to Vienna Animal Hospital's 15 case study report (1984), fifteen dogs were scheduled for euthanasia due to severe arthritis and vertebral disc complications. All were treated with bee venom injection at weekly interval for six weeks. The results showed dramatic improvement in three (20 percent), moderate improvement in seven (47 percent), no effects in five (33 percent).

Veterans Animal Hospital (1988) has reported that a case study of thirteen dogs with hip dysplasia and disc complications. Four dogs (30 percent) completely recovered; seven (55 percent) moderately improved; and two (15 percent) showed no improvement.

5.1.3.3. Horses. Von Bredow(163,164) conducted a study in which he observed the effects of bee venom injections on eight arthritic horses, ranging in age from eight to seventeen years. Six of the eight horses showed significant improvement, with three of these six demonstrating a complete recovery.

The existence of chronic obstructive pulmonary disease (COPD) in horses was thought to be between 12–25 percent of the racing population. However, with the introduction of the fiberbronchoscope, the percentage of COPD horses at the race track is now conservatively placed at 50 percent. Anytime a horse failed to race up to his potential, this disease should be considered. COPD is a big problems among the racing horses since drugs are prohibited for racing events. Holms(165) reported a controlled, randomized study of the effectiveness of bee venom for COPD of the racing horses. Horses were chosen at random from the horses that were racing at the Ontario Jockey Club Circuit. Total 3.0 ml. (60 mg. ie, 20mg./ml.) of bee venom was injected subcutaneously at the rate of 0.5 ml. (10 mg.) per site into the three acupuncture points (BL40, 41, 42) bilaterally. Seventy six percent of horses improved COPD and showed better performance.

5.2. Difficulties in the Control Studies

Hawley and Wolfe(166) have evaluated a total of 122 in clinical controlled studies and studied by observation of secondary therapy 16,071 patients who suffered from rheumatoid polyarthritis. They mention especially that most of the studies have been conducted with patients whose illness has been well established, an average of 7.6 years. Controlled studies lasted in general less than one year where studies of observations were conducted an average of 31 months. The authors said that studies of observation could furnish important information regarding the effectiveness of treatment of rheumatoid polyarthritis which are not available upon controlled studies alone. The results of their evaluation indicate that many patients rarely take their inferior medication more than one or two times even due to ineffectiveness, side effects of the treatment, or treatment started too late during the course of the illness, etc.

Hard et al.(167) have studied the relationship of clinical signs at the beginning of arthritis and x-rays of a population of 514 woman, aged 45 to 65. The x-rays have shown that the principal signs or clinical studies show a strong specific feature (87–99 percent) and a weak sensitiveness (20–49 percent). They have concluded that certain physical signs of arthritis can be reproduced and can be utilized to identify clinical illness. This same idea, reported by Bagge et al.(168), states that there is a correlation between clinical signs and x-ray results. Clinical exam has proven to be more credible and perceptible to detect the onset of arthritis.

Gordon(169) disusses the importance of therapeutic experiments, particularly the necessity of further controlled random studies before a medication could be introduced into clinical application. He also states most of the studies are short term studies, only conducted over a period of several weeks or months, while polyarthritis and rheumatoid arthritis can take on unforeseen and variable course, lasting several years. The author confirms that the measure with which the results of the controlled studies are assessed at random can be generalized, for the case of long-term arthritis is disputable.

According to the author, decisive proof for all new medication is its clinical effectiveness, and its absence of serious side-effects upon all patients regardless of their age and general health.

The proof of the effectiveness of a treatment is generally obtained with the aide of controlled random studies versus first a placebo, and followed by a related treatment. Such an approach proves difficult where bee vemom is concerend, as it is administered by intradermal injection and results in a certain pain at the site of the injection. To counter this difficulty, Kim(154) has proceeded with an injection of a solution of phosphate histamine instead of the placebo. This approach has an objective to rendering the placebo treatment credible for the patient. The author has even included an evaluation of this credibility based upon a questionnaire. As the phosphate histamine does not represent a placebo in the strict sense, it does resemble a placebo very closely.

Recently, Pincus and Callahan(26) reported that the "natural" evolution of rheumatoid arthritis cannot be characterized as independent therapy. It is not acceptable from an ethical point of harmlessness to have been proven during a period of ten years. The illness has progressed with most of his patients.

In the same vein, Wilske and Healey(170) reported that for several patients the actual treatments are not enough to prevent long term damages and the incapacitation which is the result thereof. The authors underline the need to develope new treatment and strategies, including early intervention, and to recognize the beginning of the progression of the illness.

Kushner and Dawson(171) say that the question is not only to know if one medication is superior, but rather to know at which point we are close to what we want to accomplish.

It is true that there have been only a few controlled random double blind studies on bee venom therapy. One can, nevertheless, conclude from some of the preceding observations that it seems difficult to reject all the medical literature about the clinical effectiveness of bee venom in the treatment of inflammatory illness under the pretext of such deficiencies.

5.3. Adverse Effects

No major complications or side effects were reported in this review. Transient minor complaints after bee venom injections were reported(6,107,123,133,144,157).

Yunginger et al.(172) have studied to determine the immunological and biological consequences of chronic bee venom administration to humans. In this study of a large number of beekeepers, it seems unlikely that chronic bee venom administration produces clinical disease in humans.

Forestier et al.(140) concluded, after having treated about 1,600 cases, that bee venom therapy can be considered to be relatively safe. This conclusion might be justified given that there have been few reports on severe complications.

Kim(154) reports that in publications of large series of bee venom injection, no major complications were observed. In his work, based upon total 34,600 injections (bee venom 1mg. per 1ml.) to 174 patients, it was revealed that itching constitutes the principal undesirable reaction reported by the patients; about 80 percent after initial treatment and 40 percent after 12th treatments. The side effects are secondary, they only appear at the beginning of the treatment and, including swelling (29.7 percent), headache (6.4 percent), staggering (5.8 percent) and flushing (5.6 percent).

Based on the articles in this review, it becomes very clear that bee venom therapy is not source of problems, though one has to make sure that the patient is not allergic to bee venom. Bee venom injections are being relatively well-tolerated and do not present much toxicity in animal as well as human models.

6. CONCLUSION

Apitherapy has existed since antiquity and is still being practiced in many countries. Even though doctors and scientists involved in researching bee venom as a potent anti-inflammatory agent concur that results have been promising and further research is warranted, very little research has been undertaken in the US in recent years. Reports featuring scientifically sound evidence of bee venom's effectiveness against arthritis and chronic inflammation(5,9,35,123–165,173–192) have been largely ignored in the US, and there is no monetary incentive for pursuing further investigation of this natural substance(193). Òne must acknowledge that controlled studies, and those taken randomly, especially versus placebo, are difficult to evaluate. One of the solutions would be to make a change, to include a "blind" as far as the treatment is concerned. The approach of Steigerwaldt et al.(133) and Kim(154), who used sham product as a credible irritant with his patients, seems an acceptable compromise. Comparative studies of bee venom versus standard treatment like the anti-inflammatory non-steroids would permit a statistical analysis, which would make a point for discussion.

It appears that the studies in vitro and in vivo done with animals suggest strongly that the venom could share several mechanisms with analgesic and anti-inflammatory agents. This constitutes circumstantial proof of its suggested clinical effectiveness.

Based upon all the literatures, there should be no strong objections to the use of bee venom. Actually, the great majority of authors who address themselves to the effects of apitherapy report little undesirable reactions, as long as the patient is not allergic to bee venom.

In some ways, bee venom therapy is a classic alternative therapy. It has ancient roots and, although discarded by mainstream medicine, has been carried on for thousands of years.

7. REFERENCES

1. Science News. Sweet route to heading off colon cancer. 1993, 144:p.207.
2. Fang Z. The honeybee and human health. NAAS Proc. 1982, 5:p.18.
3. Xu RX. Burn treatment with raw honey. China National Science and Technology Center 1990.
4. Rao CV, Desai D, Simi B et al. Inhibitory effect of caffeic acid esters on azoxymethane-induced biochemical changes and aberrant crypt foci formation in rat colon. Cancer Res. 1993, 53(18):4182–4188.
5. Kim CM-H. Bee venom therapy. Managing Stress and Pain 1986, 1:4:1–6.
6. Terc P. Lecture in the Monthly Assembly of Beekeepers, February 11, 1904. In: Bee venom: the natural curative for arthritis and rheumatism. G.P. Putnam's sons, New York, 1962, Appendix H, pp.183–197.
7. Langer J. Uber das Gift unserer Honigbiene. Arch.Exp.Path.Pharmk. Leipz 1897, 38:381–396.
8. Neumann W, Habermann E and Amend G. Zur papierelektrophoretischen Fraktionierung tierischer Gifte. Naturwissenschaften 1952, 39:286–287.
9. Kim CM-H. Bee venom therapy and bee acupuncture therapy (Medical Text). Published 1992 (Korean Ed.), pp515, 1000 references.
10. Habermann E and Reiz KG. Ein neues Verfahren zur Gewinnung der Komponenten von Bienengift, insbesondere des zentral wirksamen Peptids Apamin. Biochem. Z. 1965, 341:451–466.
11. Fredholm B. Studies on a mast cell degranulating factor in bee venom. Biochem. Pharmacol. 1966, 15:2037–2042.
12. Shkenderov S and Koburova K. Adolapin - A newly isolated analgesic and anti-inflammatory polypeptide from bee venom. Toxicon 1982, 20:317–321.

13. Shkenderov S. A protease inhibitor in bee venom. Identification, partial purification and some properties. FEBS Lett. 1973; 33:343–347.
14. Gauldie J. Hanson JM, Rumjanek FD, Shipolini RA and Vernon CA. The peptide components of bee venom. Eur. J. Biochem. 1976, 61:369–376.
15. Nelson DA and O'Connor R. The venom of the honey bee (Apis mellifera): Free amino acids and peptides. Can. J. Biochem. 1968, 46:1221–1226.
16. Lowy PH, Sarmiento L and Mitchell HK. Polypeptides minimine and melittin from bee venom: Effects on Drosophilia. Arch. Biochem. Biophys. 1971, 145:338–343.
17. Vick JA, Shipman WH and Brooks RB. Beta-adrenergic and anti-arrhythmic effects of cardiopep, a newly isolated substance from whole bee venom. Toxicon 1974, 12:139–144.
18. Neumann W and Habermann E. Beitrage zur Charakterisierung der Wirkstoffe des Bienengiftes. Naunyn Schmiebergs Arch. Pharmacol. 1954; 222:367–387.
19. Habermann E and Neumann WP. Reinigung der Phospholipase A des Bienengiftes. Biochem. Z. 1957, 328:465–473.
20. Shkenderov S, Ivanova I and Grigorova K. An acid monophosphatase and alpha-glucosidase enzymes newly isolated from bee venom. Toxicon 1979, 17 (Suppl. 1):169–170.
21. Ivanova I and Shkenderov S. A newly isolated enzyme with lyosphospholipase activity from bee venom. Toxicon 1982, 20:333–335.
22. Reinert M. Zur Kenutnis des Bienengiftes Festschrift. Emil Barell, Basel.
23. Owen MD. Insect venoms: Identification of dopamine and noradrenaline in wasp and bee stings. Experientia 1971, 27:544–546.
24. Owen MD, Braidwood JL and Bridges AR. Catecholamines in honey bee (Apis mellifera) and various vespid (hymenoptera) venoms. Toxicon 1982, 20:1075–1084.
25. O'Connor R, Henderson G, Nelson D, Parker R and Peck M. The venom of the honey bee (Apis mellifera). General character. In Animal Toxins, Pergamon, Oxford 1967, pp 17–22.
26. Pincus T and Callahan LF. What is the natural history of rheumatoid arthritis? Current controversies in clinical rheumatology. Rheum. Dis. Clin. N. Am. 1993, 19(1):123–151.
27. Artemov NM. The physiological proofs of apitherapy. Bee venom acts as a cholinolytic agent. XVII International Apicultural Congress, 1959.
28. Shipman WH and Cole LJ. Increased resistance of mice to X-irradiation after the injection of bee venom. Nature 1967, 215:311:2.
29. Zurier RB. Effect of bee venom on experimental arthritis. Ann. Rheu. Dis. 1973, 32:466–470.
30. Chang YH and Bliven ML. Anti-arthritic effect of bee venom. Agents and Actions 1979, 9:205–211.
31. Hyre HM and Smith RA. Immunological effects of honeybee venom using balb/c mice. Toxicon 1986, 24:5:435–440.
32. Hadjipetrou-Kourounakis L and Yiangou M. Bee venom, adjuvant induced disease and interleukin production. J. Rheum. 1988, 15:1126–1128.
33. Somerfiled SD. Bee venom and arthritis: magic, myth or medicine? New Zealand Med. J. 1986, 281–283.
34. Belliveau J. The effectiveness of bee venom an adjuvant induced colon cancer of the rats. II Am. Apitherapy Soc. Conf. 1992, Boston, USA.
35. Artemov NM. The biological bases of the therapeutic use of bee venom. University of Gorki, 1959:1–40.
36. Hammeral AM and Pitchler O. On therapy with AP Forty. Med. Clin. 1960, 55:2015–2021.
37. Vick JA and Shipman WH. Effects of whole bee venom and its fractions (apamin and melittin) on plasma cortisol levels in the dog. Toxicon 1972, 10:377–380.
38. Vick JA, Mehlman B, Brooks R, Phillips SJ and Shipman WH. Effect of bee venom and melittin on plasma cortisol in the unanesthetized monkey. Toxicon 1972, 10:581–586.
39. Hanson JM, Morley J and C. Soria-Herrera. Anti-inflammatory property of 401 (MCD-peptide), a peptide from the venom of the bee Apis mellifera (L). Br. J. Pharmacol. 1974, 50:383–392.
40. Slotta KH, Vick JA and Ginsberg NJ. Enzymatic and toxic activity of phospholipase A. In De Vries and Kochva Eds., Toxins of Animal and Plant Origin, Vol.I, Gordon and Breach, New York 1971, pp 401–418.
41. Couch TL. The effect of venom of the honeybee on the adrenocortical responses of the adult male rat. Toxicon 1972, 10:55–62.
42. Lorenzetti OJ, Fortenberry B and Busby E. The influence of bee venom in the adjuvant induced arthritic rat model. Res. Comm. Chem. Pathol. Pharmacol. 1972, 4(2):339–352.
43. Vick JA, Warren GB and Brooks RB. The effects of treatment with whole bee venom on cage avtivity and plasma cortisol levels in the arthritic dog. Inflammation 1976, 1:167–174.
44. Knepel W and Gerhards C. Stimulation by melittin of adrenocorticotrophin and beta-endophin release from rat adenohypophysis in vitro. Prostaglandins 1987, 33:3:479–490.

45. Doughert. Biochem. J. 1960, 88:599.
46. Neubould BB. Peptide 401 in adjuvant arthritis in rats. Br.J.Pharmac. 1963, 21:127.
47. Zurier RB and Quagliata F. Effect of prostaglandin E1 on adjuvant arthritis. Nature 1971, 234:304.
48. Billingham MEJ, Morley J, Hanson JM, Shipoli RA and Vernon CA. An anti-inflammatory peptide from bee venom. Nature 1973, 245:163–164.
49. Surfer R and Ballas M. Peptide 401 and prostaglandin suppression of adjuvant arthritis. Arth.Rheu. 1973, 16(2):251–157.
50. Collier HJO. A pharmacological analysis of aspirin. Adv.Phama.Chemother. 1969, 7:333–405.
51. Banks BE, Rumjanek FD, Sinclair NM and Vernon CA. Possible therapeutic use of a peptide from bee venom. Bulletin, Pasteur Institute 1976; 74; 137–144.
52. Vane JR. Inhibition of prostaglandin synthesis as a mechanism of action for aspirin-like drugs. Nature New Biology 1971, 231:232–235.
53. Mancada S, Ferreira SH and Vane JR. Pain and inflammatory mediators. Handbook of Exp. Pharm., 50/1. Inflammation. Springer Verlag, New York 1978, pp 588–616.
54. Koburova KL, Michailova SG and Shkenderov SV. Further investigation on the antiinflammatory properties of adolapin - bee venom polypeptide. Acta Physiol. Pharmacol. Bulgaria 1985, Vil.II, No.2, 50–55.
55. Shkenderov S, Koburova K and Chavdarova V. Bee venom adolapin: effect on thromboxane A2 and prostacycline plasma levels in rats with model acute inflammation. Comptes rendus de l'Academie bulgare des Sciences 1986, 39:155–157.
56. Michailova SG, Koburova KL and Shkenderov SV. Acta Phys. Pharm. Bulg. 1985, Vol. VII, 50.
57. Shkenderov S. Anti-inflammatory effect of bee venom protease inhibitor on a model system of acute inflammatory edema. Comptes rendus de l'Academie bulgare des Sciences 1986; 39:151–154.
58. Mancada S and Vane IR. NEJM 1979, 300:1142.
59. Gryglewsky R, Korbut R and Dembinska-Kiec. Prostaglandins and Thromboxanes. Plenum Press, New York 1977, 363.
60. Grady O, Warrington I and Moti M. Prostaglandins 1980, 19:319.
61. Somerfield SD, Stach JL, Mraz C, Gervais F and Skamene E. Bee venom inhibits superoxide production by human neutrophils. Inflammation 1984, 8:385–391.
62. Takeshige K and Minakami S. Involvement of calmodulin in phagocytic respiratory burst of leukocytes. Biochim. Biophys. Res. Comm. 1981, 99:484–490.
63. Smith RJ, Bowman BJ and Iden SS. Effects of trifluoperazine on human neutrophil function. Immunology 1981, 44:677–684.
64. Smolen JE and Weissman G. The roles of extracellular calcium in lysosomal enzyme release and superoxide anion generation by human polymorphonuclear leukocytes. Biochim. Biophys. Acta 1980, 677:521–530.
65. Somerfield SD, Stach JL, Mraz C, Gervais F and Skamene E. Bee venom melittin blocks neutrophil O2-production. Inflammation 1986, 10:175–182.
66. Rekka E, Kourounakis L and Kourounakis P. Antioxidant activity of and interleukin production affected by honey bee venom. Arzneimittel Forschung - Drug Research 1990, 40:912–913.
67. Chayen J, Bitensky L, Butcher RG, Poulter LW and Ubhi GS. Br. J. Derm. 1970, 82:Suppl.6, 62 & Beitr. Path. Anat. 1972, 147:6.
68. Dufourcq J. Molecular details of melittin-induced lysis of phospholipid membranes as revealed by deuterium and phosphorus NMR. Biochim. Biophys. Acta 1986, 859(1):33–48.
69. Shkenderov S and Chavdarova V. Effects of melittin on lysosomal membrane stability. Toxicon 1986, 17 (Suppl. 1), 168.
70. Schwartz G and Beschiaschvili G. Thermodynamic and kinetic studies on the association of melittin with a phospholipid bilayer. Biochim. Biophys. Acta 1989, 979(1):82–90.
71. Glenn EM and Sekbar N. In: Immunopathology of inflammation. Forscher and Houk Eds., Exc. Med. Amst. 1971. 13.
72. Di Rosa M, Papadimitriou JM and Willonghby DA. A histopathological and pharmacological analysis of the mode of action nonsteroidal antiinflammatory drugs. J. Pathol. 1971, 105:239–256.
73. Gencheva G and Shkenderov S. Inhibition of complement activity by certain bee venom components. Academy Bulgaria Science 1986, 39, No.9, 137–139.
74. Menander-Huber J. Melittin bound to calmodulin. NMR assignments and global conformation features. Exp. Biochem. 1980, 112:236.
75. Comte M, Maulte Y and Cox JA. Ca++ dependent high-affinity complex formation between calmodulin and melittin. Biochem. J. 1983, 209:269–272.
76. Jones HP, Chai G and Petrone WF. Calmodulin dependent stimulation of the NADPH oxidase of human neutrophils. Biochim. Biophys. Acta 1982, 714:152–156.

77. Barbior BM. The respiratory burst of phagocytes. J. Cin. Invest 1984, 73:599–601.
78. Banks BE, Dempsey CE, Vernon CA and Yamey J. The mast cell degranulating peptide from bee venom. Physiol. (London) 1980, 308:95–96.
79. Banks BE, Dempsey CE, Vernon CA, Warner JA and Yamey J. Anti-inflammatory activity of bee venom peptide 401 (mast cell degranulating peptide) and compound 40/80 results from mast cell degranulation in vivo. Br. J. Pharmacol. 1990, 99:350–354.
80. Habermann E and Breithaupt H. MCD-peptide, a selectively mastolytic factor isolated from bee venom. Naunyn Schmiedeberg Arch. Pharm. Exp. Path. 1968, 260:127–128.
81. Breithaupt H and Habermann E. MCD-peptide from bee venom: isolation, biochemical and pharmacological proper. Naunyn Schmiedeberg Arch. Pharm. Exp. Path. 1968, 261:252–270.
82. Zurier RB and Ballas M. Prostagladin E1 (PGE1) suppression of adjuvant arthritis: Histopathology. Arthritis Rheu. 1973, 16:251–258.
83. Ziai MR, Russek S, Wang HC, Beer B and Blume AJ. Mast cell degranulating peptide a multi-functional neurotoxin. J. Pharm. Pharmacol. 1990, 42:457–461.
84. Forster KA. Forty years of experience with bee venom therapy. Che. Med. 1950.
85. Sergeeva LI. Heparin-induced inhibition of the hemolytic activity of bee venom. Uch. Gor'k. Gos. Univ. 1974, 175:130.
86. Habermann E. Biochemistry, pharmacology and toxicology of honey bee venom. Ergeb. Physiol. 1968, 60:220–325.
87. Meszarors I. Poisoning following bee-wasp stings. Z. Gesamte Inn. Med. 1971, 26:Suppl. 193–195.
88. Hahn G and Ostermayer H. Uber das Bienengift. Ber deutsch Chem. Ges. 1936, 69B: 2407–2419.
89. Wellhoner H. Spinale Wirkungen von Apamin. Naunyn Schmiedebergs Arch. Pharmcol. 1969, 262:29–41.
90. Vladimirova IA and Shuba MF. The effect of strychnine, hydrastin and apamin on synaptic transmmission in smooth muscle cells. Neirofizologica(Kiew) 1978, 10:295–299.
91. Banks BE, Brown C, Burgess GM, Burnstock G, Claret M, Cocks TM and Jenkinson DH. Apamin blocks certain neurotransmitter-induced increases in potassium permeability. Nature 1979, 282: 415–417.
92. Maas AJ and Heertog A. The effect of apamin on smooth muscle cells of the guinea-pig Taenia cili. Eur. J. Biochem. 1979, 58:265–270.
93. Muller J and Baer HP. Apamin, a nonspecific antagonist of smooth muscle relaxants. Naunyn Schmiedebergs Arch. Exp. Pathol. Pharmacol. 1980, 311:105–107.
94. Hugues M, Duval D, Schmid H, Kitabgi P, Lazdunski M and Vincent JP. Specific binding and pharmacological interactions of apamin, the neurotoxin from bee venom, with guinea-pig colon. Life Sci. 1982, 31:437–443.
95. Lazdunski M, Romey G, Renaud J, Mourre C, Hugues M and Fosset M. The apamin-sensitive Ca++-dependent K+ channel. Handbook of Exp. Pharmacol., Springer-Verlag, 1988, Vol.83, 135–145.
96. Jenkinson DH. Peripheral actions of apamin. TIPS 1981, 2:333–335.
97. Hugues M, Schmid H, Romey G, Duval D, Frelin C and Lazdinski M. The Ca++-dependent slow K+ conductance in cultured rat muscle cells: Characterization with apamin. EMBO J. 1982, 1: 1039–1042.
98. Burgess GM, Claret M and Jenkinson DH. Effects of quinine and apamin on the calcium dependent potassium permiability of mammalian hepatocytes and red cells. J. Physiol. (London),1981, 317:67–90.
99. Renaud JF, Desnuelle C, Schmid-Antomarchi H, Hugues M, Serratrice G and Lazdunski. Expression of apamin receptor with muscles of patients with myotonic muscular dystrophy. Nature 1986, 319:678–680.
100. Hugues M, Duval D, Kitabgi P, Lazdunski M and Vincent JP. Preparation of pure monoiodo derivative of bee venom neurotoxin apamin and its binding properties to rat brain synaptosomes. J. Biol. Chem. 1982, 257:2762–2769.
101. Habermann E and Cheng-Raude. Central neurotoxicity of apamin, crotamin, phospholipase A2 and alpha-amanitin. Toxicon 1975, 13:465–467.
102. Vyatchannikov NK and Sinka AY. Effect of melittin, the major constituent of bee venom, on the central nervous system. Farmakol. Toksikol. 1973, 36:625.
103. Mourre C, Hugues M and Lazdunski M. Quantitative autoradiographic mapping in rat brain of the receptor of apamin, a polypeptide toxin specific for one class of Ca++-dependent K+ channel. Brain Res. 1986, 382:239–249.
104. Vick JA, Shipman WH, Brooks RB and Hasset CC. The beta-adrenergic and antiarrythmic effects of apamin, a component of bee venom. Am. Bee J. 1972, 112:339.
105. Vick JA and Brooks RB. Pharmacological studies of the major fractions of bee venom. Am. Bee J. 1972, 112:288.
106. Kireeva VF. Capillary permiability changes resulting from the effect of bee venom. Uch. Zap. Gor'k. Gos. Univ. 1970, 101:113.

107. Beck BF. Bee venom, its nature, and its effects on arthritic and rheumatoid conditions. D. Appleton-Century Co. New York, 1935: pp238, 341 references.
108. Schmidt-Lange W. The germicidal effect of bee venom. Muench. Med. Wchenschr., 1941, 83:935.
109. Ortel S and Markwardt F. Pharmazie 1955, 10:743.
110. Fennell JF, Shipman WH and Cole LJ. Antibacterial action of melittin, a polypeptide from bee venom. Proc. Soc. Exp. Biol. Med. 1968, 127:707–710.
111. Dorman LC and Markey LD. Solid phase synthesis and antibacterial activity of N-terminal sequences of melittin. J. Med. Chem. 1971, 14:5–9.
112. Hadjipetrou-Kourounakis L and Yiangou M. Bee venom and adjuvant induced disease. J. Rheumatol. 1984, 1(5):p.720.
113. Shipman WH and Cole LJ. Increased radiation resistance of mice injected with bee venom one day prior to exposure. Report USNRDL-TR-67–4, U.S. Naval Radiological Defense Lab., Dec. 1968, 1–10.
114. Ginsberg NJ, Dauer M and Slotta KH. Melittin used as a protective agent against X-irradiation. Nature 1968, 220:p.1334.
115. Kanno I, Ito Y and Okuyama S. Radioprotection by bee venom. J. Jap. Med. Radiat. 1970, 29:30.
116. Feldberg W and Kellaway CH. Liberation of histamine and its role in the symptomatology of bee venom poisoning. Aust.J.Exp.Biol.Med.Sci. 1937, 127:707.
117. Fredholm B and Haegermark O. Histamine release from rat mast cell granules induced by bee venom fractions. Acta Physiol. Scand. 1967, 71:357.
118. Fredholm B and Haegermark O. Studies on the histamine releasing effect of bee venom fractons and compound 48/80 on skin and lung tissue of the rat. Acta Physiol. Scand. 1969, 76:288.
119. Shipman WH and Cole LJ. Complex formation between bee venom melittin and extract of mouse skin detected by Sephadex gel filtration. Experientia 1972, 28:171.
120. Rauen HM, Schriewer H and Ferie F. Alkylans alkylandum reactons. 10. Antialkylating activity of bee venom, melittin, and apamin. Arzneim-Forsch 1972, 22:1921.
121. Artemov NM, Kireeva VF and Gudenko NA. Effects of bee venom on the blood sugar level. Uch. Zap. Gor'k. Gos. Univ. 1972, 5.
122. Kim CM-H. Apitoxin (pure honeybee venom solution) study. a. Safety test (mice, rat); b. Histamine test (cat) c. Antigenicity test (guinea pig, mice, rat); d. Hematophysiological response (human); e. AIDS treatment (human). IV American Apitherapy Conference 1994, Washington, D.C., USA.
123. Perti E. The science of apitherapy today. Dr. Philp Terc, the foremost apitherapist in Central Europe: the principles of his cures of rheumatic diseases with bees. Apitherapy Symposium at Portoroz, 1912.
124. Becker S. Treatment of rheumatic diseases with injectable bee venom. Therapie Der Gegenwart, Heft 6, 1931.
125. Burt JB. Bee venom therapy in chronic rheumatic disorders. Br. J. Phys. Med. 1937, 12:171–172.
126. Kroner J, Nicholls EE, Lintz R, Tyndall M and Anderson M. The treatment of rheumatoid arthritis with injectable form of bee venom. Ann. Int. Med. 1938, 11(7):1077–1083.
127. Fellinser E. Application of bee venom in the chronic polyarthritis. Cl. Med. 1954, 27:20–25.
128. Fishkov EL. Therapeutical use of the bee venom preparation. K.F. Clin. Med. 1954, 32:8:20–25.
129. Zaitsev GP and Poriadin VT. The use of bee venom therapy in surgical diseases. Apiculture 1958, 2:47–50.
130. Zaitsev GP and Poriadin VT. Bee venom in the treatment of arterial vessels, diseases of the spine and of circulation. XVII International Apiculture Congress, 1959.
131. Kelman IM. Application of bee venom in sanatorium conditions. Pchelovdstvo 1960, 37(3):52–54.
132. Zaitsev GP and Poriadin VT. Bee venom in the treatment of ankylosing spondylitis and polyarthritis. Moscow National Institute of Medicine, 1961.
133. Steigerwaldt F, Mathies and Damrau F. Standardized bee venom (SBV) therapy of arthritis. Controlled study of 50 cases with 84% benefit. Industrial Medicine and Surgery 1966, 35:1045–1050.
134. Hurkov S. Electrophoresis of the bee venom preparation Melivenon in the treatment of osteoarthritis. Kurort Fizioter 1971, 8:3:128–131.
135. Nikolova V. A study of the therapeutic value of electrophoresis with bee venom in children with rheumatoid arthritis. Problems in Pediatrics 1973, 16:101–106.
136. Zaitsev GP and Poriadin VT. Bee venom in the treatment of the arterial vessels of the extremities and of the diseases of the spine and joints. XVIII Apimondia Congress 1973, 1–9.
137. Serban E. Bee venom and rheumatism. Fr. Rev. Apitherapy, 1981, p.399.
138. Feldsher AS, Solodovnikox GI and Gorobets GN. Bee venom treatment of lumbosacral radiculitis. Feldsher Akush (USSR) 1981, 46(4):55–57.
139. Mund-Hoym WD. A report of the results of treating a total of 211 patients with bee venom. Medical World 1982, 33:34:1174–1177.
140. Forestier F and Palmer M. Apitherapy; rheumatology: 1600 cases investigated throughly. Fr. Rev. Apiculture 1983, p.421.

141. Lonauer G, Meyers A, Kastner D, Kalveram K, Forck G and Gerlach U. Treatment of rheumatoid arthritis with a new, purified bee venom. Abstract.
142. Moore KN. Life without arthritis. What every arthritis sufferer needs to know. R.H.K. Pub., 1987, 1–56.
143. Kim CM-H. Bee venom therapy for arthritis. Rheumatologie 1989, 41:67–72.
144. Klinghardt D. Bee venom therapy for chronic pain. J. Neuro. Ortho. Med. & Surg. 1990, 11:3: 195–197.
145. Ainlay GW. The use of bee venom in the treatment of arthritis and neuritis. Nebraska Med. J. 1939, 24:298–303.
146. Guyton FE. Bee sting therapy for arthritis and neuritis. J. Econ. Entomol. 1947, 40(4):469–472.
147. Guyton FE. Sixteen years of treating arthritis with bee stings. NAAS Proc. 1978, 1:37–41.
148. Yoirish, N. Chapter 4. Therapeutic uses of bee venom. Curative properties of honey and bee venom. New Glide Pub. 1977, pp.144–171.
149. Broadman J. Bee venom: the natural curative for arthritis and rheumatism. G.P.Putnam's sons, New York 1962, pp220, 169 references.
150. Saine J. The effectiveness of bee venom in the treatment of arthritis. NAAS Proc. 1978, 1:25–32.
151. Baker WP. The effectiveness of bee venom in the treatment of arthritis. NAAS Proc. 1980, 3:47–49.
152. Baker WP. Homeopathic bee venom therapy. NAAS Proc. 1982, 5:15–16.
153. Kazior A. Peculiar way of bee venom application in painful diseases of the spine. International Symposium on Apitherapy. Apimondia, 1985.
154. Kim CM-H. Honey bee venom therapy for arthritis (RA, OA), fibromyositis (FM) and peripheral neuritis (PN). Pain, J. of the Korean Pain Society 1992, 1(1):55–65.
155. Weeks B. Personal Communication, 1994.
156. Nicholas EE. Rheumatoid arthritis treatment with the sting of the honey bee. N.Y. St. Med. J. 1938, 38:1218.
157. Hollander JL. Bee venom in the treatment of chronic arthritis. Am. J. Med. Sci. 1941, 261:796–801.
158. Wiessmann G, Zurier RB, Mitnick D and Bloomgarden D. Effects of bee venom of experimental arthritis. Ann. Rheum. Dis. 1973, 32:466–470.
159. Eiseman JL, Bredow JV and Alvares AP. Effect of honeybee (Apis mellifera) venom on the course of adjuvant-induced arthritis and depression of drug metabolism in the rat. Biochem. Pharmacol. 1982, 31:1139–1146.
160. Tannenbaum H and Greenspoon M. Bee venom and adjuvant induced disease (Letter). J. Rheumatol. 1983, 10:522.
161. Vick JA, Warren GB and Brooks RB. The effects of whole bee venom on cage activity and plasma cortisol levels in the arthritiv dog. Inflammation 1975, Vol.1, No.2: 167–174.
162. Short T, Jackson R and Beard G. Usefulness of bee venom therapy in canine arthritis. NAAS Proc. 1979, 2:13–17.
163. Von Bredow. Treatment of equine arthritis with bee venom. NAAS Proc. 1978, 1:141–146.
164. Von Bredow J, Short T, Beard G and Reid K. Effectiveness of bee venom therapy in the treatment of canine arthritis. NAAS Proc. 1981, 4:45–48.
165. Holmes ERC. Personal Communication.
166. Hawley DJ and Wolfe F. Are the results of controlled clinical trials and observational studies of second line therapy in rheumatoid arthritis valid and generalizable as measured of rheumatoid arthritis outcome: analysis of 122 studies. J. Rheum. 1991, 18:1008–1014.
167. Hard R, Goddard P and Dieppe PA. The clinical and radiological correlations in osteoarthritis. Ann. Rheum. Dis. 1991, 50:14–19.
168. Bagge E, Bjelle A, Eden S and Svanborg A. Osteoarthritis in the elderly: clinical and radiological findings in 79- and 85-year-olds. Ann. Rheum. Dis. 1991, 50:535–539.
169. Gordon DA. The importance of therapeutic experiments. Notes on rheumatology by the French Arthritis Society, 1992.
170. Wilske KR and Healey LA. The need for aggressive therapy of rheumatoid arthritis. Current controversies in clinical rheumatology. Rheum. Dis. Clin. of N. Am. 1993, 19(1):153–161.
171. Kushner I and Dawson NV. Aggressive therapy does not substantially alter the long-term course of rheumatoid arthritis. Current controversies in clinical rheumatology. Rheum. Dis. Clin. N. Am. 1993, 19(1):163–172.
172. Yunginger KW, Jones RT, Leiferman KM, Paull BR, Welsh PW and Gleich GJ. Immunological and biological studies in beekeepers and their family members. J. Allergy Clin. Immunol. 1978, 61:2: 93–101.
173. Apitherapy. Alternative Medicine: Expanding Medical Horizons. A report to the NIH on alternative medical system and practices in the United States. NIH Pub. No. 94–06 (1994.12): 172–175,178.
174. Apivene. Monography. Laboratories H. Porcin, Paris.

175. Artemov NM, Orlov BN. New data to scientifically support the physiological use of bee venom as a medicine. Apimondia, The XXI International Congress, 1967: 348–353.
176. Artemov NM. Bee venom, its physiological properties and therapeutic uses. M.L. 188, 1941.
177. Bohmer D and Ambrus P. The effect of Forapin ointment containing bee venom on blood circulation in muscles. Therpiewoche 1981, 31: 5892–5894.
178. Broadman J. A review of the foreign literature on bee venom for the treatment of all rheumatism. General Practice 1958, 8:13, 26, 28–29.
179. Donadieu Y. Medicine and apiculture: several practical and essential comments on todays therapeutical use of bee venom, part 2. Fr. Rev. Apiculture, April 1980, p.383.
180. Forapin. Monography. Mack Lab., Allemagne, Germany.
181. Forestier F and Palmer M. Apitherapy; rheumatology: a therapy that is always young. Fr. Rev. Apiculture 1986, p.455.
182. Giza J. Bee venom therapy (Apitoxitherapy) associated with acupuncture. International Symposium on Apitherapy. Apimondia 1985, p.58.
183. Homeopathic Pharmacopeia of the U.S. Apis mellifera and apis virus. 1981 Ed., pp.88–90.
184. Kononenko IF. Bee venom preparation "Melissin" as a remedy and disease preventing agent. XVII International Congress of Apiculture, Rome, 1958.
185. Maberly FH. Brief notes on the treatment of rheumatism by bee stings. Lancet 1910, 2: p.235.
186. Parteniu A. Bee venom and arterial diseases. Treatment of peripheral arterial diseases and arterioscleroses by means of bee stings applied at lumbar trigger points. (Rumania-Abst.)
187. Pharmalgen. Monography. Pharmacia.
188. Pochinkova P. Administration of bee venom by ultrasounds. XVII International Symposium on Apitherapy. Apimondia 1973, 111–113.
189. Tadeusz O. Apipuncture (Bee Acupuncture). Fr. Rev. Apiculture 1987, 465:1–11.
190. Terc P. Report about a peculiar connection between the beestings and rheumatism. Vienna Medical Press, 1888.
191. Ven-AB. Monography. Lancet Lab., Montreal.
192. Venomil. Monography. Hollister-Stier/Miles.
193. Kim CM-H. Effectiveness of bee venom therapy, 10 years treatment. XXXIII Apimondia Congress 1993, Beijing, China.

INDEX

FSC
www.fsc.org
MIX
Papier aus verantwortungsvollen Quellen
Paper from responsible sources
FSC® C105338

If you have any concerns about our products,
you can contact us on
ProductSafety@springernature.com

In case Publisher is established outside the EU,
the EU authorized representative is:
Springer Nature Customer Service Center GmbH
Europaplatz 3, 69115 Heidelberg, Germany

Printed by Libri Plureos GmbH
in Hamburg, Germany